The Evolution of Techniques

The Evolution of Techniques

Rigidity and Flexibility in Use, Transmission, and Innovation

edited by Mathieu Charbonneau

The MIT Press
Cambridge, Massachusetts
London, England

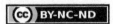
The MIT Press would like to thank the anonymous peer reviewers who provided comments on drafts of this book. The generous work of academic experts is essential for establishing the authority and quality of our publications. We acknowledge with gratitude the contributions of these otherwise uncredited readers.

This book was set in Times New Roman by Westchester Publishing Services. Printed and bound in the United States of America.

Library of Congress Cataloging-in-Publication Data is available.

ISBN: 978-0-262-54780-2

10 9 8 7 6 5 4 3 2 1

Contents

Series Foreword

Biology is a leading science in this century. As in all other sciences, progress in biology depends on the interrelations between empirical research, theory building, modeling, and societal context. But whereas molecular and experimental biology have evolved dramatically in recent years, generating a flood of highly detailed data, the integration of these results into useful theoretical frameworks has lagged behind. Driven largely by pragmatic and technical considerations, research in biology continues to be less guided by theory than seems indicated. By promoting the formulation and discussion of new theoretical concepts in the biosciences, this series intends to help fill important gaps in our understanding of some of the major open questions of biology, such as the origin and organization of organismal form, the relationship between development and evolution, and the biological bases of cognition and mind. Theoretical biology has important roots in the experimental tradition of early twentieth-century Vienna. Paul Weiss and Ludwig von Bertalanffy were among the first to use the term *theoretical biology* in its modern sense. In their understanding, the subject was not limited to mathematical formalization, as is often the case today, but extended to the conceptual foundations of biology. It is this commitment to a comprehensive and cross-disciplinary integration of theoretical concepts that the Vienna Series intends to emphasize. Today, theoretical biology has genetic, developmental, and evolutionary components, the central connective themes in modern biology, but it also includes relevant aspects of computational or systems biology and extends to the naturalistic philosophy of sciences. The Vienna Series grew out of theory-oriented workshops organized by the KLI, an international institute for the advanced study of natural complex systems. The KLI fosters research projects, workshops, book projects, and the journal *Biological Theory*, all devoted to aspects of theoretical biology, with an emphasis on—but not restriction to—integrating the developmental, evolutionary, and cognitive sciences. The series editors welcome suggestions for book projects in these domains.

Gerd B. Müller, Thomas Pradeu, Katrin Schäfer

Preface and Acknowledgments

Mathieu Charbonneau

To be culturally successful, techniques must be both effective in achieving their practical goals and efficiently transmittable from one generation to the next. To secure practical efficiency, techniques should be used flexibly, whereas, to secure cultural stability, they should, it seems, be transmitted rigidly. How are these two conflicting demands reconciled? The contributions to this volume, coming from different disciplines, address the causal factors and impacts of both flexibility and rigidity in the ways techniques are used, transformed, reconstructed, modified, and diffused through time and at varying social and timescales.

To address these issues, we invited contributors from different disciplines, including anthropology, cognitive psychology, primatology, archaeology, philosophy, and history to participate in an online five-month webinar that took place from September 2020 to January 2021. There, they presented early drafts of their contributions to be discussed online by all contributors, allowing each to gain a substantial understanding of the other contributions, to get feedback on their own ideas, and to engage in critical discussions aimed at a better overall integration of the collective project. We intended as a second step to have an in-person workshop at the Central European University in Budapest, Hungary, where all contributors would present a prefinal draft of their chapter and directly interact with one another. Unfortunately, because of the COVID-19 pandemic, this event had to be canceled. Instead, completed drafts were submitted in the summer and fall of 2021, following which philosopher Kim Sterelny wrote his final discussion chapter.

This work was part of a wider project entitled "Constructing Social Minds: Coordination, Communication, and Cultural Transmission" supported by a seven-year Synergy grant of the European Research Council (ERC Seventh Framework Program, FP7/2007–2013, grant agreement no. 609819) and aimed at better integrating the study of interindividual coordination and communication and that of cultural transmission (Brinitzer 2022).

This volume would not have been possible without the help, support, and hard work of several people. Tiffany Morisseau made the online webinar a reality and a success. Miriam Haidle and Pooja Venkatesh contributed to the online discussions. Cameron Brinitzer edited several of the chapters, making the contributions more accessible to nonspecialist but informed readers. Andrea Jenei provided friendly and efficient administrative support. Anne-Marie Bono of the MIT Press provided guidance that was crucial in bringing this project from promise to reality. Last but not least, special thanks to Dan Sperber, who offered invaluable support and mentorship throughout the making of the volume.

References

Brinitzer, Cameron Marcus. 2022. "Culturing Evolution: A History and Anthropology of a Cognitive Science of Culture in Illiberal Hungary." PhD diss., University of Pennsylvania.

Introduction

Mathieu Charbonneau and Dan Sperber

Technologies—complexes of skills, practices, and artifacts aimed at practical goals—are passed on from generation to generation, forming traditions that can persist with a high degree of stability, sometimes for millennia. These traditions have proved essential to the success of the human species in populating all but the most hostile ecosystems of our planet. They have made it possible for human populations to respond to ecological challenges in locally adapted ways—for instance, by collecting and processing foods or by transforming a variety of materials to produce artifacts such as shelters, clothing, weapons, and tools of all kinds. These traditions also play a key role in shaping social relations within and among human populations. Communication, transportation, and economic systems depend on culturally developed and transmitted technologies to properly function, persist, and evolve through time. Simpler technical traditions are also found in other species, providing the strongest evidence for the claim that culture is not uniquely human.

Technology involves both techniques and artifacts. Techniques involve complex practical actions (often employing artifacts and aimed at producing material changes in one's environment) and the skills needed to perform these actions. Yet, in the study of cumulative technological evolution, techniques have received much less attention than artifacts, notwithstanding a few programmatic discussions (Charbonneau 2015, 2016, 2018; O'Brien et al. 2010; Mokyr 2000; Manem 2020). Research on technological evolution has instead adopted artifacts as the key units of technological change (in line with Basalla 1988).

Still, the study of techniques as units of human cultures has attracted considerable scientific interest across several fields. Following in the steps of André Leroi-Gourhan and his successors, the French school of prehistory has developed the systematic study of techniques by applying the notion of *chaîne opératoire*, an analytic practice by which learned skills are decomposed into their operational and functional units. *Chaînes opératoires*, and similar means of parsing action and decision sequences involved in the production, use, refitting, and discarding of material artifacts, are now common across archaeological science (Bleed 2001). The evolutionary study of the cognitive, motor, and anatomical prerequisites for learning and deploying complex techniques is also a growing field within paleoanthropology and paleoarchaeology (Haidle 2009; Lombard and Haidle 2012; Roux and Bril 2005; Stout 2002, 2011) and cognitive science (Osiurak and Heinke 2018; Stout 2013).

In adopting artifacts as the key units of technological change, the study of cultural evolution has been primarily focusing on the study of change in artifact morphology, diversity, and function. The driving model for technological evolution has been one where individuals copy the artifacts they observe with high enough fidelity to ensure the enduring stability of the tradition (Boyd, Richerson, and Henrich 2013). Through the accumulation of small, random copying errors, the shape and functions of these artifacts would gradually change, with little involvement of individual intelligence (Boyd, Richerson, and Henrich 2011; Henrich 2016). High-fidelity copying also ensures that those modifications in artifacts that are adaptive are preserved, slowly building up increasingly sophisticated and complex technological traditions. This morphocentric model of technological evolution (Charbonneau 2015, 2018) is particularly manifest in the work of evolutionary archaeologists, focusing on morphological similarities and differences within prehistorical artifacts as bearers of phylogenetic signals (Lipo et al. 2006; O'Brien and Lyman 2000; cf. Manem 2020). Most historical case studies of technological evolution also emphasize change in artifact form, such as the evolution of sound holes in European violin design from the tenth to the eighteenth century (Nia et al. 2015), Turkmen carpet design up to the eighteenth century (Tehrani and Collard 2002), or the evolution of valved cornets in the last two centuries (Těmkin and Eldredge 2007). Laboratory experiments on cumulative cultural evolution have also followed this trend, being nearly exclusively focused on artifact evolution (Miton and Charbonneau 2018; cf. Strachan et al. 2021).

Of course, cumulative changes in artifacts are central to the process of technological evolution, but so are changes in the technical traditions that are necessary for producing and using those artifacts. Unlike artifacts, most techniques cannot be rigidly copied generation after generation. Technical actions must flexibly vary, if only to allow the successful implementation of the technique in different contexts of production (e.g., starting a fire under varying conditions of sun, rain, or wind; building a shelter on flat, steep, dry, or wet land). As complex and structured (hierarchically organized) actions, techniques often offer various degrees of freedom to their users, who can adapt them to solve novel problems and adjust to changing social and ecological circumstances, often within their individual lifetime but also over intergenerational timescales. Moreover, while actions can be observed and copied, the tacit skills and expertise that make these actions possible and effective cannot be directly copied but only inferred and reconstructed anew (with or without the help of pedagogic demonstrations and teaching), a process that calls for cognitive flexibility. How does this flexibility in the use, transmission, acquisition, and adjustment of techniques affect their evolution?

The Tension between Rigidity and Flexibility

The very existence of long-lasting technical traditions implies the persistence of the relevant knowledge content behind those traditions—knowledge that needs to be learned faithfully enough and carried on stably enough to endure as the same technique (Charbonneau 2020). For technical traditions to persist, those using the techniques must do so according to the defining features of the techniques. There is no point in transmitting specific means to carry out some practical action if users simply decide to do it their own way; a tradition reinvented at every generation is no tradition at all. On the other hand, there may be a range of possibilities between strictly copying and fully reinventing at each generation. In some cases, at least, it may be that what is transmitted is a range of indications sufficient for the full skill to be faithfully "re-produced" (with a hyphen and in the sense of produced again) rather than strictly speaking "reproduced" (in the sense of copied). Copying is not the only way to secure recurrence (Sperber 1996; Claidière, Scott-Phillips, and Sperber 2014; Miton, forthcoming).

For technical traditions to be culturally successful, they must be transmitted in ways that secure their effectiveness. While techniques play several roles beyond the functional—they can serve as social or ethnic markers, sources of prestige, tools of power, to name a few examples—failure in properly transmitting key functional aspects of a technique can lead to its functional disruption, decreased performance, or even the transmission of unusable "practical knowledge." Users of techniques must not only learn the specific functional features of the techniques, but they must deploy them appropriately for the technique to be useful at all. Here it seems that techniques effective enough to be worth transmitting demand rigidity, this time in their use.

However, if techniques are to serve as adaptive means for a population to survive and thrive, they must also allow for some amount of flexibility. For one thing, environments are not static. Populations have to adapt to novel circumstances in ways that both take advantage of their technical traditions and adjust them to a changing world (Pope-Caldwell, this volume). Such changes can be the result of exogenous factors—changes in the natural environment (for instance, desertification) or in the human environment (for instance, intergroup conflicts)—or of endogenous factors (for instance, increase in population size, changes in social organization, or new dietary norms). Such changes may call for technological invention, modification, and innovation.

A popular model of technological evolution is that of the accumulated error model (Eerkens 2000; Hamilton and Buchanan 2009). In this model, transmission and use remain quite rigid, but some slight modifications are unintentionally introduced in the traditions through miscopying, allowing just enough difference to occasionally produce novel technical variants that can track ecological and social transformations in adaptive ways, with some form of selection process ensuring the successful diffusion of the novel adaptive means (Boyd and Richerson 2005). When invention is achieved in this way, as an effect of copying errors introducing variants and selection of variants, transmission and use of techniques remain mostly rigid.

Populations can also add novel technical variants to their repertoire, either by inventing completely novel techniques from scratch or by recombining existing techniques in order to produce new ones (Charbonneau 2016; Lewis and Laland 2012). In the case of combinatorial invention, the new tradition preserves aspects of the earlier tradition (at least within the composite invention; see Manem, this volume), and to that extent it relies on some piecemeal rigidity and flexibility. In the invention from scratch case, novel technical variants are the output of individual discovery, but the transmission process based on social learning may itself remain rigidly faithful.

Studying technological innovation and evolution as iterations of a process of high-fidelity copying followed by selection works rather well for some artifact traditions, especially morphologically standardized artifacts that remain inert until used. In contrast, techniques are, by their very nature, interactive processes involving both their users and the dynamical physical world that surrounds them. In order to be effective, techniques must be adaptable to the local circumstances of their use, sometimes on the fly, sometimes from one day, one month, or one year to the next, and this means, in large part, that they need to be flexible enough to deal with the heterogeneity of the materials found in the environment and the various demands and constraints imposed by the contexts within which they are deployed.

If artifact forms can be standardized, the raw materials used to produce those forms are themselves rarely homogeneous. When pressure-flaking an obsidian core, lumbering a tree, or treating ore to form an alloy, the user must cope with the heterogeneous physical and chemical properties of the materials, which demands that the user adapt to the density, shape, mixture, quantity, and grain of the materials to be transformed. When cooking, one must usually adapt to variations in ingredients of the same type, such as the ripeness of vegetables or the tenderness of meats. When building a shelter, one often needs to balance constraints imposed by the ground's density, hardness, and slope, together with the local wind stream, precipitation patterns, light direction, and shading patterns, and so on. This is also true of contexts where the natural resources used are not themselves physical objects, such as adapting to fast-changing meteorological circumstances (e.g., the sailing of a vessel in a context of changing winds and water currents; see Astuti, this volume). The availability of the materials can also fluctuate, causing the user to adopt strategies dealing with limited, absent, or alternative materials in fruitful ways, such as using some ingredient that can serve as a "good enough" replacement for an original recipe.

Materials themselves are often reactive to users' actions. This is especially true of organic materials. Milk has several properties that vary with its source, age, and handling history, making the production of traditional dairy products as much the implementation of an ancestral recipe as an active process of contingent adjustments (e.g., needs for controlling its microbiota). Techniques involving animals, such as hunting or horseback riding (Miton, this volume) must deal with additional dynamically changing degrees of freedom and constraints. Animals (even those of a same species) often have different body shapes and sizes, muscular capacities, and more importantly, mental dispositions and capacities, both as individual members of their species (e.g., intelligence or temperament) and as abilities developed over time (e.g., mood or developmental stages).

Techniques, in their use, must be plastic enough to allow their users to flexibly implement them in the face of these varying and dynamic circumstances. Successfully using a technique therefore requires more than stereotypically following a predefined, standardized recipe. It requires, on the one hand, the user's capacity to recognize and adapt to local contingencies, and on the other, for the technique itself to allow enough degrees of freedom to accommodate the varying production contexts within which it is deployed. This, then, suggests that any two uses of a very same technique may vary in terms of the actions used, their order, and their function, depending on context. This intrinsic entanglement of the expression of a technique with the materials and ecological circumstances in which it is deployed makes the causal contribution of differences in behavior resulting from culture, individual learning, and ecology difficult to disentangle (Tenpas, Schweinfurth, and Call, this volume).

The requirement of flexible expression of techniques when used challenges the idea that techniques can be transmitted through rigid copying processes. Rigidly copying an observed action sequence may be of limited use since next-generation users (or even users of the same generation) will themselves rarely encounter the exact same conditions that shaped the specific expression of the technique they observed when learning it from others. Effective learning requires the learner to move beyond the observed sequence of action in order to parse what is contingent to the local learning situation and what serves as the key, traditional features of the technique. When the technique consists of a single action, or very few simple actions, copying might plausibly do the work, considering that there is a limited number of ways a single action can vary and remain functional. However, this assumes that single actions are limited in their dimensionality—an assumption that relies more on the coarseness of the grain of analysis at which actions are described than on the range of variation any given action offers (Csibra 2008; Charbonneau and Bourrat 2021). In any case, learners must be capable of moving past specific actions and understand their functional meaning within the whole sequence to which they belong and grasp which aspects of the action ensure its functional relevance to the overall functional economy of the action sequence (Gergely and Király, this volume). In other words, learning techniques involves not copying a stereotypical action sequence but rather capturing its hierarchically organized decision structure and learning to recognize the often unobservable, tacit cues used by the expert models when deploying the technique. This suggests that the cultural transmission of techniques is a reconstructive process working with incomplete informational inputs because the decision and perceptual cues key to mastering the skill are not themselves directly observable but must instead be inferred from, for instance, context or learned by trial and error. Reconstruction appears to be an inevitable part of the transmission process of technical knowledge (Strachan, Curioni, and McEllin, this volume) and, as a consequence, of the successful long-term stabilization of technical traditions (Stout, this volume).

Of course, some variation in technical actions, especially among novices, may be an effect of imperfect, low-fidelity copying. Much more importantly, variation is a necessary component of the acquisition and use of techniques: through observed variation in an expert model's behavior, the novice gains a better understanding of the range of actions that can yield the expected effect of the technical behavior. Instead of approaching variation as a form of deviation from what ought to have been learned (e.g., as a form of copying error or noise in transmission), variation in the expression of techniques should, in most cases, be viewed as what makes them useful and learnable in the first place (Roux et al., this volume). Moreover, the fact that variation is an intrinsic part of the expert use of a technique and of learners' acquisition of technical expertise suggests that this intrinsic flexibility may in itself be a promoter of innovation. For instance, while experts may be capable of mastering a technical skill very precisely by flexibly adapting to the context of production, they are also those who can potentially deviate from the common use of the techniques and exploit their mastery to flexibly produce novelties (Roux et al., this volume; De Munck, this volume). If so, then this poses the question of what brings experts to decide to exploit variation in innovative rather than conservative ways. Novel ecological challenges can serve as a stimulation for innovation (Pope-Caldwell, this volume; Tenpas, Schweinfurth, and Call, this volume); so can new economic demands, such as the growth of a market for new products (De Munck, this volume; Manem, this volume; Roux et al., this volume).

Finally, the social and interpersonal setups mobilized for the transmission of technical knowledge must also be flexible (Boyette, this volume; De Munck, this volume; Ongaro, this volume). Ethnographic research has documented a broad diversity of ways in which technical skills are acquired within populations (Lancy, Bock, and Gaskins 2010; Rogoff

2003; Lew-Levy et al. 2019). While some technical skills can be acquired through observational learning with minimal engagement of the model, in many societies these skills are learned through the direct engagement of the learner in the context of use of the technique or of peripheral activities (Lave and Wenger 1991; Paradise and Rogoff 2009). Peer-play also offers many opportunities for learners of a younger age to progressively build their technical expertise (Boyette 2016, this volume; Chick 2010).

Technical skills can also be passed on through various institutional forms of education, from informal and *in situ* activities (Lancy, Bock, and Gaskins 2010) to well-organized hands-on learning as in apprenticeship (De Munck, this volume; Sterelny, this volume). Decontextualized learning contexts such as those found in the classroom add to this diversity, with classroom learning itself varying in its organization and cognitive demands from one cultural (and historical) context to the next (Sternberg and Grigorenko 2004; Sterelny, this volume). This diversity of learning contexts and coordinated interactions illustrates the fact that humans are adaptable learners who are able to flexibly exploit different social configurations to ensure the successful transmission of vital skills, with each context often imposing different cognitive demands on the part of both the learner and the model or teacher (Charbonneau et al. 2023; Strachan, Curioni, and McEllin, et al., this volume).

If transmission, use, and innovation are intimately intertwined in this way, the model of invention as a process of solitary discovery seems unfit, giving way instead to a more socially situated alternative where novel ideas emerge in individuals who are themselves entrenched in deeper social relations and interactions (Cutting, this volume; Ongaro, this volume). To spread, an innovation must be accepted within a population, and for this, individuals must make sense of the novelty and flexibly integrate it within their own practices without this being too much of a challenge to, for instance, their sense of ethnic identity (Astuti, this volume; Ongaro, this volume). Producing and adopting an innovation is rarely if ever a strictly individual process (Cutting, this volume). For instance, while all individuals in a community may rigidly be reluctant to adopt on their own a novel technical variant, the community as a whole may be more flexible and deliberately adopt a novel way of doing things, overriding and overcoming individual reluctance (Ongaro, this volume). In other words, rigid individuals may, nevertheless, form flexibly open societies.

To sum up, the evolution of technical traditions appears to involve both rigidity and flexibility. On the one hand, for technical knowledge to be stabilized in the form of long-lasting traditions, it seems that it should be used according to the specific features characterizing that technique and that those features be transmitted with high fidelity, with some but not too much space for change if the traditions are to remain of functional value. On the other hand, the very plastic nature of techniques ensures they can be adapted to the specifics of their context of use, the circumstances under which they are transmitted, and the specific demands for adaptively producing and adopting novelty.

How is the tension between these two contrary demands—rigidity and flexibility in the use, transmission, and innovation of techniques—reconciled? While this question is key to our species' success, little scientific attention has been paid to it so far. This volume aims to foster a better understanding of ways in which this tension is solved and how these solutions differentially affect the evolution of technical traditions.

How the Volume Is Organized

To better understand this tension between rigidity and flexibility, we asked experts from a wide variety of disciplines and with different perspectives to provide both theoretical discussions and empirical case studies. While the contributors have been developing the notions of technical rigidity and flexibility in several ways and their individual chapters can be read independently of one another, they have all made original contribution to the common theme, often addressing each other's ideas. In a concluding discussion, the philosopher Kim Sterelny considers these contributions and their common theme in a simultaneously broad and precise evolutionary perspective.

The volume has three thematic parts:

1. Timescales of Technical Rigidity and Flexibility
2. From Rigid Copying to Flexible Reconstruction
3. Exogenous Factors of Technical Rigidity and Flexibility

There is some unavoidable arbitrariness in any such organization. While every chapter makes its main contribution to the theme of the part to which it belongs, they all also contribute to the themes of the other parts. They reflect the discussions that took place online among all the authors. We hope that this volume will foster a wider conversation and a greater interdisciplinary integration in the study of rigidity and flexibility in the evolution of techniques.

Part I, "Timescales of Technical Rigidity and Flexibility," reflects a consensus among the contributors that the tension between technical rigidity and flexibility must be simultaneously addressed at different spatiotemporal scales. As one contributor put it during our online discussions, "Larger scale processes are instantiated by the accumulation of smaller scale processes, while the smaller scale processes are constrained by the persistent context of the larger scale processes." For instance, flexibility in use and learning at the scale of an individual lifetime may result in rigid and stable traditions at the population level over the longer term. Similarly, rigidly preserved traditions may fuel flexible innovations at different timescales. Addressing this complexity demands an integrative interdisciplinarity in expertise and methods. The study of micro-interactions between individual learners and expert models and the examination of long-term cultural and technological evolution require quite different tools and methods, but they are mutually relevant and must be linked. This first part illustrates this methodological diversity and the challenges it raises for the development of an integrated picture.

The four first chapters forming part I have been ordered in terms of the timescale they study, from the shortest to the longest: technical change within a generation (Roux and colleagues), change following a single generational overturn (Astuti), and change over several centuries (De Munck) and millennia (Manem).

In an original piece of methodological interdisciplinarity, Valentine Roux, Blandine Bril, Anne-Lise Goujon, and Catherine Lara (chapter 1) investigate the relation between skill, expertise, and innovation by running a field experiment among potters inspired by laboratory experiments in cognitive psychology. They show that the most skillful individuals, who produced less variation in their pots thanks to their finer motor skills, were also better able to adapt their skills in order to produce novel pot forms. The authors argue that flexibility—defined as "the ability to cope with changing circumstances and unexpected variations and to find a motor solution for any situation and in any condition"—is acquired by experienced potters through extensive training with familiar production activities. This gives experts a much greater understanding of the task and therefore allows them to adapt to novel and varying situations better than less skillful individuals. At the same, this very experience with socially identified and shared techniques for making familiar pots helps explain the enduring stability of technical traditions.

In an ethnographic study spanning some 30 years, Rita Astuti (chapter 2) describes the change in the way Vezo fishermen of the village of Betania in Madagascar rig their canoes. Her chapter provides a case study of the adoption of a novel technique (and sail implement) within the timespan of one generation. Central to her case study is the fact that the Vezo, as a group, define themselves not by common ancestry but by their way of living, which involves sailing, fishing, and maritime coastal trade. This might suggest that techniques related to these activities would be rigidly maintained because they are key to the group's identity. Yet, after only a few years, the common sprit sail, a novelty to the Vezo, diffused within the population, eventually becoming the universally shared technique. Astuti reports how the people involved weighed the pros and cons of the novel technique compared to the traditional one, and she examines what narratives the Vezo deployed to explain the change while keeping their social identity intact.

In his contribution, Bert De Munck (chapter 3) traces the transformations of craftsmanship over several centuries from the late medieval to the early modern period in Europe. One view of craft guilds is that of a journeyman going through years of training and learning on the shop floor, where craft apprenticeship serves to ensure the faithful acquisition of a specialized and esoteric knowledge defined by rigid standards and procedures. In this view, incorporated craft guilds enforce highly rigid forms of learning and through their exclusivity serve as obstacles to individual creativity and the free diffusion of innovations. Yet the guild artisan is also associated with the idea of individual creativity and authenticity, in contrast to the image of the alienated factory worker's mechanically repetitive work. Rejecting the dichotomizing view opposing artisanal crafts and mass production, De Munck delves into "the complexity and hybridity of the artisan's manifold histories" and shows how craft guilds managed to balance the needs for rigidity and flexibility. His historical analysis focuses on the evolving ways by which apprenticeship was organized and institutionalized within craft guilds and how it adapted to cultural, political, and epistemological transformations as well as to economic and technological ones.

In a methodologically innovative contribution, Sébastien Manem (chapter 4) borrows cladistics methods of phylogenetic analysis originally developed by evolutionary biologists and studies the long-term tensions between rigidity and flexibility in the use and innovation of ceramic techniques. He deploys these methods with a twist: rather than focusing on artifact morphologies, as evolutionary archaeologists typically do, Manem instead adapts those methods to track phylogenetic signals in *chaînes opératoires*—the series of operations that transform raw material into finished product. Taking the evolution of ceramics of the Duffaits Bronze Age culture over several millennia, Manem argues that technical invention and innovation is made possible by a form of "slippery flexibility," where a novel technical variant is first used to manufacture some parts of ceramics while traditional methods are rigidly maintained for the other parts. By decomposing the *chaînes opératoires* in fine-grained, modular technical parameters, his analysis shows that we can reconstruct several important processes involved in technical innovation, including the pressures involved for flexible individual inventions and those stymieing the evolvability of technical traditions.

Part II, "From Rigid Copying to Flexible Reconstruction," contains contributions that take as a central question the nature and impact of different forms of cultural transmission. Each contribution, in its own way, moves away from the idea that high-fidelity copying is the proper general model of skill transmission, let alone of cultural learning altogether. Instead, each chapter approaches the question of skill transmission by arguing for more flexible and versatile forms of learning, where skills are not less copied-as-seen than reconstructed through a multiplicity of processes. While reconstructive, these processes result in the rigid stabilization of technical traditions in some contexts and fuel innovations in others. A common assumption of the chapters grouped in part II is that cultural learning is not merely social because the knowledge acquired is provided by others, with expert models pouring information into naive learners serving as receptables. Rather, cultural transmission should be understood primarily as an interactive process where experts materially and socially construct a rich learning environment (as Boyette and Stout emphasize in their respective chapters) or where teachers and learners are strongly involved in directing and participating in skill acquisition (as Gergely and Király argue, as do Strachan and colleagues). Each of these contributions ends up, in its own way, rejecting the common analogy between genetic and cultural inheritance (where inheritance is secured by copying) and the framework it imposes on the evolution of techniques—for instance, by treating cultural learning as a strictly unidirectional form of information transmission (Boyette; Strachan and colleagues), by adopting a dichotomy between inheritance and innovation or between individual and social learning (Boyette; Gergely and Király; Stout), or by assuming clear one-to-one lineage relationships between learners and those they learn from (Boyette; Stout).

György Gergely and Ildikó Király (chapter 5) challenge the widespread view that children acquire technical skills through passive observational learning and imitative copying and that, moreover, these rigid, high-fidelity learning mechanisms are specially adapted for capturing causally opaque actions—instrumental actions the function of which is not apparent to the learner—displayed by expert models. They argue against the dichotomy between high-fidelity transmission and innovation, which are often referred to as the two engines unique to the human species of the cumulative improvement of technical and technological knowledge. They develop a series of laboratory experiments showing that human cultural learning is a process of selective, inferential, and relevance-guided emulation (i.e., that children are specifically geared to learn on the basis of ostensive cues given to them by their models and are thereby capable of flexibly choosing, as they learn, alternative instrumental actions to those exhibited by the models). In this sense, both transmission and innovation are built in and part of the very same cognitive mechanism—relevance-based emulation—which is a learning mechanism that can be flexibly deployed in various contexts making possible both instrumental improvement of the learned actions and their faithful reconstruction.

James W. A. Strachan, Arianna Curioni, and Luke McEllin (chapter 6) take on and challenge the idea of observational learning (or copying) as a paradigm case and instead recast social learning in terms of action coordination. They argue that while observational learning may be a useful *minimal working example* for transmission experiments, the methodological gains offered by such experiments impose significant costs on the generalizability of their results. While transmission experiments generally force a unidirectional information flow from expert model to learner, many (if not most) learning interactions involve bidirectional information flow, where learners and models adapt to one another. This bidirectionality introduces a coordination dimension to social learning, which must be explained in terms of cognitive mechanisms of action coordination and joint action. These mechanisms, they argue, provide a better understanding of both technical flexibility and rigidity. Coordination mechanisms are versatile in that they allow individuals to adapt to various situational contexts and interactional demands such as role assignment or online error corrections, providing the ability to learn technical know-how in various learning setups. Coordination mechanisms, they argue, can also lead to rigidity because technical traditions depending on the coordination of individuals (e.g., driving in the left or right lane) can break down when some individuals start acting differently than expected. Instead, larger shifts at the population level may be required for effective change (e.g., governmental decree).

Noting that high-fidelity imitation (copying) is at risk of transmitting maladaptive behaviors in changing environments, Adam Boyette (chapter 7) argues that populations may flexibly adapt to such changes by *culturally* shaping their physical, social, and learning environment. Adopting a cultural niche construction approach—according to which we transform the environment in which we develop and learn from one another through the making of artifacts and the shaping of physical spaces—Boyette argues that humans can guide their children's developmental trajectories and acquisition of complex, vital skills with minimal direct, person-to-person social learning interactions. By flexibly constructing the contexts in which naive learners acquire key technical skills, we ensure the "faithful" reconstruction of technical traditions. This allows children to explore the range of tasks they need to solve freely and safely and to flexibly learn the necessary technical skills. Boyette explores this idea by studying how the BaYaka forest foragers from the Congo Basin encourage their children to autonomously explore blade tools. At the same time, they palliate the costs (such as risks of injuries) of individual exploration by providing their children with reliable and responsive caregiving. While BaYaka cultural niche construction may ensure low variation and a more rigid preservation of skills across generations, it also encourages innovations through this form of flexible and autonomous learning.

Is the prolonged technological stasis marking the Paleolithic really so puzzling? Dietrich Stout (chapter 8) rejects the assumption common in the field of cultural evolution that stability and convergence are anomalous and require special explanation. He characterizes technology as a biocultural reproductive strategy. It is the means for our species to support its distinctive life history and reproductive strategies through the investment of its surplus energy into the production of technical skills, knowledge, and equipment. Technology, therefore, is understood not as the mere production and use of artifacts by individuals or as the result of rigid, high-fidelity transmission of technological information among them. Instead, Stout understands technology as the result of collaborative, life history–oriented activities through which human collectives reconstruct practices and objects. In this perspective, technological stability is not stagnation caused by societies unable to innovate. Rather, achieving stability is the evolutionary challenge addressed by the technological niche in the first place: ensuring the persistence of a population in a changing world. According to Stout, we overcame this challenge by relying on flexible coordination and communication between individuals and the exploitation of cognitive, interactive, and ecological factors in order to reconstruct our technical way of life, from one generation to the next.

Part III, "Exogenous Factors of Technical Rigidity and Flexibility," challenges the view that both stability and change in technical traditions are the effect of endogenous factors inherent to the rigid use and transmission of technical traditions. According to that view, technical stability is best explained by the high-fidelity transmission mechanisms such as imitation while technical changes are best explained as effects of individual innovations resulting from error, serendipity, or insight and selected on the basis of their functional efficiency. While nobody denies that such endogenous factors are important, the contributions in part III highlight the importance of *exogenous* factors. These include cultural factors such as ethnic identity (Ongaro), social influences on use and invention (Cutting), the role of the materials acted on by the technique (which, especially in the case of live materials, can be quite dynamic and reactive) in shaping the technical tradition (Miton), and the impact of a changing environment (Pope-Caldwell). Disentangling the role endogenous and exogenous factors in human and nonhuman technical traditions is a major challenge (Tenpas and colleagues). Contributions in this part make clear the importance of studying exogenous factors to achieve a comprehensive understanding of technical evolution.

While many contributors focus on the capacities and factors that make technical behaviors rigid or flexible, Sarah Pope-Caldwell (chapter 9) takes issue with the human capacity to measure and choose in the face of changing ecological circumstances when it is adaptive to rigidly stick to familiar behaviors, when it is better to switch to different strategies found in the cultural repertoire of the population, and when it is preferable to invent new strategies. Starting from the assumption that humans are capable of being rigid in some circumstances and flexible in others, she examines *in which contexts* rigidity and flexibility are beneficial or detrimental and *by which mechanisms* we make the decision to pursue familiar strategies or move on to new ones. To do so, she proposes the *constrained flexibility framework*, extending existing research on cognitive flexibility in the face of environmental change. She discusses the effects of exogenous variability, predictability, and harshness on adaptive behavior and balances predictive and reactive strategies in the face of fluctuating ecological circumstances.

Central to Giulio Ongaro's ethnographical case study (chapter 10) is the role of exogenous factors in stabilizing technical traditions. He describes two kinds of technical traditions among the Akha of the Laos highlands: esoteric ethnopharmacological knowledge that is flexibly transmitted under secrecy by a few experts, and customs and practices that are widely shared, rigidly transmitted, and stabilized because of their ethnic-defining role. Because ethnopharmacological knowledge is esoteric, restricted, and not seen as constitutive of Akha identity, expertise in this domain is quite flexible and individually variable. Maintaining its secret character takes precedence over ascertaining its instrumental functionality. In the second case, techniques such as house building are transmitted in the open, closely monitored, and stabilized by the community. Yet the Akha, while they individually reject innovations that go against their identity-defining customs, remain flexible as a community and are capable of integrating novel techniques in their repertoire through group decisions.

Echoing a theme developed in the first and second parts of this volume, Nicola Cutting (chapter 11) argues against the standard dichotomy between individual learning and social learning. She reviews the growing experimental literature on children's ability to use tools in innovative ways (to which she has herself made important contributions) and argues that this dichotomy has led experimentalists to deprive children subjects of the social scaffolding necessary to solve tasks that use tools in innovative ways. In the real world, innovations are rarely achieved by individuals in isolation. It therefore remains unclear to what extent standard laboratory experiments paint a proper picture of children's capacity for flexible innovation, of the cognitive mechanisms underlying such capacity, and of their ontogeny. For her part, Cutting approaches the capacity to modify and combine existing solutions into novel ones as a form of innovative flexibility, where innovations are socially embedded and supported by social influences. She argues that experiments on innovative tool use by children remains a promising area of research when social influences are allowed to play a role, with children turning out to be much more flexible than what current research has seemed to demonstrate.

Techniques involving nonhuman animals such as horses face the difficulty of having to deal with living, complex, and changing materials. In her contribution, Helena Miton (chapter 12) asks how technical activities involving two species dynamically interacting with one another can be transmitted as stable, long-lasting traditions. In a tradition such as horse dressage, riders need to adapt expertly and flexibly to the horse's idiosyncrasies, such as its body constraints and temperament, in order to lead the horse in producing rigidly defined figures stabilized in long-standing traditions. Miton combines historical-cultural data, cognitive and veterinary research on horses, and an autoethnography of her own experience as a horse rider and teacher to understand how both riders and horses need to comply to rigidly defined codes and goals of traditional riding practices. For this, teaching traditions guide the riders in flexibly adapting to the idiosyncrasies of the horse's cognition and experience. As a result, she argues, riders and horses come to produce an integrated system where both parties are mutually adapting through the development of an interspecific form of haptic communication channel that is unnatural to both living systems.

Technical traditions are also found in nonhuman species, such as our closest relatives, the chimpanzees. Like humans, chimpanzees must strike a balance between rigidly preserving useful behaviors when beneficial and flexibly changing them when necessary. Sadie Tenpas, Manon Schweinfurth, and Josep Call (chapter 13) underline the fact that most research in comparative psychology has focused on high-fidelity social learning as the main mechanism behind the stability of nonhuman animal traditions. Too little attention has been paid to the means by which these traditions are updated in the face of novel circumstances or remain resilient in the face of disruptions. By excluding ecological and individual factors as noncultural, comparative psychologists are depriving themselves of important explanatory factors that can provide a deeper understanding of the origins and stability of nonhuman technical traditions. The authors propose complementing the social learning view by reintegrating these factors and by adapting cultural attraction theory, initially devised to account for human cultures, in order to enrich the study of chimpanzee technical traditions. In so doing, they develop a multilevel approach to technical rigidity and flexibility, examining how these are expressed at the individual, group, and population levels.

In his concluding discussion, "The Cumulative Culture Mosaic," Kim Sterelny reflects on the contributions in this volume, on the general issues these contributions address, and on the broader theme of the cultural evolution of techniques (the study of which Sterelny himself has made major theoretical contributions, in particular in his 2012 book *The Evolved Apprentice*). The factors favoring rigidity or flexibility, he underscores, have themselves evolved. Early technical skills in hominin evolution were more specific, context-dependent, and rigid. Expertise-based skills emerged progressively, making the preservation of skills more and more compatible with their improvement. Techniques with different goals develop under different ecological and cognitive constraints. Techniques used to act on the physical environment (world-facing traits) are likely to involve a great deal of trial-and-error learning, generating variation that can produce useful innovations when recognized as such and even lead to cumulative change when those techniques are used with technologies depending on maintenance for their reliability. In contrast, practices used as coordination devices within a community (community-facing traits) are less likely to vary once established, given that they act within a complex of norms allowing greater predictability of the behaviors of the members of the group by the members of the group. Because of the entwinement of use, transmission, and innovation in world-facing techniques, differences in learning regimes will affect how innovation is itself practiced. Small-scale pre-state societies favoring a face-to-face interactive form of learning contrast with large-scale state societies securing the reconstruction of expertise mostly through top-down, institutionalized schooling systems.

All in all, Sterelny's concluding discussion brings us back to the initial motivation of this volume: rethinking the use, transmission, and innovation in techniques by examining their contextual, ecological, and temporal variability. We seem to arrive at a more fluid and diversified image of technical evolution, where rigidity and flexibility are not construed as dichotomous opposites but instead define a space of possibility much larger than what is usually envisaged, a space in need of further and deeper interdisciplinary theoretical and empirical investigations.

References

Basalla, G. 1988. *The Evolution of Technology*. Cambridge: Cambridge University Press.

Bleed, P. 2001. "Trees or Chains, Links or Branches: Conceptual Alternatives for Consideration of Stone Tool Production and Other Sequential Activities." *Journal of Archaeological Method and Theory* 8 (1): 101–127.

Boyd, R., and P. J. Richerson. 2005. *The Origin and Evolution of Cultures*. Oxford: Oxford University Press.

Boyd, R., P. J. Richerson, and J. Henrich. 2011. "The Cultural Niche: Why Social Learning Is Essential for Human Adaptation." *Proceedings of the National Academy of Sciences* 108:10918–10925.

Boyd, R., P. J. Richerson, and J. Henrich. 2013. "The Cultural Evolution of Technology: Facts and Theories." In *Cultural Evolution: Society, Technology, Language, and Religion*, edited by P. J. Richerson and M. H. Christiansen, 119–142. Cambridge, MA: MIT Press.

Boyette, A. H. 2016. "Children's Play and the Integration of Social and Individual Learning: A Cultural Niche Construction Perspective." In *Social Learning and Innovation in Contemporary Hunter-Gatherers*, edited by H. Terashima and B. S. Hewlett, 159–169. Dordrecht: Springer.

Charbonneau, M. 2015. "Mapping Complex Social Transmission: Technical Constraints on the Evolution Cultures." *Biology & Philosophy* 30:527–546.

Charbonneau, M. 2016. "Modularity and Recombination in Technological Evolution." *Philosophy & Technology* 29:373–392.

Charbonneau, M. 2018. "Technical Constraints on the Convergent Evolution of Technologies." In *Convergent Evolution in Stone-Tool Technology*, edited by M. J. O'Brien, B. Buchanan, and M. I. Eren, 73–89. Cambridge, MA: MIT Press.

Charbonneau, M. 2020. "Understanding Cultural Fidelity." *British Journal for the Philosophy of Science* 71 (4): 1209–1233.

Charbonneau, M., and P. Bourrat. 2021. "Fidelity and the Grain Problem in Cultural Evolution." *Synthese* 199:5815–5836.

Charbonneau, M., A. Curioni, L. McEllin, and J. W. A. Strachan. "Flexible Cultural Learning Through Action Coordination." *Perspectives on Psychological Science*, July 17, 2023.

Chick, G. 2010. "Work, Play, and Learning." In *The Anthropology of Learning in Childhood*, edited by D. F. Lancy, J. Bock, and S. Gaskins, 119–143. Lanham, MD: AltaMira Press.

Claidière, N., T. C. Scott-Phillips, and D. Sperber. 2014. "How Darwinian Is Cultural Evolution?" *Philosophical Transactions of the Royal Society B: Biological Sciences* 369 (1642), 20130368.

Csibra, G. 2008. "Action Mirroring and Action Understanding: An Alternative Account." *Sensorimotor Foundations of Higher Cognition: Attention and Performance* 22:435–459.

Eerkens, J. 2000. "Practice Makes within 5% of Perfect: The Role of Visual Perception, Motor Skills, and Human Memory in Artifact Variation and Standardization." *Current Anthropology* 41:663–668.

Haidle, M. N. 2009. "How to Think a Simple Spear." In *Cognitive Archaeology and Human Evolution*, edited by S. A. de Beaune, F. L. Coolidge, and T. Wynn, 57–73. Cambridge: Cambridge University Press.

Hamilton, M. J., and B. Buchanan. 2009. "The Accumulation of Stochastic Copying Errors Causes Drift in Culturally Transmitted Technologies: Quantifying Clovis Evolutionary Dynamics." *Journal of Anthropological Archaeology* 28:55–69.

Henrich, J. 2016. *The Secret of Our Success: How Culture is Driving Human Evolution, Domesticating Our Species, and Making Us Smarter*. Princeton, NJ: Princeton University Press.

Lancy, D. F., J. Bock, and S. Gaskins. 2010. *The Anthropology of Learning in Childhood*. Lanham, MD: Rowman Altamira.

Lave, J., and E. Wenger. 1991. *Situated Learning: Legitimate Peripheral Participation*. Cambridge: Cambridge University Press.

Lewis, H. M., and K. N. Laland. 2012. "Transmission Fidelity Is the Key to the Build-up of Cumulative Culture." *Philosophical Transactions of the Royal Society B* 367:2171–2180.

Lew-Levy, S., A. N. Crittenden, A. H. Boyette, I. A. Mabulla, B. S. Hewlett, and M. E. Lamb. 2019. "Inter-and Intra-Cultural Variation in Learning-through-Participation among Hadza and BaYaka Forager Children and Adolescents from Tanzania and the Republic of Congo." *Journal of Psychology in Africa* 29 (4): 309–318.

Lipo, C. P., M. J. O'Brien, M. Collard, and S. Shennan. 2006. *Mapping Our Ancestors: Phylogenetic Approaches in Anthropology and Prehistory*. New Brunswick, NJ: Aldine Transaction.

Lombard, M., and M. N. Haidle. 2012. "Thinking a Bow-and-Arrow Set: Cognitive Implications of Middle Stone Age Bow and Stone-Tipped Arrow Technology." *Cambridge Archaeological Journal* 22 (2): 237–264.

Manem, S. 2020. "Modeling the Evolution of Ceramic Traditions through a Phylogenetic Analysis of the *Chaînes Opératoires*: The European Bronze Age as a Case Study." *Journal of Archaeological Method and Theory* 27 (4): 992–1039.

Miton, H. Forthcoming. "Cultural Attraction." In *Oxford Handbook of Cultural Evolution*, edited by R. Kendal, J. Tehrani, and J. Kendal. Oxford: Oxford University Press.

Miton, H., and M. Charbonneau. 2018. "Cumulative Culture in the Laboratory: Methodological and Theoretical Challenges." *Proceedings of the Royal Society B* 285:20180677.

Mokyr, J. 2000. "Evolutionary Phenomena in Technological Change." In *Technological Innovation as an Evolutionary Process*, edited by J. Ziman, 52–65. Cambridge: Cambridge University Press.

Nia, H. T., A. D. Jain, Y. Liu, M.-R. Alam, R. Barnas, and N. C. Makris. 2015. "The Evolution of Air Resonance Power Efficiency in the Violin and Its Ancestors." *Proceedings of the Royal Society A: Mathematical, Physical and Engineering Sciences* 471 (2175): 20140905.

O'Brien, M. J., and R. L. Lyman. 2000. *Applying Evolutionary Archaeology: A Systematic Approach*. New York: Kluwer Academic Publishers.

O'Brien, M. J., R. L. Lyman, A. Mesoudi, and T. L. VanPool. 2010. "Cultural Traits as Units of Analysis." *Philosophical Transactions of the Royal Society B* 365:3797–3806.

Osiurak, F., and D. Heinke. 2018. "Looking for Intoolligence: A Unified Framework for the Cognitive Study of Human Tool Use and Technology." *American Psychologist* 73 (2): 169-185.

Paradise, R., and B. Rogoff. 2009. "Side by Side: Learning by Observing and Pitching In." *Ethos* 37 (1): 102–138.

Rogoff, B. 2003. *The Cultural Nature of Human Development*. Oxford: Oxford University Press.

Roux, V., and B. Bril. 2005. *Stone Knapping: The Necessary Conditions for a Uniquely Hominin Behaviour*. Cambridge: McDonald Institute for Archaeological Research.

Sperber, D. 1996. *Explaining Culture: A Naturalistic Approach*. Oxford: Blackwell Publishers.

Sternberg, R. J., and E. L. Grigorenko. 2004. "Intelligence and Culture: How Culture Shapes What Intelligence Means, and the Implications for a Science of Well–Being." *Philosophical Transactions of the Royal Society of London. Series B: Biological Sciences* 359 (1449): 1427–1434.

Stout, D. 2002. "Skill and Cognition in Stone Tool Production: An Ethnographic Case Study from Irian Jaya." *Current Anthropology* 43 (5): 693–722.

Stout, D. 2011. "Stone Toolmaking and the Evolution of Human Culture and Cognition." *Philosophical Transactions of the Royal Society B* 366:1050–1059.

Stout, D. 2013. "Neuroscience of Technology." In *Cultural Evolution: Society, Technology, Language, and Religion*, edited by P. J. Richerson and M. H. Christiansen, 157–173. Cambridge, MA: MIT Press.

Strachan, J. W. A., A. Curioni, M. D. Constable, G. Knoblich, and M. Charbonneau. 2021. "Evaluating the Relative Contributions of Copying and Reconstruction Processes in Cultural Transmission Episodes." *PLOS One* 16 (9): e0256901.

Tehrani, J., and M. Collard. 2002. "Investigating Cultural Evolution through Biological Phylogenetic Analyses of Turkmen Textiles." *Journal of Anthropological Archaeology* 21 (4): 443–463.

Tëmkin, I., and N. Eldredge. 2007. "Phylogenetics and Material Cultural Evolution." *Current Anthropology* 48 (1): 146–154.

I TIMESCALES OF TECHNICAL RIGIDITY AND FLEXIBILITY

1 Adaptive Behavior within Technological Stability: Field Experiments with Potters from Five Cultures

Valentine Roux, Blandine Bril, Anne-Lise Goujon, and Catherine Lara

Introduction

The notion of technical choices was introduced in the 1970s and 1980s to describe the great diversity of ways of doing things and their cultural dimension (e.g., Cresswell 1994, 1993; Latour and Lemonnier 1994; Lemonnier 1993, 1992). This diversity of ways of doing has given rise to multiple technical traditions—namely, "patterned ways of doing things that exist in identifiable form over extended periods of time" (O'Brien et al. 2010, 3797).

Technical traditions can last for millennia, as shown by archaeology (Manem, this volume). Intergenerational cultural transmission (from trainer to learner) and high-fidelity copying by the learner have been considered to be the mechanisms maintaining these traditions and thus participating in cumulative cultural evolution (references in Miton and Charbonneau 2018; Kempe, Lycett, and Mesoudi 2014; Dean et al. 2014; Henrich 2015). They describe the necessary social learning processes by which perceptual-motor skills are learned and mastered through interactions between learners and trainers within a given field of promoted action (Reed and Bril 1996; Reed 1993)—that is, within an environment offering differential opportunities (cultural, social, material) for action and resultant experience. This necessary interaction between learners and trainers results in the learners learning the same way of doing things as the trainers, albeit movements (to be distinguished from actions) are not imitated (Rein, Nonaka, and Bril 2014; Bril 2021), leading to so-called copying errors and variation in the finished products (Gandon, Roux, and Coyle 2014; Harush et al. 2020; Gandon et al. 2021). From this point of view, the learning process, whatever the teaching modalities, acts as a stabilizer of the prevailing cultural way of doing that contributes directly to the formation and maintenance of traditions. Furthermore, the trainer is usually chosen from the learner's social circle, which directly contributes to the superimposition of technological and sociological boundaries with the effect of possible polarization and non-adoption of technical traits between geographically close social groups (examples are given in Roux et al. 2017; Flache 2018).

On the other hand, technical traditions keep evolving through invention or adoption of exogenous traits and following different tempo (see examples in O'Brien and Shennan 2010). Transgressing technical traditions (i.e., changing ways of doing things) raises questions at both the individual and the collective scale. At the individual scale, these questions concern the capacities individuals rely on when inventing novel traits (see references in Burdett, Reed, and Ronfard 2020; Cutting, this volume; Gergely and Király, this volume), how they react to changing exogenous environmental pressures (see Pope-Caldwell, this volume) or, once a new trait is introduced, which learning strategies are involved in promoting their diffusion—for instance, biased selection of cultural traits such as prestige or conformist biases, as described in cultural evolutionary theory (O'Brien and Bentley 2011; Shennan 2011; Lycett 2015). In the domain of technology, field experiment studies (e.g., Roux, Bril, and Karasik 2018) have shown that only high-level experts tend to select new exogenous traits for their properties because they are the only ones able to aptly explore the properties of the task (Nonaka, Bril, and Rein 2010) and properly assess the benefits of the new technique. In other words, they are the only ones to have a causal understanding defined as "the ability to predict the effect of an intentional modification of a system" (Harris, Boyd, and Wood 2021, 1798). This causal understanding, also called technical reasoning—defined by Osiurak and Reynaud (2020, 3) as "the ability to reason about physical object properties"—enables the experts to overcome the constraints exerted by tradition and related cultural representations on how objects should be produced.

At the collective scale, anthropological and archaeological studies have shown that the social conditions required for technical innovations (i.e., for inventions to be adopted at the population scale) depend on whether they are mere technical variants or major technical inventions at the origin of new lineages of objects, characterized by new sources of energy and skills (Cresswell 1996). Technical variants are not necessarily determined by social motivations. Major inventions, however, occur along social mutations and related demands for new products (Cresswell 1994; Roux 2010). Sociological and anthropological studies also underline the role of the social network structure (relationships between communities and between individuals within a community) in the spreading of new technical traits (Astuti this volume; Manzo et al. 2018).

This being said, the *stability* of technical traditions also raises many questions at both the individual and collective scale. At the individual scale, the propensity to stick to learned ways of doing things deserves further investigation. This propensity would correspond to a "cognitive conservatism" demonstrated to prevent us from changing strategies as recalled by Pope-Caldwell (this volume), who suggests that under stable conditions, the individual assessment of costs and benefits is hampered by the known benefits of a socially acquired strategy (see also Tenpas, Schweinfurth, and Call, this volume). At the collective scale, further empirical studies are needed to better assess the role of a social network's structure in promoting the stability of technical traits (e.g., Ongaro, this volume; De Munck, this volume).

In this chapter, we argue that, when facing a demand for new objects, in all communities, there are experts capable of flexibly developing efficient adaptive strategies. However, this ability does not necessarily lead to changes at the collective scale if there is no demand for new objects. To test this hypothesis, we conducted field experiments with potters[1] in five different countries. The experiments were designed to evaluate individuals' ability to cope with new challenges (i.e., new shapes of vessels).

As we shall see, the results highlight individual adaptive behavior in the five communities. This adaptive behavior may take different forms, depending on the perceptual-motor workspace—involving the task, the environment (physical and social), and the actor (as defined by Newell 1986), together with the experience of the individual. Furthermore, the different achievements of the task reveal different degrees of expertise characterized by flexible *versus* more rigid skills. Flexibility is defined here as the ability to cope with changing circumstances and unexpected variations and to find "a motor solution for any situation and in any condition" (Bernstein 1967, 398). It involves motor resourcefulness. It allows for "adaptation, maneuverability, interchangeability" as discussed in detail by Bernstein (1967, 1996) and his followers (among others, Biryukova and Sirotkina 2020; Biryukova and Bril 2012; Biryukova and Bril 2008; Newell 1996). The flexibility developed in familiar situations during learning allows the high-level experts to successfully cope with new circumstances. In contrast, artisans with lower levels of expertise have more rigid skills (i.e., skills that do not allow them to adapt to varying situations). The difference in flexibility reflects different degrees in the understanding of task properties, as shown by research on expertise (technical skills in stonemasonry, hammering, music, dance, even walking) that have indicated a direct correlation between mastery of technical task constraints and understanding of task properties (Nonaka, Bril, and Rein 2010; Parry, Dietrich, and Bril 2014).

Highlighting adaptive capacities and variable levels of skills among any artisan community is a major result because there is very little empirical evidence of adaptive/inventive skills in real life. Furthermore, skill variability between individuals who have all undergone social learning within the same technological tradition has hardly been investigated, even though this interindividual variability has been acknowledged and recognized as important, in particular, in the debates about the dormant technical potential of innovative capacities waiting to be activated (Tennie et al. 2016; De Oliveira, Reynaud, and Osiurak 2019; Osiurak and Badets 2016; Osiurak and Reynaud 2020).

The role of the learning process in promoting technical stability is another major result. This process favors technological and social biases that causally associate, on the one hand, ways of doing things and finished products and, on the other hand, techniques and social identity. As we shall see, because of these two biases, as long as the same type of object is in demand, the individual only practices the technical strategy learned to make that type of object. Hence, periods of long technical stability correspond to periods during which no new demand emerged.

Methodology: Field Experiments

Field experiments constitute a compromise between laboratory experimentation and the observation of daily life situations. It involves the construction of an experimental situation that is based on tasks and environments that are familiar to the subject. The methodology, which is inspired by experimental psychology, must allow rigorous control of the parameters involved. It must permit a resolution of the dilemma presented by the combination of laboratory analysis and natural context. In the first case, the question that is raised is as follows: To what degree can we generalize the results obtained from simple tasks that are completely devoid of all cultural meaning to real situations in daily life? In the second case, the daily life situations are characterized by the great diversity of factors involved. This makes it difficult, if not impossible, to individualize the different underlying mechanisms through observation alone. The goal of field experimentation is thus to associate the advantages of the two types of situations (field and experiments) while trying to minimize the disadvantages and biases (Bril, Roux, and Dietrich 2005; Roux, Bril, and Dietrich 1995; Bril 1986).

The field experiments were conducted with potters from five countries: India, Ecuador, Kenya, Ethiopia, and Cameroon. The diversity of technological traditions marked by differences in manufacturing processes, learning regimes, and ceramic products should allow us to highlight behavioral invariants.

In the five countries, pottery is a specialized activity conducted on a domestic scale. The experiments took place in the potters' family yard, where the pots for sale are produced daily. During the experiments, the potters could talk with anyone or stop for a moment when someone called them on their phone or asked them something. The situation was as far as possible similar to an ordinary working day. The experience lasted more or less the whole day, and sometimes people from outside the family came to watch and say a few words. The potters were paid a little more than they would have earned working all day, so they considered the experiments as regular work that did not conflict with their daily income. Furthermore, as a general rule, all potters make pots for a living, and sometimes they have new shapes to make (when a new shape comes on the market and is in high demand). When this happens, each potter tries out the new form within the household. In this respect, the experiments did not seem strange to them. On the contrary, they made sense because they were related to their work and practice. What may have surprised them was the fact that the pots were not intended to be fired but cut in half, measured, and then recycled.

All the potters from the five countries followed the same protocol. This protocol was designed at first to assess the level of skill of the potters. The potters had to form one usual shape (the most popular vessel they usually made) and two new shapes, labeled E (for Easy) and D (for Difficult), considered as expressing increasing difficulties based on a taxonomy elaborated by Roux (1989). This taxonomy proposes a classification of shapes, depending on their throwing difficulty. The level of difficulty was further measured by an index computing the risks of collapse, depending on the geometrical properties of the pot (Gandon et al. 2011). Based on this index, shape E is moderately difficult and shape D is very difficult (figure 1.1). The rationale was that the higher the level of skill, the better the results in producing new shapes and adjusting the forming sequences.

We consider that these experiments are relevant, not only to assess the potters' level of skill but also to assess the adaptive/inventive behavior of potters when facing new situations.

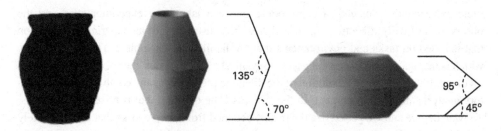

Figure 1.1
From left to right: shape of the usual pot made by the potters (may vary from one group to the other) and new shapes given as models: shape E (middle) and shape D (right). *Source*: Roux et al. 2018.

Participants

In each country, the number of participants was 16. They were selected to be representative of different levels of skill defined as such by the potters themselves, based on both technical and social criteria.

In India (Rajasthan, Jodhpur region), potters are male. They are distributed between 10 villages and two communities, Hindus and Muslims. Traditionally, Hindu potters specialized in storage jars of different sizes, while the Muslim potters specialized in kitchenware. Sometimes, older potters continue to make these traditional forms on request. Nowadays, they all specialize in the making of one type of jar only (a water jar) and use the same technical system, throwing their pots on the wheel. Transmission is vertical, from father to son. Rate of production is high. The potters participating in the experiments are between 40 and 66 years old.

In Ecuador, potters are women. They live in the Andes, in the villages of San Miguel and Taqil. San Miguel potters produce utilitarian vessels (mainly jars of different sizes, bowls, and tortilla dishes). Taqil includes two districts: Cachipamba and Cera (Lara 2016). The potters of Cachipamba tend to manufacture only culinary tableware. In contrast, the potters of Cera make a wider range of ceramics that are utilitarian and decorative. All the potters from San Miguel and Taqil make vessels using the same modeling/beating technique. Transmission is vertical, before marriage, from mother to daughter. Rate of production is quite high. The potters participating in the experiments are between 18 and 67 years old.

In Kenya, potters are women from the same ethnic group. They live in two villages separated by a muddy road, Kiriri and Ngararigeri (municipality of Ishiara, Tharaka Nithi County). The production includes mainly cooking pots and storage pots of different sizes (mainly for storing water and porridge). In Kiriri, they use the coiling by drawing technique following different processes; depending on the shape of the pot, the technique involves either starting with the upper part (the traditional method) or with the lower part (a new method). In Ngararigeri, they use only one way of making a pot, starting with the upper part. Transmission is vertical or horizontal, before or after marriage (with mother and sister-in-law). Rate of production is rather low. The potters participating in the experiments are between 40 and 79 years old.

In Cameroon, potters are men or women. They live in the north (Tikar plain). They are distributed between different ethnic groups and nine villages, all located along the main road over a 30-kilometer distance. All the potters make kitchenware. They use four main techniques—modeling by pinching, coiling by pinching, coiling by spreading, and modeling by drawing—depending on their ethnic group (Yamba, Mambila, Tikar, and Tumu). Transmission is vertical or horizontal, before or after marriage. The range of pots is mostly limited to storage jars. The rate of production is very low, and the craft is dying. The potters participating in the experiments are between 40 and 77 years old.

In Ethiopia, potters are women distributed between two ethnic groups, the Woloyta and the Oromo who live, respectively, in Goljjota and Qarsa (Oromiya region). Woloyta and Oromo make a wide range of utilitarian vessels. Certain types of vessels are common to both groups, but the manufacturing process differs between the two groups. The Woloyta use the modeling by drawing technique. The Oromo use the modeling by drawing technique or the coiling by drawing technique, depending on the vessels' shapes. Transmission is vertical or horizontal, before or after marriage. Rate of production is rather high. The potters participating in the experiments are between 18 and 47 years old.

The different pottery traditions are illustrated in figure 1.2.

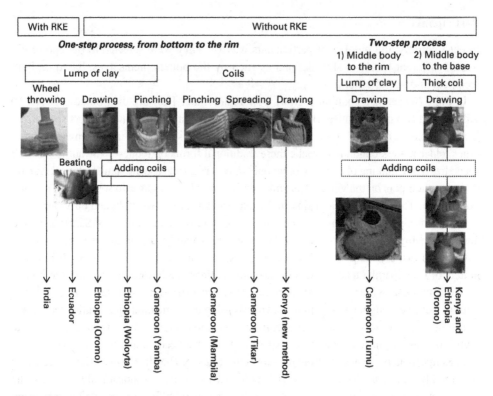

Figure 1.2
Forming techniques used by the potters from India, Ecuador, Ethiopia (Oromo and Woloyta), Cameroon (Yamba, Mambila, Tikar, and Tumu), and Kenya. RKE = Rotational kinetic energy. *Source*: ANR DIFFCERAM.

Experimental Protocol

Describing the Forming Strategies

The experimental protocol was designed to assess both the quality of the production and the strategy—described in terms of course of action, product dynamics, methods, and techniques—followed to achieve the new shapes.

The course of action describes the chaining of the elementary actions (i.e., the action dynamics). This is the scale at which it is possible to study how a global project is achieved through the execution of a succession of elementary actions.

Product dynamics describes how the raw material is transformed to make a finished product. It shows the strategy of the potters to reach the goal of their action.

The method is defined as an ordered sequence of functional operations for which different techniques can be used. Method comprises phases, stages, and operations, each of which can be achieved through different techniques. In the domain of pottery, there are three main forming phases for fashioning the body (lower part, upper part), the orifice (neck and rim), and the base. The fashioning of the body can be divided into two stages: the forming of the roughout (making a hollow volume) and of the preform (shaping the roughout in order to obtain the final geometric characteristics of the recipient). Fashioning methods are theoretically countless and more likely to be specific.

Techniques are defined as the physical modalities according to which raw material is transformed. Techniques are limited in number. In the domain of pottery, there are eight

Figure 1.3
Experimental setting: the model is placed next to the potter. *Source*: ANR DIFFCERAM.

roughing-out techniques without rotary kinetic energy (RKE) and one with RKE. There are 11 preforming techniques by pressure and percussion. These techniques are implemented according to methods, gestures, and tools whose description accounts for the diversity of the fashioning *chaînes opératoires* (Roux 2019).

The Experimental Procedure
The experiment started with the preparation of the clay and ended with the fashioning of the pot. The potters had to start with 30 kilograms of dry clay to make 18 pots (six pots per shape). Then they were free to add any amount of other necessary components, including water and temper. The different components were weighed before wedging. The potters were asked to prepare 18 lumps of equal weight for the 18 vessels to be made. Each of them was weighed. Fashioning then started.

The potters had to make six pots of similar size per shape. The fashioning sequence started with the usual shape, followed by shape *E* and ending with shape *D*. For shape *E* and *D*, the paperboard of the model was placed next to the potter so that they could refer to it while fashioning the pot (figure 1.3). Once fashioned, the vessels were put to dry. The whole manufacturing process—preparation of the clay paste, forming sequence—was videotaped (sampling rate 50 hertz) with fix plans. A calibration was used to allow for measuring the development of the shape of the pot during the shaping process.

Once the pots reached leather-hard consistency, they were photographed one after the other with a Canon camera from a few meters' distance. To get a clear silhouette, the pot was placed in front of a blue curtain. A calibration grid was used as a scale to measure the dimensions of the pots. Each pot was then cut in two parts, and measurements of the thickness of both sides of the profile were performed.

Data Analysis

The potters' performances were analyzed along two sets of data: the finished products and the manufacturing process.

For the finished products, general dimensions (absolute and relative) and section dimensions (thickness regularity) were recorded. The presence or absence of the carination and its angle, the presence or absence of a neck or a bulging rim, and the shape of the basis (flat or rounded) were noted.

The manufacturing process was analyzed from the videos in terms of forming strategies—namely, methods, techniques, course of action, and product dynamics (evolution of the carination angle). Changes in the manufacturing process (methods and techniques) were highlighted by reference to that used in regular production.

Results

Within the framework of this paper, we will focus on the forming strategies followed to make the new shapes and the quality of the products (i.e., degree of success in achieving the new shapes).

Forming Strategies

The forming strategies used to make the regular and new shapes are described for each country.

India

In India, the results show that the potters either carried out the same strategy for the regular shape and shapes *E* and *D* or adapted their strategy to the new shapes *E* and *D* (figure 1.4). In the first case, the carination was made only after the shaping of a rim, which implied numerous inefficient shaping operations before producing the carinated shape. In the second case, the carination was obtained immediately after a short thinning operation. This change of strategy is particularly relevant for successfully throwing the shape *D* whose ratio of height to maximum diameter is 0.5 against 1.35 for shape *E*.

Ecuador

In Taqil, out of nine potters (table 1.1), three potters used the traditional strategy for making the common shape and the new shapes *E* and *D* (by modeling and beating). Three potters used another strategy for shape *E*. It involves first making two open truncated cone shapes by modeling and beating them and then joining them together. It is regularly used by two potters in Cera when making high-necked jars. One potter of Cachipamba used this strategy as well but did not obtain two tronconic shapes. She obtained two convex shapes that, once joined, resulted in a carination turned inward. Six potters used a new strategy for shape *D* that involves making the lower part and the base by modeling and beating and the upper part by coiling (with various sizes of coils, depending on potters). This strategy is used regularly by three potters in Cera when making the neck of cooking pots or pitchers.

In San Miguel, all the potters used the traditional strategy for the regular shape and the new shapes *E* and *D* (by modeling and beating); see table 1.1.

Potters (age)	Usual shape	Shape E	Shape D
DJI (50), BHA (66), OMP (44), RAM (40), DAO (60), BOL (45), SAD (50)	Rim-body	Rim-body	
CHA (56), IQB (51), HAN (50), ASK (62), AHM (66), BIE (60)	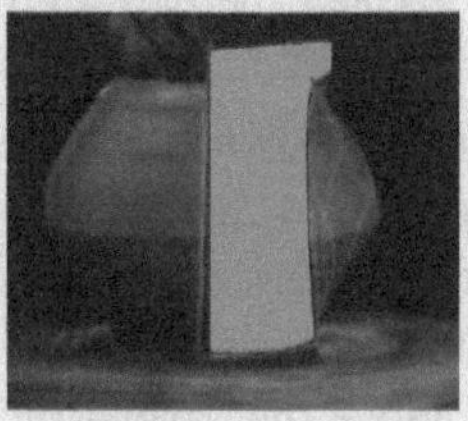	Body-rim	Body-rim 
JAG (54), BIK (60), GOV (50)		Rim-body	

Figure 1.4
Strategies followed by the Indian potters to make the usual shape and the shapes E and D. The carination is made either after forming a rim (rim-body), shown left, or immediately after a short thinning operation (body-rim), shown right. Potters are identified by their initials followed by their age. *Source*: ANR DIFFCERAM.

Table 1.1
Strategies followed by the potters of hamlets of Cera (CERA) and Cachipamba (CACH) in the village of Taqil, and the potters of hamlets of Pacchapamba (PAC) and Chico Ingapirca (CHICO) in the village of San Miguel, to make the usual shape and the shapes E and D

Potters (age)	Usual shape	Shape E	Shape D
Ecuador—Taqil			
CERA- DIPa (36), SOPa (18), ESPa (-)	Modeling by drawing-beating	Modeling-beating the lower part and coiling the upper part	Modeling-beating the lower part and coiling the upper part
CERA- CAL (45), LGU (62) **CACH-** SIN (45)		Separate modeling by drawing-beating of the lower and upper parts, joining them and piercing the opening	
CACH- REPa (35), RAPa (35) **CACH-** LIPa (65)		Modeling by drawing-beating	Modeling by drawing-beating
Ecuador—San Miguel			
PAC- JOP (67), SIM (45), MFE (66), AFE (58)	Modeling by drawing-beating		
CHICO- RH (50), CIN (39), JMO (57)			

Potters' identities are formatted as LOCATION- Potter's initials (age of potter). (-) is used when age is unknown.

Kenya

The traditional forming method, here called the "two-step process," consists of making round base pots in two steps, starting with the upper part from the maximum diameter toward the rim and, once the pot is turned upside down, finishing the base by drawing a thick coil placed on the maximum diameter. In 1997, a new method was introduced to make flat-base pots. It is called the "one-step process" method because the pots are made in one step from the base to the top.

Out of 16 potters (table 1.2), eight potters followed the same strategy for making the regular shape and the new shapes E and D; five of these potters used the two-step process and three potters used the one-step process. Six potters used a different strategy for shape E: they used either the one-step process (four potters), the one-step process method with the addition of a thick coil above the carination (one potter), or the two-step process method but with the addition of a disc to close the base (one potter). Eight potters followed a different strategy for shape D: either they used the one-step process (six), or they used the one-step process method but added a thick coil above the carination (two).

Table 1.2
Strategies followed by the potters of Ngararigeri (NG) and Kiriri (KI) to make the usual shape and the shapes E and D

Potters (age)	Usual shape	Shape E	Shape D
NG- ROS (40), ANN (76), EMI* (63), VEN (46), MAR* (40)	**Two-step process** Coiling by drawing (rim/base)	**Two-step process** Coiling by drawing (rim/base)	**Two-step process** Coiling by drawing (rim/base)
NG- LIL (79)			**One-step process**
NG- MAD (52)		**One-step process** Drawing one thick coil on a flat base and one thick coil above the carination	Drawing one thick coil on a flat base and one thick coil above the carination
NG- EST (50)		**Two-step process** Coiling by drawing (rim/base)	**One-step process** Drawing one thick coil on a flat base
KI- JAN* (56)		**Two-step process** Coiling by drawing, making the base by adding a disc	
KI- JAT* (67), ROS* (61), ONA*(50), MAR* (67)		**Two-step process** Drawing one thick coil on a flat base	
KI- AGN* (62), VID* (77), JUL* (54)	**One-step process** Drawing one thick coil on a flat base		

Note: * = potters knowing the new method ("the one-step process").

Potters' identities are formatted as LOCATION- Potter's initials (age of potter).

Cameroon

In Cameroon, out of 15 potters, 14 potters used the same strategy for the regular shape and the shapes E and D (table 1.3). Only one potter (HAB) changed strategy. She made the lower part of the shape E by modeling and the upper part by coiling; she made the lower and upper parts of shape D separately by modeling, and then she joined the two parts.

Ethiopia

The Oromo and the Woloyta use the modeling by drawing technique, from base to rim (i.e., the one-step process), but the Oromo only use this strategy to make small vessels. For bigger vessels, the Oromo use the coiling by drawing technique and, like the potters in Kenya, follow a two-step process, starting with the rim and finishing with the base. The Woloyta potters (eight) followed the same strategy for the regular shape and the new shapes E and D (table 1.4). Among the Oromo potters (seven), for shape E, three potters used the one-step process modeling technique, and one potter used the modeling technique for the lower part and the coiling technique for the upper part. For shape D, one potter followed the same strategy for the regular shape (coiling by drawing), two potters used the modeling technique, and four potters used the modeling technique for the lower part and the coiling technique for the upper part.

Explaining Variability of the Forming Strategies

When asked to make new shapes, the potters used strategies known to them, or else they used original strategies by combining known separate techniques.

In India, the potters who changed strategies for the new shapes and adopted an efficient strategy are those who knew and used this strategy to make old types of containers that are rarely in demand nowadays and whose shapes had a carination.

Similarly, in Ecuador, the potters of Taqil who changed strategies are from the hamlet of Cera, where modern forms are made (such as high-necked pitchers), for which new strategies were developed. They combined traditional and recent techniques for shapes E and D whose

Table 1.3
Strategies followed by the potters of Cameroon to make the usual shape and the shapes E and D

Potters (age)	Usual shape	Shape E	Shape D
FAN (63), SAB (73), LEC (71), FAA (62), FAT (56), MAE (46), HAD (59), ELH (76)	Coiling by pinching	Idem	Idem
MAM (65), YVO (62), CHR (40)	Coiling by spreading	Idem	Idem
MAR (55)	Modeling by drawing	Idem	Idem
MOU (77), HAM (68)	Modeling by pinching the lower part, coiling by pinching the upper part	Idem	Idem
HAB (63)	Modeling by drawing	Modeling by drawing the lower part, coiling the upper part	Modeling by drawing the lower and upper parts, joining them, piercing the opening

Potters' identities are formatted as Potter's initials (age of potter).

Table 1.4
Strategies followed by the Oromo and Woloyta Ethiopian potters to make the usual shape and the shapes E and D

Potters (age)	Usual shape	Shape E	Shape D
OR- YES (45)	**Two-step process** Coiling by drawing (rim/base)	**One-step process** Modeling by drawing (base/rim), fingerprinted carination	**Two-step process** Coiling by drawing (rim/base)
OR- SUK (29) OR- HIR (34)	**Two-step process** Coiling by drawing (rim/ base), plus coiling the neck		**One-step process** Modeling by drawing (base/rim), fingerprinted carination
OR- NAD (47), ARA (40), ENU (16), ALI (18)		**One-step process** Modeling by drawing the lower part, coiling the upper part	
WO- BUZ (26), SIS (34), ABE (42), ALM (32), ENA (30), TAR (29), TSI (30), YET (27)	**Two-step process** Modeling by drawing (base/neck)		

Potters' identities are formatted as LOCATION- Potter's initials (age of potter).

upper parts were assimilated to the high necks of the recently produced pitchers. The only potter from the hamlet of Cachipamba who changed strategies, SINCH, has family ties to one of the potters from Cera, and it can be assumed that they talked to each other about the experimentation without SINCH having seen them use the new strategies. Hence the surprising result: an inverted carination for shape E, testifying that oral instructions are not sufficient for learning new techniques. In San Miguel, none of the potters changed strategies, knowing that in this village, only one forming strategy is known.

In Kenya, all the potters of Kiriri changed their strategy to make shapes E and D, applying the new method used by all the potters to make flat-base pots (one-step process). The potter who added a disc to close the base during the forming of shape E actually combines the new method (making a flat bottom with a disc) with the traditional method (using a two-step process and closing the base when the pot is upside down). In Ngararigeri, where the new method is not practiced but is known through contacts with Kiriri potters, only three potters out of eight changed their strategy for making shapes E and D. They used a one-step method, but not the one used in Kiriri. Instead, they added a coil above the maximum diameter as in the traditional two-step method where the walls are made from a drawn coil.

In Cameroon, only one potter, HAB, changed her strategy for forming pots E and D. She is the only one who is familiar with different forming techniques, in this case modeling and coiling, having learned the former from her mother and the latter from her father, who was also a potter but from a different ethnic group.

In Ethiopia, only the Oromo changed their strategy for forming pots E and D. The Woloyta did not. Unlike the Woloyta, the Oromo use different forming techniques depending on the shape and size of the containers. The techniques used to make pots E and D (modeling, coiling) are those used for morphological types perceived as comparable.

Finished Products

The experimental products show significant variability across and within each community, although the two models to be shaped were the same. This variability is illustrated in figures 1.5 and 1.6 by two pots of types *E* and *D* made by two potters from each community. The pots shown in figure 1.5 were made by potters who are socially viewed as being highly skilled; the pots shown in figure 1.6 were made by potters who are socially viewed as having low-level skills. This variability applies to many aspects of the pots, such as the overall shape, the dimensions, the presence of a flat or a round base, the presence (or absence) of a carination or of a neck, the linearity of the profiles, the relative dimensions of the rim to the base or the maximum diameter to the height, the relative position of the carination to

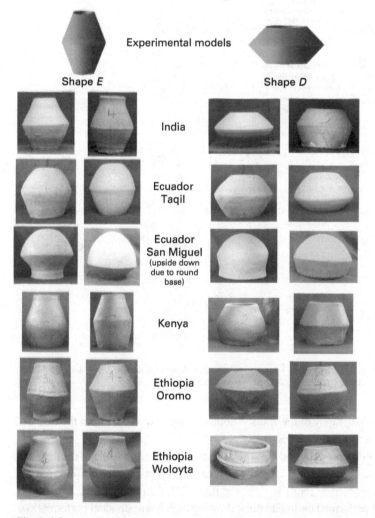

Figure 1.5
Examples of finished products made by potters who are socially considered highly skilled (to photograph the pots from Ecuador, these have been balanced upside down). *Source*: ANR DIFFCERAM.

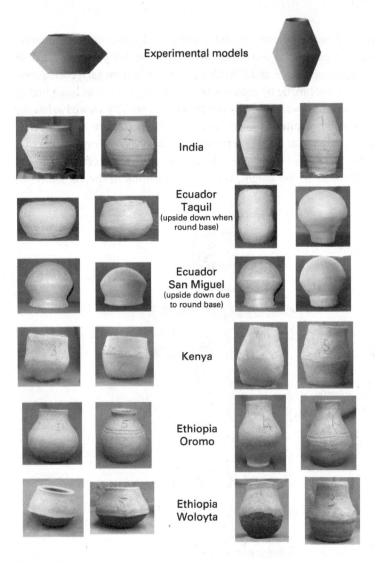

Figure 1.6
Examples of finished products made by potters who are socially considered to have low-level skills (to photograph the pots from Ecuador, these have been balanced upside down).

the height, or even the topography of the walls. Another feature, not visible from the outside, is the thickness of the walls and their regularity.

This important variability between and amid each community is also visible in the distance to the model, which has been analyzed in terms of relative proportions and profiles; for the Indian, Kenyan, and Ecuadorian (Taqil) vessels, see figure 1.7. A measure of the linearity of the walls has also been performed on Indian pots, showing high interindividual performances (Roux, Bril, and Karasik 2018).

Explaining Variability of the Finished Products

The important variability of the finished products can be explained by two main factors. The first factor has its origin in the diversity of technological niches (Stout, this volume),

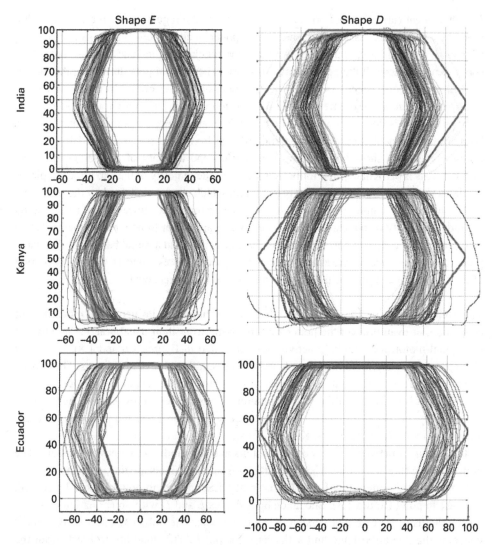

Figure 1.7
Normalized profiles of shapes E and D for comparison of relative proportions and profiles with the model (experimental pots from India, Kenya, and Taqil in Ecuador). *Source*: Figure provided by A. Karasik.

which here is defined in terms of technological traditions and their cultural, social, and material context. The second factor lies in differences in skill levels.

Variability explained by the diversity of technological niches The technological niche affects the performances of potters, either positively or negatively. Depending on what they have learned as well as their everyday practice, the potters testify to different perceptions of the salient properties of the models E and D (i.e., the striking carination especially for model D, the linearity of the profiles, the relative proportions, and the shape of the base).

The impact of the technological niche on the potters' performances is visible through the influence of the regular production on the experimental one. Thus, among the Woloyta, the regular pots are carinated and without neck. As a consequence, all the potters produced shape

D with a patent carination and no neck. In San Miguel, the regular pots are carinated with a round base and a neck. As a consequence, all the potters made experimental pots with a patent carination, a round base, and a neck. In Taqil, the regular pots include flat-base pots. As a consequence, almost all the potters made flat-base experimental pots. In Kenya, the regular pots have a globular shape. As a consequence, only one potter out of 16 made a carinated linear profile. Another extreme example may be found with DEJ, an Oromo potter whose daily production is limited to one specific type of container (incense burner). This potter has been absolutely unable to produce anything different from what she produces daily.

These examples show how cultural habits affect the capacity of adaptation to a new demand. This is particularly true for pots of shape *E*. The majority of the experimental pieces display a neck or a rim, more or less pronounced, which can be explained by the proximity of this shape to the common shapes found in most communities. However, this result should not hide the ability of potters to modify their habits and adapt to new requirements. While almost all of the common pots produced in the experiment had a round base, overall, a large proportion of the potters were able to produce flat bases for pots *E* and *D*, with the exception of the significant example of the potters from San Miguel (Ecuador).

Variability explained by skill levels   If everyday habits do influence the modalities of adaptation, or even prevent the potter from breaking from their usual practice in terms of perceptual-motor and cognitive activity (as illustrated in figures 1.6 and 1.8), skill level also appears as a key factor in the capacity to successfully achieve new forms. The experimental production of the potters does indeed show different levels of success testifying to different levels of skills within each community. These different levels of success can be measured by the distance to the model, the linearity of the profiles, and the thickness regularity (details in Roux, Bril, and Karasik 2018).

The distances between the experimental pots and the model vary between potters. The potters whose finished products are the farthest from the models are the ones who are used to producing only one type of pot. The case of YES, an Oromo potter of low-level skill, is a case in point (figure 1.8). She made both shapes *E* and *D* with no carination and with a neck that is just a different version of the usual pots. The only adaptation to the model concerns the overall volume and a flat base for pot *E*. This inability to break from the regular repertoire and develop adaptive skills is a sign of low-level skills and appears often with shape *E*, given its less salient features compared to the regular production.

The linearity of the profile of the pots is another significant parameter. Indeed, in India, this parameter shows significantly different values between high-level and low-level skilled potters (see Roux et al. 2018). Figures 1.5 and 1.6 suggest that such a variability in the linearity of the profiles also applies to the other communities.

The index of thickness regularity also shows interindividual variation within each community. The smaller the value, the more regular the wall of the pots and the more skilled the potter (table 1.5; see also Roux et al. 2018). At the scale of the community, the smaller the value, the more homogeneous the production; the larger the value, the more variation between the individuals of the community. The values obtained are quite large, which confirms the differences in skill levels between individuals and communities.

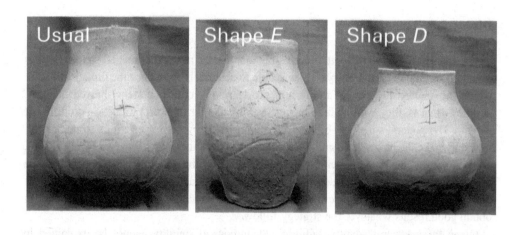

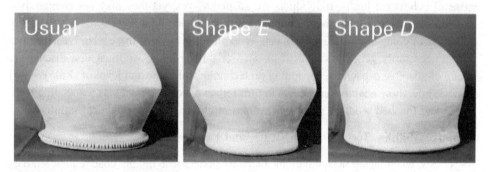

Figure 1.8
Examples of experimental pots produced by potters who have not been able to break from their technological traditions. Top row: YES, an Oromo potter of low-level skill, has not been able to carry out the carination. Bottom row: RH, a highly skilled potter from San Miguel, adapted the experimental shapes by lengthening or eliminating the neck (to photograph the pots from Ecuador, these have been balanced upside down).

Table 1.5
Index of regularity of wall thickness (mean, standard deviation, and coefficient of variation)

Country	Shape E				Shape D			
	nb of pots	Mean	σ	CV (%)	nb of pots	Mean	σ	CV
India	95	1,72	0,93	54%	96	1,61	0,55	34%
Ecuador-Taqil	54	2,14	0,77	36%	54	2,1	0,8	40%
Ecuador- San Miguel	42	2,05	0,83	53%	42	1,5	0,7	48%
Kenya	96	1,45	0,65	45%	96	1,65	0,78	47%
Cameroon	90	1,97	0.71	35%	84	1,93	0.94	49%
Ethiopia-Oromo	42	2,43	1,13	47%	42	2,30	1,01	44%
Ethiopia-Woloyta	48	1,76	0,71	40%	48	1,78	0,87	49%

Note: This index is computed as the square root of the sum of the squared differences of the thicknesses between the two profiles taken at different heights (rim, maximum diameter, base, and between each segment) and divided by the number of measures. A small value indicates small differences between the thicknesses of the right and left walls; a higher value indicates walls of irregular thickness. CV= coefficient of variation.

Discussion

Our results show that all the potters were able to make the new experimental shapes (shapes *E* and *D*), even though not everybody was able to accurately reproduce the models. In other words, no matter the way of doing and the learning context, potters showed adaptive abilities—that is, they were able to change task goals (the finished products)—to different extents. The potters' adaptive abilities are demonstrated by both the various strategies followed to make the new shapes and the resultant products, indicating different levels of skill, with the most skilled potters being more successful in being able to tackle change in shape independent of the adopted strategies. Indeed, it is not enough to change strategy to face the new situation to obtain good results. It also depends on skills. A highly skilled potter can obtain good results whatever the strategy followed.

Given these general adaptive abilities, technological stability cannot be explained in terms of behavioral limits. Individuals are adaptable, yes, yet they may not change their way of doing things during their lifetime; they can develop high-level skills but not adopt new ways of doing things. How, then, can we explain the fact that these adaptive capacities do not necessarily generate changes? To answer this question, we must first discuss the nature of adaptive capacities, depending on technological niche and level of skill. We will then see how these adaptive capacities in turn reflect a technological bias associating techniques and finished products and how this bias might play on technological stability.

Adaptive Abilities, Technological Niches, and Experience

As we saw, when facing new situations, there are potters who did not change their strategy and others who did. The former are the ones who knew only one strategy. The latter are those who are (or were) exposed to different strategies in their daily practice. As a result, either the potters used strategies known to them to produce shapes they perceived as similar to the new models we presented them with, or the potters used original strategies by combining known separate techniques (e.g., modeling a pot in two parts instead of one or combining modeling and coiling). In either case, a change in strategy depended on both the potter's technological niche (use of one or several strategies for making different types of pot) and experience (e.g., HAB in Cameroon, the daughter with parents who each practiced a different forming technique, or in Kenya, the social links between potters living close by). It also depended on the nature of the technique: with the wheel throwing technique, it is theoretically possible to change the method and the course of action but not the technique. On the contrary, with the techniques not using the rotary kinetic energy (i.e., modeling, coiling), it is possible to change the technique, the method, and the course of action.

These results are in line with the "variability of practice hypothesis" that shows that performance improves more with variable practice than with constant practice and enhances capacity of transfer (Huet et al. 2011; Pacheco and Newell 2015). Learning different shaping strategies for making various forms of pots develops the capacity to detect the information useful to differentiate the new shape from the usual one and to incorporate this information into the requirements for the new shaping task. It thus leads to an increased capacity to change the forming process when facing the production of a new form, which in turn explains that the original strategies combining different techniques were developed by two potters who use more than one strategy (in Cameroon, HAB; in Kenya, JAN).

Levels of Skill

Contrasting the strategies with the quality of the finished products shows that there is no direct correlation between a potter's level of skill and the choice to change of strategy. In each community, whatever the technical tradition and the strategy adopted, potters have succeeded or failed. This means that the way of doing things is not determinant in the success of the response to a new demand. On the contrary, the degree of skill is determinant. Whatever the strategy adopted, the more skilled the potter, the better the end-result. Those who succeeded in approaching the models can be qualified as highly skilled. Those who are less successful in making the new shapes can be qualified as low skilled. The former testifies to flexible skills. The latter testify to rigid skills (i.e., skills that do not allow adaptation to new situations).

The highly skilled potters have developed a level of dexterity that can be defined as the capacity to modulate one's behavior depending on shifting environments (here new forms). In this regard, they show adaptive flexibility, requiring perceptual attunement to key information sources, and a capacity to search for more varied and efficient movement solutions to fit the task dynamics.

The hypothesis of the necessity of flexible skills to reorganize the "skill structure" already acquired is supported by previous results obtained by field experiments with Indian potters in which the researchers analyzed the potters' hand position repertoire (Gandon and Roux 2019). The results suggest that the cost of motor skill adaptation to novel shapes depends on the potters' skill level (defined by the quality of the experimental products). This cost is high for potters with low-level skills and low for highly skilled potters. The latter mastered a wide range of hand positions and the flexible motor skills required to explore, through individual learning, how to make novel shapes. In comparison, the potters with low-level skills had a low range of hand positions and rigid motor skills. This hypothesis is also supported by the previous results obtained with Indian stone-bead knappers (Bril, Roux, and Dietrich 2005). Field experiments showed that low-level skilled artisans (characterized by low-quality production) had great difficulties in adapting to new raw material (glass instead of stone), revealing rigid skills as compared to high-level skilled artisans (characterized by high-quality production). Elementary movements were critical, arguing in favor of a rigidity found at the motor level (Biryukova and Bril 2008; Biryukova et al. 2015). It is this rigidity that led most of the highly skilled knappers not to train the low-skilled knappers (they are working in different workshops), considering that it would take at least two years to teach them how to handle the hammer correctly (i.e., mastering the elementary movements required to make thin flakes).

In sum, artisan communities are made up of individuals with different levels of skill. Accordingly, the cost of motor skill adaptation varies. Differences in skill levels can be explained by the dynamics of the system that depends on the characteristics of three sets of constraints: the person (the characteristics of the organism), the environment (the technological niche), and the properties of the task. Variation in any element of the system may change the dynamics, hence the development of different skill levels within any community.

Technological Stability

Technical traditions can last for millennia (Manem, this volume). The question is, therefore, how do we explain why technical traditions are stable, knowing that any community includes individuals who have adaptive/inventive capacities and have high-level skills?

Our experiments showed that the solutions found to fulfill the task involve selecting known forming strategies associated with shapes perceived as close to the models, whatever the skill level, which has then played a role in the success of the task. This selection depends on the shape of the pot and reveals a "technological bias" that may partially explain the tendency toward technological stability.

We define "technological bias" as a causal inference linking techniques and objects—that is, explaining the choice of a technique to make an object based on the properties of the object rather than on the properties of the technique: "I use this technique because I make this finished product" (Roux et al. 2017). It falls under the "covariation principle," which states that "if event A accompanies outcome B, and if event A is absent when outcome B is absent, then people tend to attribute A as the cause of B" (Carley 2001, 181; Kelley 1973, 1967). This principle is visible in the strategies followed by the different communities to make the experimental pots depending on the shape of the objects they knew—when a strategy is used for a given type of pot, then people attribute the type of pot as the cause of the strategy (e.g., Ecuador, Ethiopia). It can also be illustrated by several cases of reverse-engineering. Thus, in Ecuador, a potter who used to fire big jars in open firing married a man who made small objects and used a kiln. During the lifetime of her husband, this potter then used the kiln. When he died, she returned home, produced big jars again, and fired them in open firing technique, associating big jars with open firings and small objects with kilns (attributing the type of pot as the reason for the firing technique). In Cameroon, a potter of the Tikar ethnic group learned to make her vessels using the modeling technique. She then married a potter of the Mambila ethnic group who used the coiling technique. In order to help him, she learned the coiling technique, which she practiced during her husband's lifetime. When he died, she returned to her traditional production and her initial way of doing things, the one she first learned (attributing the type of pot as the reason for using the forming technique).

The association between finished products and forming strategies occurs in the course of the learning process. Let us first recall that learners construct their skills following a trainer's model. The model provides the method to follow. On the contrary, the mastery of the technique, which cannot be achieved through imitation (Byrne and Russon 1998; Bril 2021), requires lengthy repetition over several years for achieving tasks specific to each cultural niche (Ericson and Lehman 1996; Bril et al. 2010; Nonaka, Bril, and Rein 2010). This necessarily progressive acquisition of skills then creates a cognitive association between the finished products and the technical strategy.

This covariation principle between technique and finished product gets combined with another covariation linking techniques and social identity, which also gets created in the course of practice through the dyadic relationship involved in training (Lave 1991; Lave and Wenger 1991). The technological bias gets thus combined with a social bias ("I use this technique because I am from this social group"), reinforcing technical stability (Roux et al. 2017; see also Astuti, this volume; Ongaro, this volume).

At last, by reference to longue durée archaeological studies (Manem 2020), there is a covariation between the stability time of a trait and its learning duration: the longer the social learning of motor skills, the more stable the technical trait (Manem 2020). The phylogenetic tree shows that forming techniques are more stable than finishing techniques that require less complex motor skills. This can be understood in the light of, on the one hand, the cost of motor skill adaptation, which is much higher for forming techniques than for the other techniques involved in the manufacturing process, and on the other hand, the necessarily dyadic interactions involved in learning new forming techniques and that contrast with the indirect interactions that may suffice for learning the other techniques. (For examples of learning new clay recipes or new decorative techniques through indirect contacts, see Gosselain 2000 and Roux 2015).

In sum, the constraints imposed by the learning of motor skills, including both the repetition of elementary gestures over many years and a dyadic relationship between the learner and the trainer, create a technological bias linking techniques and finished products and a social bias linking techniques and community. These two biases lead the individuals to do the same things the same way. They explain that techniques have been repeatedly transmitted over multiple generations to form stable technical traditions, all social conditions being equal. In other words, they explain not high-fidelity copying—as we said in the introduction, studies have shown how copying errors can occur during the transmission process—but why once a way of doing things is learned, it is repeated over time, according to a "conformist bias" (as elaborated in Henrich 2015). To break with technical traditions and innovate, several conditions have to be met—among them, having highly skilled people capable of transgressing from their usual ways of doing/thinking.

Conclusion

The field experiments carried out in five different cultural contexts highlight that individuals are able to find new ways of doing objects whatever their technological niche (including modalities of learning). Their adaptive capacities are expressed both in the strategies followed and their level of skill. They reveal a technological bias linking strategies (techniques and methods) and finished products regardless of skill level. This technological bias favors technological stability, despite inventive abilities at the individual scale. It is all the stronger that it is combined with a social bias linking technology to social identity. These two biases, technological and social, explain the conformist bias that allows the reproduction of traditions without necessarily causal understanding (Harris, Boyd, and Wood 2021; Henrich 2021).

To question the conditions for change is therefore to question both the person capable of breaking with tradition and the social conditions favorable to the adoption of a new trait. In the field of ceramic technology, let us recall that only a few techniques were invented. For example, the potter's wheel was invented a few millennia after the appearance of ceramics, and the wheel throwing technique a few millennia later, testifying to both exceptional individual inventors and particular social conditions favorable to their adoption (Roux 2003, 2010).

Acknowledgments

This work was supported by ANR (French National Agency for Research) within the framework of the program known as CULT (Metamorphosis of societies—"Emergence and evolution of cultures and cultural phenomena"), and the DIFFCERAM project (Dynamics of spreading of ceramic techniques and style: actualist comparative data and agent-based modeling) (n°ANR-12-CULT-0001–01) (PI, V. Roux). The data were collected in Cameroon by G. Delebarre and E. Zangato, in Ethiopia and Kenya by A.-L. Goujon, in Ecuador by C. Lara, and in India by B. Bril and V. Roux. We thank all the potters in India, Ecuador, Kenya, Ethiopia, and Cameroon for their availability and their unfailing kindness.

Notes

1. For contingent reasons during fieldwork, the pictures of the pots produced by the potters from Cameroon could not be used.

References

Bernstein, N. A. 1967. *The Coordination and Regulation of Movements*. London: Pergamon.

Bernstein, N. A. 1996. "On Dexterity and Its Development." In *Dexterity and Its Development*, edited by M. L. Latash and M. T. Turvey, 1–244. New York: Psychology Press.

Biryukova, E. V., and B. Bril. 2008. "Organization of Goal-Directed Action at a High Level of Motor Skill: The Case of Stone Knapping in India." *Motor Control* 12 (3): 181–209.

Biryukova, E. V., and B. Bril. 2012. "Biomechanical Analysis of Tool Use: A Return to Bernstein's Tradition." *Zeitschrift für Psychologie* 220 (1): 53–54.

Biryukova, E. V., B. Bril, A. A. Frolov, and M. A. Koulikov. 2015. "Movement Kinematics as an Index of the Level of Motor Skill: The Case of Indian Craftsmen Stone Knapping." *Motor Control* 19 (1): 34–59.

Biryukova, E. V., and I. Sirotkina. 2020. "Forward to Bernstein: Movement Complexity as a New Frontier." *Frontiers in Neuroscience* 14 (553): 1–17.

Bril, B. 1986. "The Acquisition of an Everyday Technical Motor Skill: The Pounding of Cereals in Mali (Africa)." In *Themes in Motor Development*, edited by M. G. Wade and H. T. A. Whiting, 315–326. Dordrecht: Martinus Nijhoff.

Bril, B. 2021. "Cognition Demonstrated by Artifacts: Tool-Use Expertise and Tool-Use Learning." In *Psychology and Cognitive Archaeology: An Interdisciplinary Approach to the Study of the Human Mind*, edited by T. B. Henly and M. J. Rossano, 97–112. New York: Routledge.

Bril, B., R. Rein, T. Nonaka, F. Wenban-Smith, and G. Dietrich. 2010. "The Role of Expertise in Tool Use: Skill Differences in Functional Action Adaptations to Task Constraints." *Journal of Experimental Psychology: Human Perception and Performance* 36 (4): 825–839.

Bril, B., V. Roux, and G. Dietrich. 2005. "Stone Knapping: Khambhat (India), a Unique Opportunity?" In *Stone Knapping: The Necessary Conditions for a Uniquely Hominin Behaviour*, edited by V. Roux and B. Bril, 53–72. Cambridge: McDonald Institute for Archaeological Research.

Burdett, E., R. Reed, and S. Ronfard. 2020. "A Cognitive Developmental Approach Is Essential to Understanding Cumulative Technological Culture." *Behavioral and Brain Sciences* 43:20–21.

Byrne, R. W., and A. E. Russon. 1998. "Learning by Imitation: A Hierarchical Approach." *Behavioral and Brain Sciences* 21 (5): 667–684.

Carley, K. M. 2001. "Learning and Using New Ideas: A Sociocognitive Perspective." In *Diffusion Processes and Fertility Transition: Selected Perspectives*, edited by J. B. Casterline, 179–207. Washington, DC: National Academy Press.

Cresswell, R. 1993. "Of Mills and Waterwheels." In *Technological Choices: Transformation in Material Cultures since the Neolithic*, edited by P. Lemonnier, 181–213. New York: Routledge.

Cresswell, R. 1994. "La nature cyclique des relations entre le technique et le social. Approche technologique de la chaîne opératoire." In *De la Préhistoire aux Missiles Balistiques. L'Intelligence Sociale des Techniques*, edited by B. Latour and P. Lemonnier, 275–289. Paris: Editions La découverte.

Cresswell, R. 1996. *Prométhée ou Pandore? Propos de Technologie Culturelle*. Paris: Editions Kimé.

Dean, L. G., G. L. Vale, K. N. Laland, E. Flynn, and R. L. Kendal. 2014. "Human Cumulative Culture: A Comparative Perspective." *Biological Reviews* 89 (2): 284–301.

De Oliveira, E., E. Reynaud, and F. Osiurak. 2019. "Roles of Technical Reasoning, Theory of Mind, Creativity, and Fluid Cognition in Cumulative Technological Culture." *Human Nature* 30 (3): 326–340.

Ericson, K. A., and A. C. Lehman. 1996. "Expert and Exceptional Performance: Evidence from Maximal Adaptation to Task Constraints." *Annual Review of Psychology* 47:273–305.

Flache, A. 2018. "Between Monoculture and Cultural Polarization: Agent-Based Models of the Interplay of Social Influence and Cultural Diversity." *Journal of Archaeological Method and Theory* 25 (4): 996–1023.

Gandon, E., R. Casanova, P. Sainton, T. Coyle, V. Roux, B. Bril, and R. J. Bootsma. 2011. "A Proxy of Potters' Throwing Skill: Ceramic Vessels Considered in Terms of Mechanical Stress." *Journal of Archaeological Science* 38 (5): 1080–1089.

Gandon, E., T. Nonaka, T. Coyle, E. Coyle, R. Sonabend, C. Ogbonnaya, J. Endler, and V. Roux. 2021. "Cultural Transmission and Perception of Vessel Shapes among Hebron Potters." *Journal of Anthropological Archaeology* 63: 101334.

Gandon, E., and V. Roux. 2019. "Cost of Motor Skill Adaptation to New Craft Traits: Experiments with Expert Potters Facing Unfamiliar Vessel Shapes and Wheels." *Journal of Anthropological Archaeology* 53:229–239.

Gandon, E., V. Roux, and T. Coyle. 2014. "Copying Errors of Potters from Three Cultures: Predictable Directions for a So-Called Random Phenomenon." *Journal of Anthropological Archaeology* 33:99–107.

Gosselain, Olivier P. 2000. "Materializing Identities: An African Perspective." *Journal of Archaeological Method and Theory* 7, no. 3 (2000): 187–217.

Harris, J. A., R. Boyd, and B. M. Wood. 2021. "The Role of Causal Knowledge in the Evolution of Traditional Technology." *Current Biology* 31 (8): 1798–1803.e3.

Harush, O., V. Roux, A. Karasik, and L. Grosman. 2020. "Social Signatures in Standardized Ceramic Production: A 3-D Approach to Ethnographic Data." *Journal of Anthropological Archaeology* 60: 101208.

Henrich, J. 2015. *The Secret of Our Success: How Culture Is Driving Human Evolution, Domesticating Our Species, and Making Us Smarter*. Princeton, NJ, NJ: Princeton University Press.

Henrich, J. 2021. "Cultural Evolution: Is Causal Inference the Secret of Our Success?" *Current Biology* 31 (8): R381–383.

Huet, M., D. M. Jacobs, C. Camachon, O. Missenard, R. Gray, and G. Montagne. 2011. "The Education of Attention as Explanation of Variability of Practice Effects: Learning the Final Approach Phase in a Flight Simulator." *Journal of Experimental Psychology: Human Perception and Performance* 37 (6): 1841.

Kelley, H. H. 1967. "Attribution Theory in Social Psychology." *Nebraska Symposium on Motivation* 15:192–238.

Kelley, H. H. 1973. "The Processes of Causal Attribution." *American Psychologist* 28 (2): 107–128.

Kempe, M., S. J. Lycett, and A. Mesoudi. 2014. "From Cultural Traditions to Cumulative Culture: Parameterizing the Differences between Human and Nonhuman Culture." *Journal of Theoretical Biology* 359:29–36.

Lara, C. 2016. "Traditions céramiques et occupation précolombienne du piémont oriental des Andes équatoriennes: le cas de la vallée du fleuve Cuyes." PhD diss., Université de Paris Nanterre.Latour, B., and P. Lemonnier. 1994. *De la Préhistoire aux Missiles Balistiques. L'Intelligence Sociale des Techniques*. Paris: Editions La découverte.

Lave, J. 1991. "Situating Learning in Communities of Practice." In *Perspectives on Socially Shared Cognition*, edited by L. B. Resnick, J. M. Levine, and S. D. Teasley, 63–82. Washington, DC: American Psychological Association.

Lave, J., and E. Wenger. 1991. *Situated Learning: Legitimate Peripheral Participation*. Cambridge: Cambridge University Press.

Lemonnier, P. 1992. *Elements for an Anthropology of Technology*. Ann Arbor: Museum of Anthropology, University of Michigan.

Lemonnier, P. 1993. *Technological Choices: Transformation in Material Cultures since the Neolithic*. New York: Routledge.

Lycett, S. J. 2015. "Cultural Evolutionary Approaches to Artifact Variation over Time and Space: Basis, Progress, and Prospects." *Journal of Archaeological Science* 56:21–31.

Manem, S. 2020. "Modeling the Evolution of Ceramic Traditions through a Phylogenetic Analysis of the *Chaînes Opératoires*: The European Bronze Age as a Case Study." *Journal of Archaeological Method and Theory* 27:992–1039.

Manzo, G., S. Gabbriellini, V. Roux, and F. Nkirote M'Mbogori. 2018. "Complex Contagions and the Diffusion of Innovations: Evidence from a Small-N Study." *Journal of Archaeological Method and Theory* 25 (4): 1109–1154.

Miton, H., and M. Charbonneau. 2018. "Cumulative Culture in the Laboratory: Methodological and Theoretical Challenges." *Proceedings of the Royal Society B: Biological Sciences* 285 (1879): 20180677.

Newell, K. M. 1986. "Constraints on the Development of Coordination." In *Motor Development in Children: Aspects of Coordination and Control*, edited by M. G. Wade and H. T. A. Whitingm, 341–360. Dordrecht: Martinus Nijhoff.

Newell, K. M. 1996. "Changes in Movement and Skills: Learning Retention and Transfer." In *Dexterity and Its Development*, edited by M. L. Latash and M. T. Turvey, 393–429. New York: Psychology Press.

Nonaka, T., B. Bril, and R. Rein. 2010. "How Do Stone Knappers Predict and Control the Outcome of Flaking? Implications for Understanding Early Stone Tool Technology." *Journal of Human Evolution* 59 (2): 155–167.

O'Brien, M. J., and R. A. Bentley. 2011. "Stimulated Variation and Cascades: Two Processes in the Evolution of Complex Technological Systems." *Journal of Archaeological Method and Theory* 18:309–337.

O'Brien, M. J., R. L. Lyman, A. Mesoudi, and T. L. VanPool. 2010. "Cultural Traits as Units of Analysis." *Philosophical Transactions of the Royal Society B: Biological Sciences* 365 (1559): 3797–3806.

O'Brien, M. J., and S. J. Shennan. 2010. "Issues in Anthropological Studies of Innovation." In *Innovation in Cultural Systems: Contributions from Evolutionary Anthropology*, edited by M. J. O'Brien and S. J. Shennan, 3–17. Cambridge, MA: MIT Press.

Osiurak, F., and A. Badets. 2016. "Tool Use and Affordance: Manipulation-Based versus Reasoning-Based Approaches." *Psychological Review* 123 (5): 534.

Osiurak, F., and E. Reynaud. 2020. "The Elephant in the Room: What Matters Cognitively in Cumulative Technological Culture." *Behavioral and Brain Sciences* 43:1–66.

Pacheco, M. M., and K. M. Newell. 2015. "Transfer as a Function of Exploration and Stabilization in Original Practice." *Human Movement Science* 44:258–269.

Parry, R., G. Dietrich, and B. Bril. 2014. "Tool Use Ability Depends on Understanding of Functional Dynamics and not Specific Joint Contribution Profiles." *Frontiers in Psychology* 5 (306): 1–15.

Reed, E. S. 1993. "The Intention to Use a Specific Affordance: A Conceptual Framework for Psychology." In *Development in Context: Acting and Thinking in Specific Environments*, edited by R. Wosniak and K. Fisher, 45–76. New York: Psychology Press.

Reed, E. S., and B. Bril. 1996. "The Primacy of Action in Development. A Commentary of N. Bernstein." In *Dexterity and Its Development*, edited by M. Latash and M. T. Turvey, 431–451. New York: Psychology Press.

Rein, R., T. Nonaka, and B. Bril. 2014. "Movement Pattern Variability in Stone Knapping: Implications for the Development of Percussive Traditions." *PLOS One* 9 (11): e113567.

Roux, V. 1989. "Development of a Taxinomy to Measure Throwing Difficulties of Prehistorical and Protohistorical Ceramic Vessels." In *The Potter's Wheel: Craft Specialization and Technical Competence*, edited by V. Roux, 93–145. New Delhi: Oxford and IBH Publishing.

Roux, V. 2003. "A Dynamic Systems Framework for Studying Technological Change: Application to the Emergence of the Potter's Wheel in the Southern Levant." *Journal of Archaeological Method and Theory* 10:1–30.

Roux, V. 2010. "Technological Innovations and Developmental Trajectories: Social Factors as Evolutionary Forces." In *Innovation in Cultural Systems: Contributions from Evolutionary Anthropology*, edited by M. J. O'Brien and S. J. Shennan, 217–234. Cambridge, MA: MIT Press.

Roux, V. 2015. "Standardization of Ceramic Assemblages: Transmission Mechanisms and Diffusion of Morpho-Functional Traits across Social Boundaries." *Journal of Anthropological Archaeology* 40:1–9.

Roux, V. 2019. *Ceramics and Society: A Technological Approach to Archaeological Assemblages*. Cham: Springer Nature Switzerland.

Roux, V., B. Bril, J. Cauliez, A.-L. Goujon, C. Lara, C. Manen, G. de Saulieu, and E. Zangato. 2017. "Persisting Technological Boundaries: Social Interactions, Cognitive Correlations and Polarization." *Journal of Anthropological Archaeology* 48 (4): 320–335.

Roux, V., B. Bril, and G. Dietrich. 1995. "Skills and Learning Difficulties Involved in Stone Knapping: The Case of Stone-Bead Knapping in Khambhat, India." *World Archaeology* 27 (1): 63–87.

Roux, V., B. Bril, and A. Karasik. 2018. "Weak Ties and Expertise: Crossing Technological Boundaries." *Journal of Archaeological Method and Theory* 25 (4): 1024–1050.

Shennan, S. J. 2011. "Descent with Modification and the Archaeological Record." *Philosophical Transactions of the Royal Society B* 366:1070–1079.

Tennie, C., D. R. Braun, L. S. Premo, and S. P. McPherron. 2016. "The Island Test for Cumulative Culture in the Paleolithic." In *The Nature of Culture*, edited by M. Haidle, N. J. Conard, and M. Bolus, 121–133. Dordrecht: Springer Netherlands.

2 Innovation and Social Identity in Madagascar

Rita Astuti

Introduction

This chapter is about a change in the way the Vezo of Betania, a fishing village in Madagascar, rig their canoes—that is, the way they set the sail. I describe the change, and I report how villagers assessed its pros and cons. I then discuss a conjecture, put forward by one individual, as to how this change first came about, as well as how several villagers described the mechanism by which it became widespread.

The chapter is ethnographic. It documents something that happened over the course of my repeated fieldwork visits and that I initially documented almost by chance, by taking a photo and then realizing that what I had photographed was a piece of canoe equipment that I had not seen before. At the time, I was pursuing other research leads and didn't follow up on this accidental observation. Only much later, when almost the entire village had adopted the new way of rigging, did I join the conversation about the pros and cons of the new technique and ask questions about its origins. I mention this to draw the reader's attention to the fact that, unlike the authors of most of the other chapters, I will be addressing questions that were of interest to my interlocutors, using their language and concepts. This matters because what I document in this chapter is not just a technological change but how this change was explained, discussed, and evaluated by the agents concerned. These explanations, discussions, and evaluations are part of the cultural and social environment that determines whether a technical innovation is accepted or rejected, and thus they have a bearing on the questions addressed by this volume.

In what follows, I shall refer to the change that took place in the rigging of the canoe as an innovation. I do so because the people who adopted it came to regard the new way as an improvement on the old way. And I situate the discussion of this technical innovation in the context of local ideas about social identity because Vezo of Betania say that the canoe is the very "root of their Vezo-ness" (Astuti 1995).[1] In other words, unlike so many other *ethnic* groups of people who define themselves by reference to their ancestry and the traits that are inherited through it, my Vezo interlocutors define who they are (their being Vezo) by reference to the most important tool for their livelihoods and what they do with it: sail, fish, and trade along the coast. Given this pivotal role in determining group identity, one might have expected Vezo people to want to protect the "root" from significant change—that is, to display rigidity in the way they maintain and pass on their sailing techniques. As we shall see, this is not the case (compare with Ongaro, this volume). I will return to the significance of this in the conclusion after presenting the details of the case.

Technical Innovation: The Sequence of Observed Events

During my first period of research—between November 1987 and June 1989—people in Betania rigged their canoes with *tehy mitsanga* (literally: standing poles). I will refer to this type of rig as a *double sprit sail* (figure 2.1a): a rectangular sail held up by two sprits (poles)

(a)

(b)

Figure 2.1
(a) Canoes with double sprit sails (1988). 1: sprits; 2: ties at the outrigger boom. (b) Canoe with common sprit sail (1994). 1: mast; 2: sprit; 3: rope sling. *Source*: Photo courtesy of Maurice Bloch, 1988.

that were tied to the outrigger boom and rested inside a holder at the bottom of the hull (which had six holes, four along the long axis and two along the short axis). Changing the position of the sail required undoing the knot, lifting one or both of the sprits, repositioning them in the appropriate hole, and retying them to the outrigger boom. Quite a laborious operation.

In January 1989, when I visited Belo-sur-Mer, a village about 60 kilometers south of Betania, people were using a different rig, which they referred to as *tehy mihanto* (literally hanging pole). I will refer to this type of rig as a *common sprit sail*.[2]

This sail is rigged on one movable mast that is tied to the boom of the outrigger and rests inside a holder at the bottom of the hull (which has two holes along the long axis) and one sprit (which, suspended in a sling made of rope, pivots around the stationary mast). This rig makes it possible to change the position of the sail without any unknotting, lifting, and reknotting—a much easier operation that, as we shall see, affords new maneuvers (tacking) that were not possible with the double sprit sail.

When I returned in 1994, I noticed that some villagers in Betania had adopted the common sprit sail; my host father, Gramera, was one of them (figure 2.1b). My visit was short and my ethnographic attention was elsewhere, and I therefore didn't investigate the issue. By the time of my visit in 1998, virtually all Betania villagers had abandoned the double sprit sail and had adopted the new rig, whether for fishing or local transportation. But the fact that some were still using the double sprit sail meant that I was able to discuss with people the reasons for their diverging preferences.

By 2004, the topic was no longer of interest as the new rig had become universal.

Pros and Cons (as Discussed in 1998)

According to those who had adopted it in 1998, the main advantage of the common sprit sail is that it makes sailing very easy (*mora mare*). This is because, as noted above, it does away with the need to reposition the sprits in response to either a change in the wind or in the desired direction of travel. In both cases, the sailors only have to pull on the ropes and let the pivoting sprit move to the new required position. As people put it (using the French word), the new technique makes the process of sailing *automatique*.

Because of this automaticity, the common sprit sail makes tacking (what Vezo describe as turning this way and that, zigzagging into the wind) a realistic option. With the old rig, tacking was not possible because each change in direction would have required manually and arduously moving the sprits. Instead, people were forced to take the sail down and paddle.

The new rig has another advantage, which is that the sail can be kept almost parallel with the hull (i.e., close-hauled), which allows sailing, as much as is physically possible, against the wind. This was not an option with the old system because the two sprits, both tied to the outrigger boom, could not be aligned along the axis of the hull. Notably, this is not an advantage that my interlocutors pointed out explicitly, probably because it offers improved tacking, which, under the old system, was not even attempted.

So, what are the disadvantages? The main reason some people resisted the new rig was that while it makes sailing *easier*, it also makes it more *dangerous*. If there is a sudden change in the wind direction, the suspended sprit is out of the sailors' control. This means that the canoe is at a higher risk of capsizing or breaking up. In addition, in strong winds, the sling

Figure 2.2
A close-up of the pivoting sprit and its supporting sling (1998). *Source*: Photo taken by Rita Astuti, 1998.

that holds the sprit (see figure 2.2) comes under a lot of pressure; if it breaks the consequences are catastrophic. The hull cracks, and in one instance, a man was reported to have been hit by the sprit and to have died as a result. The sprit itself can also break under the increased pressure, and so can the mast (but see next paragraph for a different view). People complained that the increase in the theft of poles was due to all these breakages.

Those who use the new rig disagreed. Badiga, the man who was first to adopt it in Betania, remarked that, like death, capsizing is something that just happens ("capsizing, like death, is not selective"). In more technical fashion, he also suggested that responding to a sudden change in wind is far more dangerous with the old rig than the new one. This is because one has to quickly lift and reposition the two sprits, with the risk of losing control over them, with equally catastrophic consequences. When making this point, he showed me his foot. Many years back, when he was adjusting the position of an original, double sprit rigging, one of the sprits had slipped from his hands and landed on his foot and crushed it. His damaged foot is proof, he said, that the new system is better than the old one. Another man pointed out that the reason some people think that the new rig causes the mast and the sprit to break is that they recycle the old poles instead of using new purpose-built ones appropriate for the new system.

As I mentioned, by the time I was having these conversations in 1998, the majority of people had adopted the new system. Those who hadn't seemed keen to inflect their decision with a moral valence. For example, one elder remarked that with the new technique, the wind becomes the "owner" of the canoe, and he doesn't like that—he would much rather be the one in control. Another remarked that the reason the new technique is popular is that it requires less effort; he tried it but went back to the old rigging because, when he sails, he wants to use his own strength. Another common remark was that the new rigging technique was particularly trendy with young men who like to show off their fast sailing, even if it means taking reckless risks.

As far as I could tell, such personal preferences were the main factors in determining whether people decided to drop the old rig for the new one. When I inquired about the costs of transitioning from one system to the other, some reported that there were no costs since they had used the old equipment (sail, sprits, and sprit holder) and that they only had to add a few ropes. Others argued that to use the new rig properly and safely, a new set of poles was needed and that the sail had to be repurposed, rotating it by 90 degrees.

All in all, my overall impression was that the new system did not have insurmountable entry costs that would have stopped people transitioning from old to new. Having said this, in at least one case—the case of Badiga who, as mentioned, was the first to adopt the new rigging technique in Betania—the impetus for change came when an external event disrupted his equipment. I will have more to say about this man's reflections on how the new system came about, but here I am interested in what he said about the moment when he decided to make the transition, having had plenty of time to observe it and discuss it with people who were using it already. The moment came when a cyclone (in his recollection in 1989) buried his canoe into the sand. At that point, as he set out to replace it, he decided to cut a new mast and a sprit and to start sailing in the new way.

Where Did the New Idea Come From?

In an article on "The Evolution of Pacific Canoe Rigs," Adrian Horridge writes that all variations on the common sprit sail scattered across the Pacific region (e.g., China, Philippines, Indonesia) were "certainly introduced independently to many places by western colonists and I know of no case that can be shown to be indigenous" (1986, 92).

When asked where the new rigging technique came from, most Betania villagers pointed to the south. No one named a particular person, but they all agreed that the people they had copied the technique from had themselves copied it from other people further to the south. I should note here that all along the coast, villagers are connected by extensive kinship links. This means that they have many reasons to travel north or south for funerals, ancestral rituals, and celebrations of one kind or another, in addition to fishing and trading. They thus have plenty of opportunities to observe localized "ways of doing things" (*fomba*), including sailing.

While the southern origin of the new rig is universally recognized by the people of Betania, Badiga (the man we encountered earlier) offered a more complex account of how he believes it came about, which very much aligns with Horridge's claim about its Western origins. As a prelude to the story, I need to briefly introduce the schooners that sail up and down the western coast of Madagascar. Schooners (known as *botsy*) are built locally and are used to transport cargo and people (figure 2.3). They were introduced by a Breton family of shipbuilders, the Joachims, who came to Madagascar via Réunion at the request of the Merina King Radama II (in power from 1861 to 1863) as part of his pro-European policies and treaties. After Radama's assassination in 1863, the Joachim family made their way to the west coast and eventually established themselves in Morondava, Morombe, and Belo-sur-Mer, where they opened boatyards and taught the locals how to build, rig, and sail European-style schooners.

The history of how this knowledge came to be owned and passed on by Vezo villagers is outside the scope of this chapter; all that matters here is that schooners, a Western introduction, are fore-and-aft rigged; that is, the sails are set along the direction of the boat's hull. With this in mind, let me return to Badiga's account (which I paraphrase). He speculated that the person who first thought of changing the way the canoe was rigged must have reasoned as follows: schooners cannot be propelled by paddling, and yet they manage to move even if the wind doesn't blow in their favor; they do this by zigzagging until they get to their destination. By contrast, in such unfavorable wind conditions, a canoe doesn't move at all, and people have to paddle to get to where they want to go. He thus suggested that the rig of the canoe was changed to mimic that of a schooner—that is, by turning one of the sprits into a mast and adopting a fore-and-aft rig. He concluded that with the new rig, the canoe

Figure 2.3
A schooner partly rigged (2017). *Source*: Photo taken by Rita Astuti, 2017.

became "like" a schooner: it can zigzag just like one, with the added advantage that, unlike a schooner, if there is no wind at all, one can still paddle and move on.

Badiga did not offer any detail of how this person went from his "thought" to the implementation of the new rigging system—like many others, Badiga asserted that there is no "history" about how this happened. But he suggested that the new technique must have developed in the stretch of coast near Andavadohaka (about 250 kilometers south of Betania), where there are lots of small islands scattered along the coast. Unlike in Betania, where people can sail in a straight line out in the open sea, in the Andavadohaka region, people have to make their way around these islands. And to do this by sailing, rather than paddling, they have to be able to use, and easily switch between, all available points of sail: run, reach, and close-hauled. According to Badiga, these local conditions prompted the idea of rigging the canoe in the new way, turning it into something of a schooner.

How Did It Become Widespread?

Badiga admitted that when he first saw the new rig, he didn't like it. He was put off by the fact that there were far too many ropes compared with the old system. But then, after observing people using the new rig, he started to see "what makes it a good thing," adding, "What's good becomes something that people are okay doing, and it becomes something that people will learn." Thus, while the old system was good for "keeping oneself alive" (i.e., to support one's livelihood), the new system is much better than that: no need to struggle; one just sits at the back of the canoe, steering it along. Badiga's story was one of progress (although he didn't use that term).

As we saw earlier, by 1998, most people in Betania had adopted the new rig. Badiga had led the way, and everyone recognized him as the local origin point of the new system. However, when I shared Badiga's speculation about the origin of the new rig with Gramera, my host father, he strongly disagreed (in part because he misunderstood it as a claim on Badiga's part to have been the one who came up with the idea, which wasn't the case). Gramera insisted that the people of Betania, including Badiga, had just copied what other people were doing elsewhere. This is not something that "comes out of somebody's head," he added; it is not anybody's thought but is just something that people "copy" from other people.

In this, Gramera articulated a view that was shared by all the people I talked to. They all stressed the ease with which they were able to copy by "just watching." Badiga agreed with this account. He told me that after overcoming his initial resistance and having come to appreciate the advantages of the new rig, all he did was to discuss it with a relative who was already using it. And when I inquired whether he had asked the relative to take him out sailing on his canoe so that he could learn the new technique, he replied no, that it was enough for him to just watch the way the canoe was rigged. Another elder who, in 1998, was still contemplating whether to adopt the new system, told me that he didn't need anyone to teach him how to use it; all he needed was to watch, and he would know.

The claim that people learned the new technique simply by watching others resonates with the local folk-sociological theory that anyone, irrespective of their ancestry, can become Vezo by learning the knowledge that makes people Vezo—namely, sailing, fishing, trading fish, building a canoe, and so on. My adult interlocutors say that learning to be Vezo, which is what children must do since they are not born Vezo, is not something that requires Western-style schooling. Instead, children are taken out to sea; they initially get seasick and throw up; they then get used to it; and when they are used to it, they watch and just learn to do things. This shared narrative, which is also applied to newcomers from the interior who have never before been anywhere near the sea, is a cornerstone of the local folk-sociology: it is only because people emphatically say that Vezo knowledge is so easy to acquire that they can sustain the idea that anyone can become Vezo.

In reality, things are more complicated than the narrative suggests. Children do learn a great deal by watching, but they also learn by playing with toy canoes and schooners (figure 2.4), by listening to stories about sailing and fishing, by being told what to do, and by being told off when they make mistakes. And while newcomers do become Vezo, they require guidance and mentorship, typically by Vezo kin or in-laws who are prepared to take them out to sea with them.

Still, in the case of the acquisition of the new rigging technique, my impression was that people did in fact copy it from others without much guidance. This is plausible since what they were copying was not something radically new but more like a variation on what they already knew about sailing and about the wind, the sail, the ropes, the masts, the hull, the outrigger, and so on. Thus, for example, when Badiga discussed the new rig with his relative, he will have used a shared vocabulary and tapped background knowledge that made the conversation meaningful and helpful. While I did not witness this particular exchange,

Figure 2.4
Vezo children playing with a model schooner (2017). *Source*: Photo taken by Rita Astuti, 2017

I heard countless stories (which often went over my head) about sailing and fishing expeditions—stories full of details about unexpected wind conditions and what the sailing team had done about it. It is this shared knowledge that must have allowed Badiga to ask just the right questions about the new rig and to receive just the right answers.

The point here is that the introduction of the new rigging technique required people to incorporate new features into a well-known technical system, even if that involved changing that system quite radically (e.g., transforming one sprit into a mast). I suspect that this is why I didn't encounter much anxiety about the loss of the old system. Some mentioned that it would be a good idea to show children how the old system works, in case they have problems with the new way and want to revert back; by contrast, Badiga thought that there is no reason to teach children about the old way, since the new one is so much better.

Innovation and Social Group Identity

If one asks Vezo people why they do a certain thing in a certain way, one is likely to receive the following answer: it's because this is the *fomba*, the way of doing things. Only superficially this is a circular answer (I am doing X in this way because doing X in this way is the way of doing it). In fact, the term *fomba* signals that the way of doing something is customary (rigid, not flexible) and that it is not open to much negotiation, if at all.

Still, depending on the context, invoking *fomba* may be a lazy way of stopping the conversation and discouraging any further questions, while in fact there are reasons that people can easily spell out (if they are so minded) to explain why they do what they do in that way. For example, while it is undoubtedly the case that painting the bottom part of the canoe's hull with tar (see figures 2.1a and 2.1b) is *fomba*, in the sense that it is a well-established and shared practice that has been passed down from previous generations, there is a perfectly obvious reason that people can easily articulate to explain why that particular *fomba* exists—that is, it is a way of making the hull waterproof.

By contrast, there are contexts in which invoking *fomba* is really all that people can muster. This is because the only reason they do that thing in that particular way *is* that it is *fomba*. This applies to the details of all ritual practices that have been laid out by the ancestors, as well as to the numerous taboos the ancestors imposed on their descendants. In such cases, people just defer to whatever the ancestors prescribed or proscribed and leave it at that. Trying to explain why during an ancestral offering the pot where the rice is cooked must be placed in this and not that position is a futile exercise: the only reason is that the ancestors stipulated it, and because of this, it is what their descendants will do (see Astuti and Bloch 2013 for a fuller discussion).

The contrast between *fomba* that can be easily explained and those that cannot maps onto the distinction between bodies of practices that can be flexibly adapted and innovated and those that must be rigidly observed and obeyed. And although the contrast should not be overdrawn, it is certainly the case that changing the details of a ritual or redefining an ancestral taboo would require a completely different process of innovation than what I have described in this chapter. Central to it would be a delicate and potentially dangerous process of negotiation with the ancestors involving sacrificial offerings and a certain amount of pleading (see Astuti 2007 and Cole 2001).

None of the people with whom I discussed the new rigging method thought that changing this *fomba* required the involvement of the ancestors. When I asked whether, before adopting the new technique, they had called on the ancestors to inform them of the change and seek their blessing, my interlocutors told me that "sailing doesn't have ancestral laws" (hence no need to seek ancestral approval and blessing if one changes the way one sails) or that both old and new sailing techniques are just ways of "dealing with the sea," so it is a matter of individual (as opposed to ancestral) preference whether one uses one or the other.

Thus, even if the canoe is said to be "the root of Vezo-ness," the people who are made Vezo by it can flexibly transform it into something that sails like a European schooner! This openness to change and innovation, so different from the rigidity that comes from deferring to the ancestors, dovetails with an observation I made at the time of my first period of fieldwork—namely, that there were no origin stories about the canoe. When asked where the canoe came from and what its origin was, my interlocutors would refer to the fact that the people of the past already knew how to make canoes, and they would leave it at that. Only once did an old man offer a more elaborate narrative. The first time that people tried to make a canoe, he said, they used a tree that was too heavy, and the canoe sank. They returned to the forest and saw another type of tree that they thought might be suitable, and indeed it was. This time the canoe floated. That is the tree that people still use today, but the names of the people who made the discovery are not remembered, nor the time or place where it happened.

I read this lack of interest in the moment of discovery as a manifestation of a more general disregard for the past, which lies at the core of how my Vezo interlocutors construct their identity (see Astuti 1995, chap. 2 and 3). Specifically, what is striking about the story of how the canoe was invented is that it fails to transform a rather fortuitous moment of human discovery into an ancestral event. Free from the control of the ancestors and their *fomba*, the canoe remains open to exploration. Like those people who first discovered how to make it float, present-day Vezo continue experimenting with it.

In this sense, the canoe truly is the root of Vezo-ness in that, like the identity that it creates, it is not anchored to the past. And it is thanks to this cultural elaboration of what it means to be Vezo that the canoe remains open to innovation: innovation without a named inventor that spreads easily as people watch, discuss, and copy what looks like a better way of doing things.

Acknowledgments

Many thanks to Mathieu Charbonneau and Dan Sperber for inviting me to participate in this project and for their patience throughout. The work presented in this chapter was made possible by the generosity of the people of Betania and Belo-sur-Mer, Madagascar, to whom I am forever grateful.

Notes

1. Sailing is almost exclusively a male domain, and in this chapter I will only refer to the testimonies of men. Women have other ways of rooting their Vezo-ness.

2. The same rig is illustrated in Les Vezo du Sud-Ouest de Madagascar by Koechlin (1975: 78-79; 87-88), who carried out his research in the late 1960s and early 1970s in Bevato, a village about 100 kilometers south of Belo.

References

Astuti, R. 1995. *People of the Sea: Identity and Descent among the Vezo of Madagascar*. Cambridge: Cambridge University Press.

Astuti, R. 2007. "La moralité des conventions: Tabous ancestraux à Madagascar." *Terrain* 48:101–112.

Astuti, R., and M. Bloch. 2013. "Are Ancestors Dead?" In *Companion to the Anthropology of Religion*, edited by J. Boody and M. Lambek, 103–117. London: Wiley-Blackwell.

Cole, J. 2001. *Forget Colonialism? Sacrifice and the Art of Memory in Madagascar*. Berkeley: University of California Press.

Horridge, A. 1986. "The Evolution of Pacific Canoe Rigs." *Journal of Pacific History* 21 (2): 83–99.

Koechlin, Bernard. *Les Vezo du sud-ouest de Madagascar: Contribution à l'étude de l'éco-système de semi-nomades marins*. Berlin, New York: De Gruyter Mouton, 1975. https://doi.org/10.1515/9783111330112

3 Apprenticeship, Flexibility, and Rigidity: A Long-Term Perspective

Bert De Munck

Introduction

Our present-day views on craft and craftsmanship are very much informed by nineteenth- and twentieth-century visions in which craft was increasingly seen as the antithesis of technology-driven mass production and standardized production processes. Karl Marx's famous distinction between concrete labor and abstract labor was a distinction between artisanal labor and labor applied in the context of manufactories, with artisanal labor being likened to art. Marx (1973, 297) noted that "this economic relation—the character which capitalist and worker have as the extremes of a single relation of production—therefore develops more purely and adequately in proportion as labor loses all the characteristics of art; as its particular skill becomes something more and more abstract and irrelevant, and as it becomes more and more a purely abstract activity, a purely mechanical activity, hence indifferent to its particular form." In this regard, his views, which he voiced around 1857 to 1858, did not differ much from more conservative opinions, like those articulated in the context of the famous Arts and Crafts Movement, in which a return to the medieval crafts was advocated. The most famous representative of this movement, William Morris (1882, 9), wrote, "These arts, I have said, are part of a great system invented for the expression of a man's delight in beauty: all peoples and times have used them; they have been the joy of free nations, and the solace of oppressed nations; religion has used and elevated them, has abused and degraded them; they are connected with all history, and are clear teachers of it; and, best of all, they are the sweeteners of human labor, both to the handicraftsman, whose life is spent in working in them, and to people in general who are influenced by the sight of them at every turn of the day's work: they make our toil happy, our rest fruitful."

Notwithstanding all their ideological and scientific differences, both thinkers fell back on a dichotomous view in which arts and crafts were at once closely aligned with and in opposition to alienated factory labor. Craftsmanship is likened to a genuine and authentic type of labor, which was lost during modernity.

Such views inform our understandings up to the present day. In his bestseller *The Craftsman*, the famous sociologist Richard Sennett (2009, 9) considered craftsmanship the antithesis of modern types of alienated labor, defining craftsmanship as something intrinsic to human nature—that is, as "an enduring, basic human impulse, the desire to do a job well for its own sake." In this definition, the autonomy and independency of the artisan takes center stage, although Sennett, not unlike Marx and Morris, also attributes a profoundly social dimension to craftmanship. While medieval workshops were considered to be characterized by a communal atmosphere, for Sennett, *craftsmanship* was "joined skill in community" (2009, 51). Medieval workshops would have provided a social structure for the development and transfer of skills, not with the help of codified instructions on paper but through the daily incorporation of the legitimate standards. In the words of Sennett (2009, 54), "the successful workshop will establish legitimate authority in the flesh, not in rights or duties set down on paper."

This clearly brings the question of rigidity versus flexibility to a head. While premodern crafts are likened to art—which invokes an autonomous and creative artisan—craftsmanship is also seen as a result of the incorporation of collective standards and procedures. As will be shown below, this ambivalence is actually the result of our modern, dichotomous view, which eclipses the complexity and hybridity of the artisan's manifold histories. What is particularly unclear is the range of technical knowledge an artisan masters and the level of individuality and specialization the artisan is able to deploy. Related to that, to what extent is an artisan to be distinguished from an artist? And last but not least, how did this all transform in the long run? What happened in between the medieval period and nineteenth-century industrialization?

Most approaches to apprenticeship postulate a field of tension between, on one hand, the need to pass on techniques in a structured and more or less standardized way and, on the other, the need for a certain flexibility when it comes to applying the techniques in variegated and ever-changing circumstances. In a way, the medieval and early modern apprenticeship system is considered to provide the mechanisms for balancing these somewhat contradictory requirements. The principle of learning on the shop floor by imitating a master amounts to passing on a certain technical tradition from generation to generation in a certain social context while at the same time enabling apprentices to adapt the techniques to the ever-changing economic and cultural contexts. However, the way in which rigidity and flexibility was balanced drastically transformed in the long run because of economic as well as cultural and political transformations.

My chapter traces these transformations through a focus on the way in which apprenticeship was organized and institutionalized. The current literature emphasizes that the medieval and early modern apprenticeship system allowed for a substantial amount of flexibility and adaptation to new contexts. Most evidence for that argument is circumstantial, but it includes the finding that books and recipes were hardly used on the shop floor, that imitating the master was not standardized, that test pieces were very broadly defined and simultaneously allowed for specialization, and that the learning content was defined in an open-ended way and customized in apprentice contracts (for overviews, see De Munck, Kaplan, and Soly 2007, and Prak and Wallis 2019). Rigidity increased in the long run, however. This has already been argued with a focus on the guild system, but other factors may have been important too, including the introduction of books, the growing importance of design and novelty in the appreciation of products, and last but not least, religious and epistemological transformations.

As I will argue, the long-term evolutions cannot be explained by looking at the economic context alone. While mass production and technological innovation are considered to be accompanied by deskilling and increasing division of labor, I will show that the antithetical opposition of craft and technological innovation is in all likelihood a nineteenth-century fabrication. Our present-day views on medieval and early modern apprenticeship are very much informed by a nineteenth-century "invention of tradition," in which crafts were seen as the flipside of mass production in factories. A proper understanding of our current views on craftsmanship instead requires us to look at crucial transformations in the late medieval and early modern period. This is what I will present here, based on a review on the recent literature on craftsmanship and guilds in Northwest Europe and empirical data from my own work on the Southern Netherlands (roughly present-day Belgium).

Specialization and Division of Labor in the Late Middle Ages

Nineteenth- and twentieth-century views not only disregard evolutions that have taken place during the early modern period (roughly the fifteenth to the eighteenth century), they are even doing injustice to the complexity of medieval craftsmanship itself. On the one hand, many medieval crafts did already experience thorough division of labor and specialization, especially in the textile industries, the largest sector by far before the Renaissance. The production of woolen cloth required up to a dozen different types of workers, from wool combers, carders, spinners, warpers, overweavers, and fullers to dyers and shearers (Munro 2003a). All these professions could moreover have their own specialization, with dyers in blue, for instance, being distinguished from black dyers. Nor did specialization fail to increase. As soon as textile industries expanded from roughly the eleventh century on, competition forced entrepreneurs to specialize and focus on niche products, resulting in the use of different types of fabric (light woolens versus heavy woolens or a combination of wool and linen), different colors, and different patterns (van der Wee 1975, 1988; Thijs 1993; Munro 2003b, 2009). On the other hand, however, these types of specialization were not necessarily synonymous with deskilling—rather to the contrary. Innovation in terms of new types of products was mostly based on the development of new and additional technical knowledge. As economic historian Herman van der Wee (1975, 213) argued, "What was involved was more the deepening than the widening of human capital," and what was stressed was "the input of labor, as against capital and raw materials."

In this vein, the textile industry was not too different from the art sector. Our views on art and artists, too, are very much informed by nineteenth-century opinions in which artists are pictured as independent geniuses whose success is entirely the result of their natural talent and skill. According to ideas such as those of the nineteenth-century cultural historian Jacob Burckhardt (1860), artists would have emancipated themselves during the Renaissance, first in Italy and then in the rest of Europe. Yet contemporary views on the history of art show that sculpting, painting, and related arts could be viewed as an industry—a luxury industry targeting high-end markets (van der Wee 1975, 1988; Thijs 1993; Goldthwaite 1995). The huge importance of skills and technical knowledge in these industries does not prevent specialization; targeting niche markets was the order of the day. As art historians have shown, painters adapted their style and subject matter to the tastes and purchasing power of their target markets and customers (North and Ormrod 1998; De Marchi and van Miegroet 2006; Sluijter 2009). Moreover, a thorough division of labor has been revealed in the workshops of artists as well. Many hands were involved in the finishing of one painting, with different workers often having different specializations, be it heads or human figures, or nature or specific types of decoration (Peeters 2007). Famous artists and workshops are even shown to have collaborated on specific paintings, with each artist attributing a specific specialization (Honig 1998).

In short, medieval and late medieval craftsmanship was not incompatible with division of labor, art was not antithetical to specialization, and art and craftsmanship were not situated in different realms; rather, both were luxury industries competing in high-end markets. This did not prevent, of course, that the entrepreneurs involved faced the challenge of dealing with the distinction between general and specific skills when it comes to investing in training. According to standard human capital theory, entrepreneurs only have an incentive to invest in specific skills (i.e., skills that can be applied only in their specific firm). With respect to general skills that are useful to a wide range of employers, it is up to either a public authority or the individual apprentice to make the investment. The latter can do so by simply paying the master or by accepting a below-market wage for some time (Becker 1964). However, this theory applies in a context of highly competitive markets, which of course did not exist in the early modern period (Acemoglu and Pischke 1999). Moreover, the problem of free riding—that is, employers benefiting from the investments made in training by other masters by hiring skilled workers instead of training them themselves—could, at least in theory, be solved by craft guilds.

In one of the most seminal papers in the field, Stephan R. Epstein (1998) has argued that craft guilds potentially solved the so-called hold-up problem faced by employers. While masters typically wanted to recoup their investment in training by committing their apprentices longer than strictly needed to learn the trade—so that the trainee paid the master with free or below-market-price skilled work—apprentices also had an incentive to abscond once they had mastered the trade sufficiently in order to earn a wage elsewhere. The guild could prevent this by simply describing a minimum term to serve (as a precondition to enter the labor market as either a journeyman or a master) or related measures such as imposing a ban on masters taking on other masters' apprentices or having the apprentice pay an upfront fee. Unfortunately, empirical research on whether guilds or other political authorities installed such regulations is still inconclusive (see Prak and Wallis 2019). Other research suggests that even the rigid English apprenticeship system, which prescribed a minimum term of seven years across trades and cities, allowed for a high degree of flexibility in practice—with the terms being adapted at the discretion of master and apprentice (Minns and Wallis 2012; Schalk et al. 2017).

The history of the guilds in particular reveals that they had to deal with tensions between rigidity and flexibility whenever they wanted to regulate the transmission of techniques. This is all the more the case as the passing of techniques is not only a matter of economic efficiency but also of building a community and defining the boundaries of that community (De Munck 2011). Apprenticeship rules were not only installed with an eye toward passing on knowledge and skills but also for deciding who had entry to the community of masters. The status of master was acquired if one could prove they had mastered the trade and were able to deliver high-quality products, which was often tested with a masterpiece. Becoming a master was actually akin to becoming a member of a political community, with a collective identity that was based on the ability to honestly perform high-quality work (for a case study, see De Munck 2018). The way in which guilds navigated these different goals and contexts is very instructive of how the field of tension between rigidity and flexibility transformed in the long run.

Craft Guilds and Specialization

The field of tension between specialization and generality (and the transferability of skills) is detectable at the level of the juridical and institutional embedding of crafts. From the medieval period onward, most craftspeople were organized in craft guilds. Each guild was supposed to gather a specific group of artisans: the shoemaker's guild representing all shoemakers, the carpenter's guild all carpenters, the baker's guild all those involved in making bread, and so on (Farr 2000). However, this was far from straightforward. The shoemaking industry also encompassed old shoemakers or shoe repairers; the bakery industry gradually included pastry shops and sugar bakeries; and the wood industry was divided into sawyers, carpenters, and cabinetmakers, with the latter in turn encompassing such groups as wood inlayers, coffin makers, and panel makers. Product and process innovations within these groups, moreover, challenged the identity and boundaries of these groups constantly. The sixteenth-century wood industry in Antwerp was witness to a prolonged discussion between the carpenters and the cabinetmakers about who was allowed to produce exactly which products. In principle, the cabinetmakers were allowed to make loose furniture (with the use of glue) and the carpenters everything that was a fixed part of a house. But what to do with fixed banks once they are introduced? Among other things, the introduction of so-called panel work and the related shift from rough heavy boxes to light and attractive furniture assembled with mortise and tenon joints (in which panels could be fitted) jeopardized the existing boundaries between these two guilds (De Munck 2007a).

Within the guilds as well, specialization was a cause for concern, especially when it came to training and the assessment of skills. In most guilds, membership as a master artisan was conditional upon the finishing of an apprenticeship term and a masterpiece. The length of the term was, up to a degree at least, related to the difficulty of the trade. But what if different specializations are present within one and the same craft? That the guilds themselves struggled with this issue clearly emerges from an in-depth analysis of their masterpieces. The earliest descriptions of test pieces in the guilds' fifteenth- and sixteenth-century ordinances are mostly very vague, which suggests that a standardized range of skills needed to be demonstrated. In Antwerp, the tanners' guild in 1583 simply stipulated that prospective masters had to "skin, scrape, and sprinkle a hide" as a test. A journeyman who wanted to become a master shoemaker had to "cut and make a pair of thick leather shoes, a pair of boots, and a pair of slippers." And the cloth dressers' guild in 1696 simply prescribed that a new member first needed to shear nine ells of white cloth (De Munck 2007c, 68–74).

Yet other guilds clearly felt the need to specify exactly which product needed to be made as a trial piece or which types of skills needed to be demonstrated. The instruction of the Antwerp diamond guilds in their founding ordinance in 1582 mentions different types of stones as well as different operations, such as cutting, polishing, and finishing. At the end of the fifteenth century, the above-mentioned guild of cabinetmakers not only enacted that two cabinets and a table had to be made, but it also distinguished between two different types of cupboard ("een spenne en een tritsoer") and specified that the table could be either round or square (De Munck 2007a; 2007c, 68–74). Moreover, guilds increasingly added that the deans of the guilds could decide ad hoc which piece was to be produced exactly. The Antwerp gold and silversmiths not only enacted a November 24, 1524, ordinance that the masterpiece must comprise either a major work (e.g., a platter) or a gold ring set with a diamond, it also stipulated that prospective masters could simply make "what they were used to making"—that is, what they had learned as an apprentice (De Munck 2007c, 72). Something similar was the case among the Paris goldsmiths (Bimbenet-Privat 1995, 29).

In the nineteenth and most of the twentieth centuries, liberal and progressive thinkers—building on Enlightenment ideas—mostly considered craft guilds as conservative hindrances to technical innovation and the emergence of efficient markets. Yet research has shown that they might have accommodated, if not stimulated, economic innovation (Epstein and Prak 2008). In response to specialization and the invention of niche products, they have at least tried to prevent that such entry barriers as apprenticeship terms and masterpieces would turn into thresholds for those who had only learned part of the craft or acquired a specific set of skills. Moreover, guilds may have had a stimulating impact on product as well as process innovation, too. Maarten Prak (2003) has, for instance, argued that the Dutch Saint Luke's guilds as a structure and network were a key factor in the success of the seventeenth-century Dutch art market, which was famous for its pioneering role in the shift toward more realistic and more standardized depictions of everyday life (the so-called genre paintings and still lives) fabricated with the help of new techniques such as the use of coarse brushstrokes and a reduction in the use of colors. It all indicates that there was at least a certain flexibility in adapting to changing market circumstances, even with regards to the institutions that were once renowned for their rigidity.

Specialization and Vertical Integration in the Long Run

Especially from the sixteenth century on, European guilds struggled to accommodate product innovation and vertical integration. Both dimensions can be illustrated by looking at discussions about the introduction of products such as leather belts with silver buckles or tin pots with earthenware lids. These products implied the input of manufacturers from different guilds, which was made possible by the activities of large merchants or entrepreneurs who simply bought the separate components of the article and employed artisans to assemble the final product. Guilds often protested this practice because they considered it a violation of their rules, which prescribed that guild-based masters were to be independent and the sole warrantors of the final product's quality and standards. The guilds' attitude even hampered vertical integration in one single sector. In the manufacture of leather products, tanning and shoemaking were typically considered two different crafts, with each guild defining separate quality standards (for the leather and the shoes, respectively) and entry requirements (apprenticeship terms and master trials). Juridical litigation could arise when tanners started to subcontract to shoemakers, who then manufactured shoes on the tanners' premises, and vice versa (De Munck 2007b, 127–128; 2007c, 233–236). The guilds involved mostly tried to confirm and monitor their rules, insisting that whoever manufactured shoes needed to have finished the shoemakers' term and trial and whoever tanned hides needed to have completed the terms and trials of the tanners. In early modern Antwerp, such discussions emerged in a context in which the guilds of the tanners and the shoemakers were not sufficiently distinguished. Debate arose as shoemakers began to purchase hides themselves and hired tanners to tan for them. After decades of discussion, it was eventually decided to clearly separate the two guilds, which in practice meant that they each had their own apprenticeship term and masterpiece (De Munck 2007c, 235).

Within a single guild as well, product innovation could cause trouble. The guilds' products mostly had identifiable hallmarks that guaranteed certain quality standards. The tanners' hallmarks, for instance, guaranteed the proper origin of the hide used, distinguishing not only between different animals but also different parts of the animal. In the wood sector, the hallmark guaranteed the proper origin of the wood used, for instance, in order to distinguish genuine ebony wood from Spanish wood, which was used as a cheaper substitute. And in metal trades, a certain alloy was mostly guaranteed with a hallmark, with the tinsmiths' marks, for instance, ensuring that the tin did not contain more than 2 percent lead. What happened in the long term was that additional hallmarks were introduced to accommodate new products and quality levels. This was, for instance, the case with the Antwerp guild of pewterers who were forced to create new marks in order to accommodate new alloys introduced by the masters (De Munck 2008, 215–222; 2010a, 38–39; 2012). In addition, markets were increasingly flooded with products without quality marks or with shopkeepers' marks, which were not made according to the guild's standards. This was due to the increasing importance of wholesalers that provided shops with imported products and products made by "false masters" or outside the guild context. At the other end of the spectrum, the number of shops increased—called "faiseurs de rien vendeurs de toutes" in some of the literature. While guild-based masters had often combined manufacturing and selling, in the seventeenth and eighteenth centuries, these processes were increasingly separated (De Munck 2010a, 39–43; 2012, 10–13).

In all, textbook craftsmanship, which was based on autonomous masters who all made the same or similar products independently and with the help of a few journeymen and one or two apprentices, was very much a fiction by the end of the eighteenth century—at least in a range of trades. Vertical integration, subcontracting, wholesale, and retail all undermined the ideal of the master-housefather working on his own premises and selling products directly to customers (Lis and Soly 2006, 2008; De Munck 2010a). The guilds' efforts to accommodate product and process innovations moreover suggests that early modern craftspeople not only applied technological innovations but were on the very basis of innovations as well. This is clearly the case for product innovations, as can easily be illustrated by new "brands" such as Venetian glass, Cordoban leather, and Italian maiolica, to name only a few renowned types of products originating in early modern Europe. Nor was this simply a matter of designing a new product. These products typically implied adaption to the production process as well. One type of maiolica called "faience"—with which potters tried to imitate the renowned white Chinese porcelain—required a very sophisticated production process. First, the already-fired earthenware was covered with an opaque layer of white tin glaze. Subsequently, colored figures were typically painted on the white background with "underglaze" paint, after which the object was covered with a transparent glaze made of lead and tin oxides and fired again in order to fuse and burn the painting and the different layers. Clearly, in such innovations, new product designs were deeply entangled with new production techniques and even new procedures—the latter, in this case, even involving new insights into chemical and thermochemical processes.

New insights were often the result of the combination of preexisting techniques, including from different sectors. Spillovers could occur between, for instance, the glass and the earthenware industries, on the condition that such industries clustered together in certain cities. Such conditions were very often met because highly skilled artisans were typically very mobile in the early modern period. While journeymen were sometimes obligated to "tramp" from place to place for one or two years after their initial training in order to acquire additional skills, established artisans were often forced to migrate as dynastic and religious warfare disrupted their access to suppliers and customers (Reith 2008). The history of the early modern guilds thus fits into the long-standing argument in economic geography and the history of technology that diversity fuels the recombination of ideas, techniques, and technologies, which in turn fuels innovation (e.g., Audretsch and Feldman 2004). Yet innovation is not a straightforward affair either, and the definition and nature of innovation has transformed throughout history. These transformations, moreover, had a profound effect on craftspeople, a history that is still largely to be written.

The Separation of Art and Craftsmanship

Guilds and craftsmanship deteriorated with the advent of industrialization and technological innovation. But contrary to what is thought, craftsmanship underwent major transformations even before the mid-eighteenth century too. Only they weren't transformations of scale or technology-driven change. The major transformation was the growing distinction between artisan and artist. Before the Renaissance period, artisans and artists were hardly distinguished at all. What we consider artists today were often joined in the same guild as artisans, as was the case with sculptors, who often shared a guild with other artisans working with stone, like masons. During the Renaissance, however, sculptors and painters started to distinguish themselves from what they then referred to as "mere artisans" or mere "mechanics." The artists built on the classical distinction between the mechanical arts and the liberal arts and aligned themselves to intellectuals and those who had learned Latin and were familiar with texts and images from Greek and Roman antiquity. In the process, artisans were increasingly denied what was then called *ingenium* (ingenuity). The Antwerp sculptors argued that talent was needed to become an artist but not to become a mason, claiming that while masons could support themselves from the very first day of their training, "apprentices in sculpture did not know for four or five years whether they would be able to continue in the profession" (De Munck 2010b, 346–347; Filipczak 1987, 16).

This evolution was part of a broader and more thorough transformation in which artisanal skills were increasingly instrumentalized. By the mid-eighteenth century, the famous Enlightenment philosopher Adam Smith (1776, 151–153) distinguished inventors and artisans with ingenuity from the rank-and-file artisans who only had to manufacture the products invented and designed by others: "The arts, which are much superior to common trades, such as those of making clocks and watches, contain no such mystery as to require a long course of instruction. The first invention of such beautiful machines, indeed, and even that of some of the instruments employed in making them, must, no doubt, have been the work of deep thought and long time, and may justly be considered as among the happiest efforts of human ingenuity. But when both have been fairly invented and are well understood, to explain to any young man, in the compleatest manner, how to apply the instruments and how to construct the machines, cannot well require more than the lessons of a few weeks: perhaps those of a few days might be sufficient" (quoted in De Munck 2017, 819).

Paradoxically, artisans were actually valued highly in this period, but only as sophisticated robots of sorts. In the famous *Encyclopédie* of Diderot and D'Alembert, they were called "automatons," while entrepreneurs referred to them as "sets of hands" (Koepp 1986, 2009; Sewell 1986; Schaffer 1999; Lis and Soly 2012, 485–488, 422; De Munck 2014, 55–61). Artisans were no longer considered able to invent or design new products or to conceive and monitor a complex production process. Building on a distinction between mind and hand, this capacity was henceforth denied to most of them.

Interestingly, until at least the early sixteenth century, no distinction was made between mind and hand. Nor was embodied knowledge necessarily seen as inferior to intellectual knowledge before that time. As Michel Foucault first noted, the Renaissance epistemological context was such that access to the truth—which was at the time synonymous with God's wisdom—was possible through words as well as things. Just as finding the right words was considered getting closer to the truth in rhetoric, imitating God's creation by crafting could be seen as a way of getting nearer to God's wisdom. In order to understand this adequately, we have to free ourselves, again, from modern notions about invention and ingenuity. While imitation is seen as inferior to invention today, in the Renaissance context, it could actually be superior to manufacture imitations of, for instance, precious stones or diamonds, as you were then actually emulating God's *act* of creation (Bucklow 2009). This is in any case how the mystic humanist Nicolas Cusanus defended craftsmanship against the pretense of artists. In his famous dialogue *Idiota de mente*, Cusanus (1937, 51) features a wooden spoon maker who actually argues that craftsmanship is superior to art because artisans do not use models, just as God did not have a model when he created the world: "All finite art depends on infinite art. . . . As an example let me take the making of spoons. The spoon has no model other than the idea in our mind. The sculptor or painter takes his models from the things which he wishes to represent, but I do not do this when I produce spoons, saucers and pitchers from wood. . . . For this reason my art produces rather than reproduces natural forms and is, therefore, more like infinite art."

A proper understanding of this idea thus requires including an epistemological perspective and a reflection about the very notion of invention. Historians of science have shown that a more instrumental view on invention emerged in the beginning of the seventeenth century. The famous science theorist Francis Bacon praised practice and the mechanical arts because experience would have the capacity to reveal the secrets of nature, but theory was nevertheless superior. For Bacon, the invention of theories was the final aim of the natural sciences. This still involved practitioners, but it was based on a strict distinction between theory and practice. New insights were based on the data obtained from practical experiments and observations— for which practitioners were needed—but they were to be classified and interpreted by natural philosophers (Atkinson 2007, chap. 2). In areas such as navigation, cartography, surveying, and fortification, practical mathematics in the meantime grew more important and eventually became integrated as a source of knowledge—and even the very essence of nature—in natural philosophy (Cormack 2017a, 2, 4–5). This was all to the detriment of artisans, whose prestige dwindled as processes of codification and abstraction grew more important in science and beyond (Vérin 1998, 2002; Dubourg Glatigny and Vérin 2008; Valleriani 2017).

All this had a profound impact on the social and political status of the artisans, which was subject to transformations, too. As will be further discussed below, craftsmanship was embedded in a religious atmosphere in which work and skills were inseparable from devotion and piety. As suggested by the above-mentioned views of Cusanus, the religious context was deeply entangled with the artisans' epistemology. The idea that artisans had access to fundamental truths was conditional on a specific religious worldview—that is, one in which God was immanent or at least had left his signature in everything (P. H. Smith 2000a, 2000b; De Munck 2014, 2019). Conversely, however, this specific religious and also political atmosphere presupposed a range of collective activities, the historical transformations of which could be seen as the flip side of the transformations related to technical knowledge and epistemologies.

The Sociopolitical Dimension of Learning

In the Middle Ages, learning was clearly not reducible to the acquisition of technical knowledge and skills. It was rather embedded in a political and religious culture in which acquiring skills went hand in glove with being socialized in a corporative milieu. Apprentices not only worked with their master but lived under his roof as well. A master was typically considered to act as a surrogate father of sorts. He was not only supposed to train the youth but also to educate and discipline him. Masters were held responsible for their apprentices' morals and were to guard that they fulfilled their religious duties (Prak 2004; De Munck 2010c). Moreover, the terminology used in contemporary documents—including in apprenticeship contracts—often included references to "serving." Apprenticeship was part of a life-cycle tradition in which living and being socialized in another household was a standard feature of the life of a large part of lower and middling groups (S. R. Smith 1973, 1981; Krausman Ben-Amos 1991; Rappaport 1989, 232–238). Boarding in another household mostly took place between age 12 and 24, roughly, with 14 or 15 often being the medium age at which an apprenticeship started. The term to serve as an apprentice often ranged between two and eight years, with four to five years being the mean or medium (De Munck 2007c, 177–185). During this period, apprentices were to "serve" their master to the best of their abilities, not unlike living-in female servants. Masters, in turn, were to teach, educate, and discipline the youth as a good housefather does.

The role of the guilds entirely chimed with this. Becoming part of the guild, too, was not only a matter of being able to demonstrate the mastering of a certain range of skills. It implied a range of rituals, like having meals and attending mass together with the other masters. A guild was not only an economic institution but a religious brotherhood as well. Members were not only confronted with product standards but were supposed to take part in collective activities such as masses, processions, feasts for a patron saint, and funerals of fellow members. The ubiquity of terms referring to family, like brother and brotherhood, suggests that guilds can be seen as artificial families of sorts. Finishing an apprenticeship was, in a way, like entering a new or a broader family, and becoming a master was equivalent to assuming a sociopolitical role in which public and private aspects were profoundly entangled

(Farr 2000). A proper understanding thereof requires appreciating the deeply feudal and corporative context in which this all took place. The body politic was in this period literally imagined as a "body," with a head and members. While the prince or, in some cities, a conglomerate of nobles was the head of the body, they did not represent a range of individuals on a territory as we would imagine today. Being part of the body as an individual rather implied being part of a member, which could be a noble family, the clergy, or indeed, a guild. And the guild itself was in turn conceived of as a body with a head (the board) and members (the masters), just as a household was (with the master as the head and the children, apprentices, servants, and the wife as members) (De Munck 2018, esp. chaps. 1 and 2).

To be sure, the corporative system and the collective "guild ethic" did not fail to be challenged by an economic reality to which the guilds' logic had difficulties to adapt. In the thirteenth and fourteenth centuries, masters and apprentices concluded individual apprenticeship contracts that not only differed from economic sector to economic sector but could also be customized to each individual case. In a sample of 11 weaver apprentices contracts, the terms agreed on ranged from two to five years; in a sample of six contracts for silversmith apprentices, the terms were three to six years (Des Marez 1911; Verriest 1911; also De Munck, De Kerf, and De Bie 2019, 223). Unfortunately, the current state of the art does not permit one to chart to what extent and in what sense this tension between a sociopolitical and an economic logic increased in the long run, but at least after the mid-seventeenth century, this tension would seem to have increased. In France, the term *alloué* appears, which refers to an apprentice who agrees to learn without aspiring to become a master himself. In the literature on England, reference is made to "clubbing-out apprenticeships," in which the apprentices no longer boarded with their masters, especially from the late eighteenth century on (Snell 1985, 257–263). On the Continent, the decline of boarding had been observed even sooner—in some instances, from the mid-seventeenth century on (De Munck 2010c, 9–14).

All this suggests that guilds gradually lost their brotherhood-like characteristics and gradually transformed into modern political and economic institutions that guarded the "mysteries of the trade" (as the technical knowledge was often called at the time) in a juridical way, with apprenticeship terms and masterpieces serving as entry barriers—just like Adam Smith and later modern thinkers envisioned them to be. This is arguably the origin of the view on the guilds as conservative institutions out of tune with what was then called "commercial society," but the historical shift was not one from regulation to deregulation. The enlightened philosophers were engaged in a broader reflection on the relationship between the individual, religion, and the state, and what eventually emerged was a one-to-one relationship between the individual and the state in which each individual was supposed to have "natural rights"— like the right to life, liberty, and property, as summarized by John Locke. Collectives such as the guilds were increasingly obsolete in this context, and they were eventually abolished altogether from the end of the eighteenth century on.

The Commodification of Skills

Some historians have attributed the decline of "das ganze haus" model to the emergence of precapitalistic labor relations, which would have turned the relationship between master and apprentices into one resembling the relationship between employer and employee. In an article on apprenticeship, Reinhold Reith and Andreas Grießinger distinguished sectors in which concentration trends took place, such as textiles and the building industry, from "traditional" sectors. In the former, concentration trends and increasing numbers of journeymen and apprentices per masters would have profoundly transformed the relationship between master and apprentice (Grießinger and Reith 1986; Reith 2007). However, research has shown that in so-called traditional sectors like gold and silversmithing as well, the ratio of apprentices who boarded declined. Moreover, in such sectors, too, the relationship between masters and apprentices seems to have grown more businesslike. A small sample of juridical litigations between masters and (the representatives of) apprentices shows that apprentices often refused to do household chores and that their parents or guardians stressed that they had paid for the acquisition of technical skills. They insisted that the apprentice would learn the tricks of the trade rather than being used as a servant or a cheap workforce (De Munck 2010c, 15–16; also De Munck 2018, chaps. 4 and 5).

Even so, this is not to say that the more businesslike relationship between master and apprentice was synonymous with deskilling. The point is rather that apprenticeship commodified skills, which could just as well mean that large amounts of money were paid for the acquisition of highly coveted skills. In the Antwerp gold and silversmiths sector, genuine learning ateliers emerged, where apprentices—including immigrant apprentices—came to learn very specific and specialized types of skills, or what was referred to in the apprenticeship contracts as "advanced skills." These apprentices were no longer life-cycle servants but were present in the master's workshop only to acquire the skills paid for—and, perhaps, to work for the master in return. Nor was this type of apprenticeship still related to a guild logic of finishing an apprenticeship term in order to become a master. The Antwerp gold and silversmith guild opposed the emergence of such learning ateliers, arguing that the masters in question hired more apprentices than was allowed. In a juridical dossier filed against him, the most important such master in the late seventeenth century, Guillaume De Rijck, simply responded that they were not apprentices but journeymen who attended his atelier to work and acquire additional skills (De Munck and De Kerf 2018, 48–56).

The guild system itself profoundly transformed in the meantime. Until the fifteenth century, master status was largely inheritable; a master's sons could become masters without a great deal of formal obligations. In contrast to outsiders, they often did not have to finish an apprenticeship term or make a trial piece. So, up to the late Middle Ages, you were either born in the guild or you had to become socialized in your new family by living with and serving a master for some years and then proving that you were able to do the job. This too changed in the long run, however. By the mid-eighteenth century, master's sons often had to meet requirements very much like those of outsiders, up to and including paying high entry fees. This suggests that the guild was no longer a corporation with the master's household as "members"; rather, it was an organization external to the private household of the master and more akin to either a modern civil society organization or an economic institution enacting rules to protect the interests and privileges of its members.

Unfortunately, current research does not tell us to what extent this had an impact on the innovativeness of guilds. Most economic historians and historians of technology would probably hypothesize that the innovative capacity increased because of increased flexibility and mobility. Relatedly, guilds have often been seen as institutions that guarded the secrets of the trade (in a context in which a master's son often inherited the father's workshop), which could lead to the observation that knowledge has become more public. According to Joel Mokyr, this was a breeding ground for a more intense collaboration between practitioners and inventors, which in his view has served as a catalyst of the technological innovations driving the industrial revolution and is behind the kickoff of the so-called knowledge economy (Mokyr 2002; Hilaire-Pérez 2007). What is clear is that learning and acquiring skills were henceforth completely separate from acquiring a corporative status. It continued to be connected to a working ethos and embedded in a paternalistic culture, but skills were nevertheless instrumentalized and, to put it in Weberian terms, "disenchanted." This explains why a philosopher and economist like Adam Smith (1776, 151–153) could conceive of skills as only a factor in a larger production process: "the improved dexterity of a workman may be considered in the same light as a machine or instrument of trade which facilitates and abridges labor, and which, though it costs a certain expense, repays that expense with a profit."

Craft in the Nineteenth Century

Learning and acquiring social and cultural prestige continued to be important in the nineteenth century, of course. Under pressure of modern economic thinking based on the laissez-faire principle, guilds and other professional organizations were abolished around 1800 (often under French rule or influence) or later in the nineteenth century. In France, the D'Allarde and Le Chaplier Laws of 1791 abolished the guilds and outlawed collective organizations, and in England, the Combination Acts banned trade unions and collective bargaining in 1799 and 1800. Yet this did not prevent guildlike mechanisms like training apprentices on the shop floor from persisting. Even in large manufactories, senior workers could train apprentices as if they were masters of sorts, in a system called gang labor that allowed smaller social units to integrate into the larger whole. Nor did hands-on skills and craftsmanship disappear. In recent decades, several scholars have shown the continuing importance and persistence of artisanal skills as well as small-scale manufacturing (Samuel 1977, 1992; Berg 1980; Sabel and Zeitlin 1985).

Of course, this did not prevent mechanization and division of labor to become more important, resulting in both new types of hierarchies and new types of labor control. While more workers were reduced to their labor power, engineering and technical and technological know-how became more highly valued. Also, new managerial techniques replaced face-to-face contact and enabled the disciplining of workers with formal rules, protocols, differential wages, tables with working hours and targets, and so on. At least in large manufactories, workers were increasingly reduced to little cogs in a large productive machine (see Marglin 1974). In this context, craft experienced a return. The famous Arts and Crafts Movement was but the most famous expression of an atmosphere in which mechanized production and modernity in general were criticized in religious and conservative circles. In these circles, a revival of the Middle Ages was preached and practiced in myriad ways: the buying, selling, and imitation of old art; the invention of neo-styles like neo-Gothic and neo-Renaissance; the emergence of renovation and restoration as a discipline, and so on (Caen, De Munck, and Langouche 2008). In it, medieval craftman-ship was very much idealized, with the medieval craftsman rendered as a pious beacon of harmony as well as a proud and disinterested artist targeting high quality for its own sake.

In this context, a dichotomous view emerged in which craftmanship was seen as the antithesis of innovation and technology-driven production. Yet, as shown in this chapter, this is doing injustice to late medieval and early modern craftspeople and the small-scale production they stood for. While small commodity production has often been able to sustain economic growth and productivity, craftspeople were often at the center of innovation. Yet innovation and invention were, in this period, entirely different from our modern concep-tions of it. The ideological and epistemological context were far more favorable to embod-ied and collective types of knowledge—up to the point that craftsmanship could be seen as conducive to scientific progress (Cormack 2017b). Art historians and historians of science have argued that the seventeenth-century scientific revolution—in which observa-tion and experiment substituted for deductive philosophizing about the nature of nature—were very much indebted to the practices and experiences of craftspeople and other practitioners. While craftspeople observed how nature (raw materials) reacted to mechanical processing or such procedures as heating, the mathematical knowledge used for hydraulics and navigation was appropriated and built on by scientists such as René Descartes.

Unfortunately, this was also the period in which artisans themselves became discredited. Not unlike Bacon, Descartes eventually developed an instrumentalized view of artisanal skills. Reflecting on the rationality of artisanal work in the first decades of the seventeenth century, Descartes eventually abandoned the idea that artisanal work could give access to the mathesis of the world. While he first saw "orderly souls" (*âmes réglées*) at work when observing artisans, the dominant metaphor for the rational order gradually became the machine—which furthered the idea that the human body was devoid of talent and ingenuity and that talent and ingenuity should rather be looked for in the mind of learned philosophers. As art historian and historian of science Pamela Smith has observed, "artisanal bodily experience was absorbed into the work of the natural philosopher at the same time that the artisan himself was excised from it" (Smith 2004, 186).

References

Acemoglu, D., and J.-F. Pischke. 1999. "Beyond Becker: Training in Imperfect Labour Markets." *The Economic Journal* 109 (453): 112–142.

Atkinson, C. 2007. *Inventing Inventors in Renaissance Europe: Polydore Vergil's* De inventoribus rerum. Tübin-gen: Mohr Siebeck.

Audretsch, D. B., and M. Feldman. 2004. "Knowledge Spillovers and the Geography of Innovation." In *Hand-book of Regional and Urban Economics*, Vol. 4, edited by J. V. Henderson and J.-F. Thisse, 2713–2739. Amster-dam, Elsevier.

Becker, G. H. 1964. *Human Capital*. Chicago: University of Chicago Press.

Berg, M. 1980. *The Machinery Question and the Making of Political Economy*. Cambridge: Cambridge Univer-sity Press.

Bimbenet-Privat, M. 1995. "Goldsmiths' Apprenticeship during the First Half of the Seventeenth Century: The Situation in Paris." In *Goldsmiths, Silversmiths and Bankers: Innovation and the Transfer of Skill, 1500–1800*, edited by D. Mitchell, 23–31. Stroud: Alan Sutton Publishing.

Bucklow, S. 2009. "The Virtues of Imitation: Gems, Cameos and Glass Imitations." In *The Westminster Retable: History, Technique, Conservation*, edited by P. Binski and A. Massing, 143–149. London: Hamilton Kerr Institute.

Burckhardt, J. 1860. *Die Kultur der Renaissance in Italien*: ein Versuch. Basel: Schweighauser'schen Verlagsbuchhandlung.

Caen, J., B. De Munck, and L. Langouche. 2008. "Het verleden herscheppen. De restauratie-ethiek en -praktijk in het negentiende-eeuwse glasatelier Bethune-Verhaegen." In *Wedijveren met de middeleeuwen. Negentiende-eeuws corporatisme en de restauratiepraktijk in België en Nederland*, edited by J. Caen and B. De Munck, 145–162. Special issue, *Trajecta* 17 (2).

Cormack, L. B. 2017a. "Introduction: Practical Mathematics, Practical Mathematicians, and the Case for Transforming the Study of Nature." In *Mathematical Practitioners and the Transformation of Natural Knowledge in Early Modern Europe*, edited by L. B. Cormack, S. A. Walton, and J. A. Schuster, 1–10. Springer: Cham.

Cormack, L. B. 2017b. "Handwork and Brainwork: Beyond the Zilsel Thesis." In *Mathematical Practitioners and the Transformation of Natural Knowledge in Early Modern Europe*, edited by L. B. Cormack, S. A. Walton, and J. A. Schuster, 11–36. Springer: Cham.

Cusanus, N. 1937. *Idiota de mente*, cap II. Opera Omnia.

De Marchi, N., and H. J. van Miegroet, eds. 2006. *Mapping Markets for Paintings in Europe 1450–1750*. Turnhout: Brepols.

De Munck, B. 2007a. "Construction and Reproduction: The Training and Skills of Antwerp Cabinetmakers in the 16th and 17th Centuries." In *Learning on the Shop Floor: Historical Perspectives on Apprenticeship*, edited by B. De Munck, S. L. Kaplan, and H. Soly, 85–110. London: Berghahn Books.

De Munck, B. 2007b "La qualité du corporatisme: Stratégies économiques et symboliques des corporations anversoises du XVᵉ siècle à leur abolition." *Revue d'histoire moderne et contemporaine* 54 (1): 116–144.

De Munck, B. 2007c. *Technologies of Learning: Apprenticeship in Antwerp Guilds from the 15th Century to the End of the Old Regime*. Turnhout: Brepols.

De Munck, B. 2008. "Skills, Trust and Changing Consumer Preferences: The Decline of Antwerp's Craft Guilds from the Perspective of the Product Market, ca. 1500–ca. 1800." *International Review of Social History* 53 (2): 197–233.

De Munck, B. 2010a. "One Counter and Your Own Account: Redefining Illicit Labour in Early Modern Antwerp." *Urban History* 37 (1): 26–44.

De Munck, B. 2010b. "Corpses, Live Models, and Nature: Assessing Skills and Knowledge before the Industrial Revolution (Case: Antwerp)." *Technology and Culture* 51 (2): 332–356.

De Munck, B. 2010c. "From Brotherhood Community to Civil Society? Apprentices between Guild, Household and the Freedom of Contract in Early Modern Antwerp." *Social History* 35 (1): 1–20.

De Munck, B. 2011. "Gilding Golden Ages: Perspectives from Early Modern Antwerp on the Guild-Debate, c. 1450–c. 1650." *European Review of Economic History* 15:221–253.

De Munck, B. 2012. "The Agency of Branding and the Location of Value: Hallmarks and Monograms in Early Modern Tableware Industries." *Business History* 54 (7): 1–22.

De Munck, B. 2014. "Artisans, Products and Gifts: Rethinking the History of Material Culture in Early Modern Europe." *Past and Present* 224:39–74.

De Munck, B. 2017. "Disassembling the City: A Historical and an Epistemological View on the Agency of Cities." *Journal of Urban History* 43 (5): 811–829.

De Munck, B. 2018. *Guilds, Labour and the Urban Body Politic: Fabricating Community in the Southern Netherlands, 1300–1800*. London: Routledge.

De Munck, B. 2019. "Artisans as Knowledge Workers: Craft and Creativity in a Long-Term Perspective." *Geoforum* 99 (February): 227–237.

De Munck, B., and R. De Kerf. 2018. "Wandering about the Learning Market: Early Modern Apprenticeship in Antwerp Gold- and Silversmith Ateliers." In *Navigating History: Economy, Society, Knowledge, and Nature. Essays in Honour of Prof. Dr. C. A. Davids*, edited by P. Brandon, S. Go, and W. Verstegen, 36–63. Leiden: Brill.

De Munck, B., R. De Kerf, and A. De Bie. 2019. "Apprenticeship in the Southern Netherlands, c. 1400–c. 1800." In *Apprenticeship in Early Modern Europe*, edited by M. Prak and P. Wallis, 217–246. Cambridge: Cambridge University Press.

De Munck, B., S. L. Kaplan, and H. Soly, eds. 2007. *Learning on the Shop Floor: Historical Perspectives on Apprenticeship*. London: Berghahn Books.

Des Marez, G. 1911. "L'apprentissage à Ypres à la fin du XIIIe siècle: Contribution à l'étude des origines corporatives en Flandre." *Revue du Nord* 2 (1): 1–48.

Dubourg Glatigny, P., and H. Vérin, eds. 2008. *Réduire en art: la technologie de la Renaissance aux Lumières*. Paris: Éditions de la Maison des sciences de l'homme.

Epstein, S. R. 1998. "Craft Guilds, Apprenticeship, and Technological Change in Preindustrial Europe." *Journal of Economic History* 58 (3): 684–713.

Epstein, S. R., and M. Prak, eds. 2008. *Guilds, Innovation, and the European Economy, 1400–1800.* Cambridge: Cambridge University Press.

Farr, J. R. 2000. *Artisans in Europe, 1300–1914.* Cambridge: Cambridge University Press.

Filipczak, Z. Z. 1987. *Picturing Art in Antwerp, 1550–1700.* Princeton, NJ: Princeton University Press.

Goldthwaite, R. A. 1995. *Wealth and the Demand for Art in Italy, 1300–1600.* Baltimore: Johns Hopkins University Press.

Grießinger, A., and R. Reith. 1986. "Lehrlinge im deutschen Handwerk des ausgehenden 18. Jahrhunderts: Arbeitsorganisation, Sozialbeziehungen und alltägliche Konflikte." *Zeitschrift für Historische Forschung* 13:149–199.

Hilaire-Pérez, L. 2007. "Technology as a Public Culture in the Eighteenth Century: The Artisans' Legacy." *History of Science* 45 (2): 135–153.

Honig, E. A. 1998. *Painting and the Market in Early Modern Antwerp.* New Haven, CT: Yale University Press.

Koepp, C. J. 1986. "The Alphabetical Order: Work in Diderot's Encyclopédie." In *Work in France: Representations, Meaning, Organization, and Practice,* edited by S. L. Kaplan and C. J. Koepp, 229–257. Ithaca, NY: Cornell University Press.

Koepp, C. J. 2009. "Advocating for Artisans: The Abbé Pluche's Spectacle de la Nature (1732–51)." In *The Idea of Work in Europe from Antiquity to Modern Times,* edited by J. Ehmer and C. Lis, 245–273. Aldershot: Ashgate.

Krausman Ben-Amos, I. 1991. "Failure to Become Freemen: Urban Apprentices in Early Modern England." *Social History* 16:155–172.

Lis, C., and H. Soly. 2006. "Export Industries, Craft Guilds and Capitalist Trajectories." In *Craft Guilds in the Early Modern Low Countries: Work, Power and Representation,* edited by M. Prak, C. Lis, J. Lucassen, and H. Soly, 107–132. Aldershot: Ashgate.

Lis, C., and H. Soly. 2008. "Subcontracting in Guild-Based Export Trades, Thirteenth-Eighteenth Centuries." In *Guilds, Innovation, and the European Economy, 1400–1800,* edited by S. R. Epstein and M. Prak, 81–113. Cambridge: Cambridge University Press.

Lis, C., and H. Soly. 2012. *Worthy Efforts: Attitudes to Work and Workers in Pre-industrial Europe.* Leiden: Brill.

Marglin, S. 1974. "What Do Bosses Do? The Origins and Functions of Hierarchy in Capitalist Production. Part I." *Review of Radical Political Economics* 6 (2): 60–112.

Marx, K. 1973. *Grundrisse: Foundations of the Critique of Political Economy.* Harmondsworth: Penguin.

Minns, C., and P. Wallis. 2012. "Rules and Reality: Quantifying the Practice of Apprenticeship in Early Modern England." *Economic History Review* 65 (2): 556–579.

Mokyr, J. 2002. *The Gifts of Athena: Historical Origins of the Knowledge Economy.* Princeton, NJ: Princeton University Press.

Morris, W. 1882. *Hopes and Fears for Art: Five Lectures.* Boston: Roberts Brothers.

Munro, J. 2003a. "Medieval Woollens: Textiles, Textile Technology, and Industrial Organisation, c. 800–1500." In *The Cambridge History of Western Textiles,* edited by D. Jenkins, 181–227. Cambridge: Cambridge University Press.

Munro, J. 2003b. "Medieval Woollens: The Western European Woollen Industries and Their Struggles for International Markets, c. 1000–1500." In *The Cambridge History of Western Textiles,* edited by D. Jenkins, 228–324. Cambridge: Cambridge University Press.

Munro, J. 2009. "Three Centuries of Luxury Textile Consumption in the Low Countries and England, 1330–1570: Trends and Comparisons of Real Values of Woollen Broadcloths (Then and Now)." In *The Medieval Broadcloth: Changing Trends in Fashions, Manufacturing, and Consumption,* Vol. 6, Ancient Textile Series, edited by K. Vestergård Pedersen and M.-L. B. Nosch, 1–73. Oxford: Oxbow Books.

North, M., and D. Ormrod, eds. 1998. *Art Markets in Europe 1400–1800.* Aldershot: Ashgate.

Prak, M. 2003. "Guilds and the Development of the Art Market during the Dutch Golden Age." *Simiolus: Netherlands Quarterly for the History of Art* 30 (3/4): 236–251.

Prak, M. 2004. "Moral Order in the World of Work: Social Control and the Guilds in Europe." *Social Control in Europe.* Vol. 1, 1500–1800, edited by H. Roodenburg and P. Spierenburg, 176–199. Columbus: Ohio State University Press.

Peeters, N., ed. 2007. *Invisible Hands? The Role and Status of the Painter's Journeyman in the Low Countries c. 1450–c. 1650.* Leuven: Peeters.

Prak, M., and P. Wallis, eds. 2019. *Apprenticeship in Early Modern Europe.* Cambridge: Cambridge University Press.

Rappaport, S. 1989. *Worlds within Worlds: Structures of Life in Sixteenth-Century London.* Cambridge: Cambridge University Press.

Reith, R. 2007. "Apprentices in the German and Austrian Crafts in Early Modern Times—Apprentices as Wage Earners?" In *Learning on the Shop Floor: Historical Perspectives on Apprenticeship*, edited by B. De Munck, S. L. Kaplan, and H. Soly, 179–202. London: Berghahn Books.

Reith, R. 2008. "Circulation of Skilled Labour in Late Medieval and Early Modern Central Europe." In *Guilds, Innovation, and the European Economy, 1400–1800*, edited by S. R. Epstein and M. Prak, 114–142. Cambridge: Cambridge University Press.

Sabel, C., and J. Zeitlin. 1985. "Historical Alternatives to Mass Production: Politics, Markets and Technology in Nineteenth-Century Industrialization." *Past and Present* 108:133–176.

Samuel, R. 1977. "Workshop of the World: Steam Power and Hand Technology in Mid-Victorian Britain." *History Workshop* 3 (1): 6–72.

Samuel, R. 1992. "Mechanization and Hand Labour in Industrializing Britain." In *The Industrial Revolution and Work in the Nineteenth Century*, edited by L. R. Berlanstein, 26–43. London: Routledge.

Schaffer, S. 1999. "Enlightened Automata." In *The Sciences in Enlightened Europe*, edited by W. Clark, J. Golinski, and S. Schaffer, 126–165. Chicago: University of Chicago Press.

Schalk, R., P. Wallis, C. Crowston, and C. Lemercier. 2017. "Failure or Flexibility? Apprenticeship Training in Premodern Europe." *Journal of Interdisciplinary History* 48 (2): 131–158.

Sennett, R. 2009. *The Craftsman*. London: Penguin.

Sewell, W. H., Jr. 1986. "Visions of Labor: Illustrations of the Mechanical Arts before, in, and after Diderot's Encyclopédie." In *Work in France: Representations, Meaning, Organization, and Practice*, edited by S. L. Kaplan and C. L. Koepp, 258–286. Ithaca, NY: Cornell University Press.

Sluijter, E. J. 2009. "On Brabant Rubbish, Economic Competition, Artistic Rivalry and the Growth of the Market for Paintings in the First Decades of the Seventeenth Century." *Journal of the Historians of Netherlandish Art* 1 (2): 1-31.

Smith, A. 1776. *An Inquiry into the Nature and Causes of the Wealth of Nations*. 1st ed., bk. 1, chap. 10, part 2. London: W. Strahan, London.Smith, P. H. 2000a. "Artists as Scientists: Nature and Realism in Early Modern Europe." *Endeavour* 24 (1): 13–21.

Smith, P. H. 2000b. "Vital Spirits: Redemption, Artisanship, and the New Philosophy in Early Modern Europe." In *Rethinking the Scientific Revolution*, edited by M. J. Osler, 119–136. Cambridge: Cambridge University Press.

Smith, P. H. 2004. *The Body of the Artisan: Art and Experiment in the Scientific Revolution*. Chicago: University of Chicago Press.

Smith, S. R. 1973. "The London Apprentices as Seventeenth-Century Adolescents." *Past and Present* 61:150–151.

Smith, S. R. 1981. "The Ideal and the Reality: Apprentice–Master Relationships in Seventeenth-Century London." *History of Education Quarterly* 21:449–460.

Snell, K. D. M. 1985. *Annals of the Labouring Poor: Social Change and Agrarian England 1660–1900*. Cambridge: Cambridge University Press.

Thijs, A. K. L. 1993. "Antwerp's Luxury Industries: The Pursuit of Profit and Artistic Sensitivity." In *Antwerp: Story of a Metropolis, 16th–17th Century*, edited by J. van der Stock, 105–113. Antwerp: Martial & Snoeck.

Valleriani, M. 2017. "The Epistemology of Practical Knowledge." In *The Structures of Practical Knowledge*, edited by M. Valleriani, 1–20. Springer: Cham.

van der Wee, H. 1975. "Structural Changes and Specialization in the Industry of the Southern Netherlands." *Economic History Review* 28:203–221.

van der Wee, H. 1988. "Industrial Dynamics and the Process of Urbanization and De-urbanization in the Low Countries from the Late Middle Ages to the Eighteenth Century." In *The Rise and Decline of Urban Industries in Italy and the Low Countries (Late Middle Ages–Early Modern Times)*, edited by H. van der Wee, 307–381. Leuven: Leuven University Press.

Vérin, H. 1998. "La réduction en art et la science pratique au XVIe siècle." In *Institutions et conventions: La réflexivité de l'action économique*, edited by R. Salais, É. Chatel, and D. Rivaud-Danset, 119–145. Paris: Editions EHESS.

Vérin, H. 2002. "Généalogie de la 'réduction en art: Aux sources de la rationalité moderne." In *Les nouvelles raisons du savoir: Vers une prospective de la connaissance*, edited by T. Gaudin and A. Hatchuel, 29–41. La Tour d'Augue: Editions de l'aube.

Verriest, L. 1911. *Les luttes sociales et le contrat d'apprentissage à Tournai jusqu'en 1424*. Classe des Lettres, Deuxième série, Tome IX. Brussels: Mémoire in-8° de l'Académie Royale de Belgique.

4 When Rigidity Invents Flexibility to Preserve Some Stability in the Transmission of Pottery-Making during the European Middle Bronze Age

Sébastien Manem

Introduction

Valentine Roux and colleagues (this volume) define technical flexibility as a control over variability on an individual scale—for example, when an expert potter either masters several ways of doing different kinds of pottery or sufficiently masters some pottery-making such that they can explore novel ways of meeting the needs of consumers. In contrast, a rigid technical behavior would involve a difficulty in adopting new strategies to realize novel results, such as low-level experts being less efficient in realizing new ceramic shapes. It thus seems that flexibility and rigidity are opposites on a continuum when considering these properties at the scale of the individual.

In this chapter, I approach the question of technical flexibility and rigidity from a broader intergenerational timescale, offering a different but complementary take on their relation. I argue that flexibility and rigidity need not always be opposed to one another, and that under certain conditions they can be understood in symbiosis: it is because the traditional technical behaviors are rigid that they can be *partially* rethought and modified in a flexible way to produce new technical innovations, and this flexibility promotes, and can even ensure, the transmission of the rigid technical behaviors.

A dual synchronic and diachronic reading is adopted here to understand the evolution of "ways of doing things" when adapting to consumers' new needs and to what I refer to as "the order of development"—that is, an independent technical evolution taking place only at the level of the producers and transmission networks. This allows us to understand particular technological choices as compromises between flexibility and rigidity. To support the hypothesis of the meeting between flexibility and rigidity in both a synchronic and diachronic analysis, this topic is approached in the manner of the Swiss Cheese Model of system accidents (Reason 2000). The evolutionary trajectory leading to the technological choices made by low-level experts results from the alignment of circumstances—individual (invention) and collective (innovation)—that must be identified. Doing so, however, requires that one looks at the internal organization of technical behaviors rather than the individual performances of those behaviors. This organization concerns the specific building blocks of technical traditions, together with their degree of interdependence and dependence.

I show that adopting a broader evolutionary timescale in conjunction with a finer-grained analysis of technical traditions can help us better understand this interplay between technical rigidity and flexibility, and how the use of cladistic methods and analysis applied to technological evolution are particularly suited to this task. Building on previous work (Manem 2020), I address issues related to stability, rigidity, and flexibility in the use and transmission of techniques in the scope of the concept of descent with modifications between technical traditions. To support these theoretical claims, I examine the evolution of European Middle Bronze Age pottery-making with the *chaîne opératoire* approach, focusing on molding and coiling as the means to produce two new particular types of vessels: pots and cups.

Ceramic *Chaînes Opératoires*, Techniques, Methods, and Technical Traditions: Where to Place Rigidity and Flexibility

A Framework of Technical Behaviors

To understand how I intend to demonstrate that a technical tradition is partially rethought and modified (invention at the individual scale), and then accepted and transmitted to the next generations (innovation at the collective scale), it is necessary to define what kind of rigidity and flexibility I am referring to and "where" rigidity and flexibility operate in the way of doing things and in relation to the transgenerational evolutionary trajectory of potters' behaviors. To do this, I need to first define the features of the technical behavior that will be discussed in this chapter. This will allow me to show that the switch from one technique to another, or in a broader sense, the adoption of a new technical behavior, is not to be perceived as a replacement of a previous behavior. Moreover, I argue that we need to adopt a finer-grained understanding of technical behaviors that goes beyond the notion of technique; we need to include the parameters of the techniques and to differentiate techniques from methods. Put another way, the notion of technique refers to a complex that in fact contains many transmissible elements. Using this finer-grained framework will allow me to identify the evolutionary bridges between behaviors before and after a switch or, to be more precise, what such switches demand in terms of the motor constraints and perceptions involved in a way of doing things.

First and foremost, this study does not cover the whole ceramic *chaîne opératoire*—that is, "a series of operations that transform raw material into finished product, whether it is a consumer object or a tool" (Cresswell 1976). Instead, it is only focused on the fashioning *chaîne opératoire*, "a series of operations that transform the paste into a hollow volume" (Roux 2019, 41) and the consumer object, but not their use and reuse by consumers. The *chaîne opératoire* is deliberately limited in this study to the fashioning techniques and methods (defined below), although the data extracted to cover the subject of this chapter are based on a broader *chaîne opératoire* that also involves the operations following the fashioning, such as the finishing operations and the surface treatments (Manem 2020).

By technique, I refer to the "physical modalities used to transform the raw material" (Roux 2019, 42). These physical modalities can be sorted according to parameters (see the exhaustive list of parameters in Roux 2019, 42–43, and Rice 2015, 135). I am particularly interested in two parameters that are relevant in the archaeological context that I describe below (figure 4.1). The first parameter (A) concerns the elementary volume on which the forces

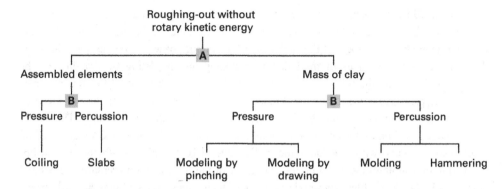

Figure 4.1
Classification chart of the roughing-out technique without rotary kinetic energy and according to two parameters. Parameter A is the elementary volume on which the forces act (assembled elements versus mass of clay). Parameter B identifies the forces used during the fashioning (pressure versus percussion). *Source*: Adapted from Roux 2019.

act (Fewkes 1940, 1944; Roux 2019, 42); it distinguishes the techniques involving a mass of clay and those involving assembled elements. The second parameter (B) concerns the forces used during the fashioning and contrasts the techniques involving pressure from those of percussion. These parameters are important because it is through the evolution of these parameters that we can truly understand how a new technique can become a technical tradition (see also Manem 2020).

A technique must be distinguished from a method (Tixier 1967, 807) or the "ordered sequence of functional operations carried out by a set of elementary gestures for which different techniques can be used" (Roux 2019, 41). A sequence comprises phases and stages. The phases describe the different parts of the pottery (e.g., base, body, neck) that are fashioned with the same or different techniques (e.g., base with technique X and body and neck with technique Y). The sequencing of the phases can vary. For instance, some potters start by the base, then move on to the body and the neck or rim, while others start the shaping with the neck or rim and then finish with the base (van der Leeuw 1993; Krause 1985). There are two successive stages—from roughing-out to preforming—intended to progressively obtain the desired form. A roughout is a hollow volume that does not present the final geometric characteristics of the vessel. A preform is a hollow volume that has the final geometric characteristics of the recipient before undergoing finishing operations, surface treatments, drying, and firing (Roux 2019).

The issues of flexibility and rigidity that will be discussed in this chapter concern two techniques—coiling and molding—and some methods that will be discussed through their fashioning *chaînes opératoires*. The coiling technique "consists in roughing-out recipients or parts of recipients using coils. The coil is a roll of paste obtained either by rolling an elementary volume of paste between the palms or on a flat surface with both palms or by modeling exerting interdigital pressure" (Roux 2019, 54). Molding consists in "roughing-out and preforming recipients by spreading a clay mass onto a convex or concave mold. The clay is progressively thinned by percussion" (Roux 2019, 61).

The diversity of manufacturing techniques can be classified (figure 4.1) according to their parameters and stages, which reveals nested "families" and "subfamilies" of techniques (Roux 2019, fig. 2.42). The techniques are therefore not seen as isolated from one another. Instead, they are linked by their shared parameters. For instance, taking the parameter of forces (percussion versus pressure) into consideration, the roughing-out techniques of molding and hammering are both members of the subfamily of techniques based on percussion. Similarly, the roughing-out techniques of modeling by pinching and modeling by drawing can be grouped in the same subfamily based on pressure. These two subfamilies of techniques (percussion and pressure) are themselves a branch of a larger family related to the parameter of an elementary volume on which the forces act, specifically on the mass of clay. The family of techniques used on masses of clay is distinguished from the family of techniques used on assembled elements, which includes the coiling and slabs techniques—themselves distinguished in two subfamilies based on the parameter of the forces, again pressure and percussion. At a higher level, these two main families of techniques (mass of clay and assembled elements) are related by another parameter—here, muscular energy without rotary kinetic energy.

Questioning Rigidity and Flexibility at the Meeting of Synchronic and Diachronic Axes

As mentioned above, technical behaviors may be complex and are constituted by an interplay of different elements that may vary. For instance, when producing a vessel, a potter can successively rough-out different parts of the ceramic with the same or different roughing-out techniques and then preform them with the same or different preforming techniques. Alternatively, a potter may rough-out and then preform each part with the same (or different) techniques before continuing with the next part. Moreover, the roughing-out techniques may depend on the same or different parameters. A potter may use a roughing-out technique involving a mass of clay for the base but another roughing-out technique involving, this time, assembled elements for the body (parameter: elementary volume on which the forces act). Therefore, by these combinatorics alone, the variability of technical behaviors subject to flexibility or rigidity is potentially tremendous, especially if one includes other variables in the *chaîne opératoire* (e.g., operating procedures, gestures, and tools; Roux 2019, 43–44).

Chaînes opératoires can also vary from one generation to another. For instance, each stage of a *chaîne opératoire* can change through time with or without other stages changing with it. To fully understand why a potter invents a new behavioral variant by flexibly altering some element of a previous behavior while keeping other elements stable, we need to track the history of the technical element that has been changed within the *chaîne opératoire* and how it relates to changes in other technical elements. It is therefore essential that we examine how the choice of transforming a technical tradition relates to the way it has been used and maintained by previous generations of potters. Indeed, as I will argue, technical traditions tend to have a "harder core" around which other elements can vary more freely from one generation to the next.

Observing whether the use and transmission of a technique appears flexible or rigid within a studied culture will depend on whether one considers this technique on its own or in the context of the evolution of the whole *chaîne opératoire* of which it is part. For instance, the use and transmission of a roughing-out technique (e.g., coiling, molding) can be perceived as stable but flexible in the method. So, while the technique is transmitted to the next generation, the method used for that technique may vary. For a generation of potters, coiling may be used only for the base, while for the next generation coiling may be used for the body only or for both the base and the body. The reasoning here is that there may exist an evolutionary independence of each step in the *chaîne opératoire*, allowing one step to change without leading the overall technical behavior to collapse by losing its functionality. Studying the flexibility and rigidity of techniques therefore requires us to understand whether such modifications in the *chaîne opératoire* are cultural, functional, or both (e.g., related to a new ceramic shape, considering the link between form and function; Shepard 1956, 224).

Approaching the relation between changes in consumer demands and the relative flexibility and rigidity of technical behaviors of producers, as well as the classification of techniques by parameters and their relative evolutionary dependence and independence, naturally brings about two groups of questions:

(1) At what level of the *chaîne opératoire* do adaptations to consumer needs mostly appear? Do those changes happen more frequently at the level of the techniques, in the parameters of the techniques, or at the level of the methods (phases and stages)? Are certain technical traits immune to such flexible changes, and if so why?

(2) How does rigidity facilitate flexibility, and what is the future of this symbiosis in terms of transmission and evolution? What is the relative contribution of the decision to adapt an existing technical trait to the needs of the consumers when compared to the order of development of the techniques taking place from generation to generation of potters?

Tracking the Evolution of Techniques and Methods with Cladistics Analysis: A Theoretical and Methodological Brief

We have seen that techniques are particularly complex to apprehend. This raises the methodological question of how to understand flexibility and rigidity in archaeology by tracking their numerous components in a process of transmission over centuries or millennia. It quickly becomes difficult to analyze the possible interactions between all these components, especially if the archaeological-cultural context with which one is working presents strong synchronic and diachronic technical variability. This is where the use of cladistic analysis comes in handy: it allows us to integrate large amounts of data and thus to track and analyze large-scale cultural evolutionary patterns without sacrificing details.

Darwinian archaeology (Shennan 2008)—where culture is conceptualized as the result of the transmission of knowledge within social contexts—appears to be a solution for addressing the intergenerational results of flexibility and rigidity in transmitted techniques and methods. The results of flexible and rigid behaviors in transmission processes may induce a mix of continuity and invention turning into innovation in the *chaînes opératoires*. Technical evolution can be understood as a form of descent with modification (Riede 2005, 2006, 2008), as technical traditions satisfy its three criteria: "(1) the existence of variation in the entities involved, (2) the presence of a mechanism by which at least some of that variation is heritable, and (3) the differential inheritance of particular patterns of variation across time and/or space" (Lycett 2015, 22). Thus, the ceramic *chaîne opératoire* approach may be fully integrated into cultural evolution (Manem 2020), "the theory that cultural change in humans and other species can be described as a Darwinian evolutionary process, and consequently that many of the concepts, tools and methods used by biologists to study biological evolution can be equally profitably applied to study cultural change" (Mesoudi 2016, 481).

Cladistics is an approach in evolutionary biology to model the hypothesis of ancestor-descendant relationship (Darlu et al. 2019; Wiley and Lieberman 2011). Cladistics generates phylogenetic trees that constitute explicit hypotheses of relatedness among the units under consideration. These trees show change within lineages and their diversification. Cladistics is based on distinguishing between kinds of similarities and using those considered to build phylogenetic relationships between species. Key traits in such analyses are derived traits shared between two or more taxa[1] that are inherited from the taxa's most recent common ancestor. Taking shared derived traits as the units of phylogenetic relationships is based on four basic principles. First, the emergence of a new taxon results from the bifurcation of a preexisting taxon as the outcome of descent with modification. Second, two taxa are considered to be more closely related to one another than either is to a third taxon if they share a common ancestor that is not also shared by the third taxon. Third, the shared derived traits represent the evidence for exclusive common ancestry between two taxa to the exclusion of a third taxon if these shared derived traits are not also shared by this third taxon. Fourth and finally, shared derived traits make it possible to define a monophyletic group—that is, "a taxon comprised of two or more species that includes the ancestral species and all and only the descendants of that ancestral species" (Wiley and Lieberman 2011, 9).

Phylogenetic methods such as cladistics have been used with great success to treat material culture since the seminal works of Mike O'Brien and R. Lee Lyman in evolutionary archaeology (O'Brien, Darwent, and Lyman 2001; O'Brien and Lyman 2003). If for biologists, the preferred taxa are biological species (e.g., Pujos 2006), then for evolutionary archaeology the taxa of choice have been groups of artifacts, often derived from morphological or stylistic features of the material culture (e.g., Cochrane 2009). In the present case, however, *chaînes opératoires* are understood to form technical traditions (Roux 2019, 6). Consequently, for the present analysis, the chosen taxa are the ceramic *chaînes opératoires*, with the techniques (and their parameters) and methods (and their phases and stages) serving as the characters (traits) used to define these taxa.

The shared derived traits are inventions that became innovations. Thus, two *chaînes opératoires* A and B sharing a new innovation I1 form, together with their hypothetical common technical tradition ancestor, a monophyletic group or clade X. This monophyletic group X excludes the *chaîne opératoire* C, which does not share the innovation I1. Three *chaînes opératoires* A, B, and C sharing an older innovation I2 (i.e., older than the innovation I1) form, with their hypothetical older common technical tradition ancestor, a more inclusive monophyletic group or clade Y. So, while evolutionary archaeologists have focused on groups of artifacts, *chaînes opératoires* can also be used as units. For a more detailed description of methods and analyses and a demonstration of the utility of *chaînes opératoires* as taxa, see Manem 2020.

Cladistics can trace in some detail lineages of potters and the successive innovations in technical traditions. In particular, cladistics makes it possible to understand if flexibility (e.g., an invention) in a technical tradition—and in a context where the result of this flexibility is transmitted at least to the next generation (innovation)—is based on more or less recent innovations (e.g., from a century or five centuries ago) or on ancestral traits never before modified by the generations of potters of the cultural group studied. This approach allows for the tracing of complex evolutionary processes in more depth. For instance, it is possible to detect whether some steps of the *chaînes opératoires* have evolved across centuries while others remained stable, but also how within those that have evolved, some may have done so independently (or not) from the other steps of the *chaînes opératoires* of which they were part.

European Bronze Age Ceramic *Chaînes Opératoires* as Case Study

The present phylogenetic analysis uses more than 15,000 data points in the matrix (based on methods and techniques of roughing-out, preforming, finishing, and surface treatment) of 60 ceramic *chaînes opératoires* identified from pottery from 14 sites and three cultural contexts in France and the United Kingdom (Duffaits, Trevisker/Deverel-Rimbury, and Norman pottery), which date to the Middle Bronze Age (1600–1350 BC) (see also Manem 2020). One result shows that ceramic technical traditions from these cultural groups were deeply dominated by a branching process (Collard, Shennan, and Tehrani 2006, 171)[2] and anchored in the Atlantic Early Bronze Age, thus showing kinship links between learning networks and, consequently, between the bearers of technical traditions not previously suspected in this part of Western Europe.

The first result resolved an issue raised in previous work (Manem 2008, 2017; Manem, Marcigny, and Talon 2013) where it had been observed that several technical behaviors were similar in the southwest of France and around the English Channel (Normandy and southwest England), without understanding whether the origin of this similarity was related to a convergence, a blending,[3] or a branching process. In the end, two distinct dynamics are observed: those related to consumer needs and the other to the producers. On the one hand, the dynamics relating to the needs of consumers (e.g., shape and decorative style, dissociated from the decorative techniques that meet other dynamics) may be homogeneous or standardized and intercultural over a vast regional area. The borrowing of shape and decoration induces contacts between contemporaneous consumers undoubtedly catalyzed by the many economic interactions between cultures that characterize this period. It results in highlighting vast stylistic provinces (e.g., Lachenal et al. 2017) that do not necessarily reflect a cultural reality, in the sense that a stylistic province can cover several cultural groups (Gelbert 2002; Roux 2019). On the other hand, the dynamic between generations of potters remains highly vertical when it implies a transmission by acquisition of motor habits with a tutor. The evolution of technical traditions was therefore endogenous. The meeting of these two dynamics generates a stylistic province regrouping several social groups that can be distinguished by their technical traditions—as around the English Channel (Manem 2017; Marcigny, Bourgeois, and Talon 2017).

On a broader scale, this phylogenetic tree supports the global tendency (observed by Collard, Shennan, and Tehrani 2006) toward branching processes, which have been more important in the evolution of cultural similarities and differences among human populations than blending processes. However, our result is also atypical when compared to other studies not based on the *chaîne opératoire* concept (for a non-exhaustive list, see Manem 2020). Indeed, the branching rates of our tree (here based on the retention index, which allows comparison between trees without the same quantity of data) are exceptionally high, close to the highest rates recorded in biology as compiled by Collard, Shennan, and Tehrani (2006). One study (Pardo-Gordó, Rivero, and Aubán 2019) based on decorative techniques in

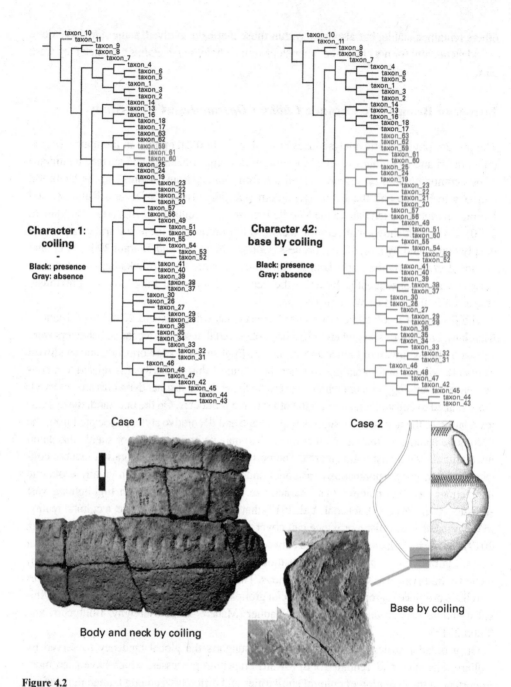

Figure 4.2
Phylogenetic tree of 60 ceramic *chaînes opératoires*. Characters 1 (coiling) and 42 (base by coiling) are mapped.
Source: Adapted from Manem 2020.

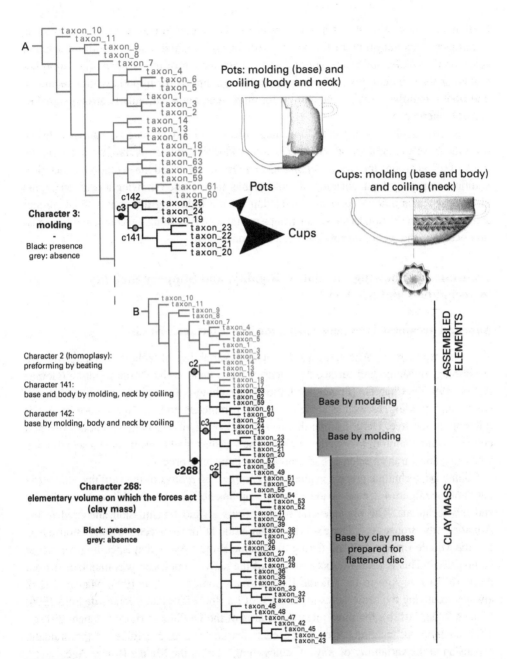

Figure 4.3

Phylogenetic tree of 60 ceramic *chaînes opératoires*. The main characters discussed are mapped in the tree as well as the monophyletic groups involving the techniques on clay mass, including the molding tradition. *Source:* Adapted from Manem 2020.

Neolithic ceramics also shows a much lower rate compared to that of ceramic fashioning techniques. This important difference between our matrix and matrices built on data not based on the *chaîne opératoire* concept and the acquisition of motor habits (i.e., morphological or stylistic features of the material culture) supports the hypothesis that technical traditions—mainly roughing-out and preforming—contain a particularly strong signal of cultural inheritance.

These two results—one, a deep branching process and differentiated dynamics between consumer needs (vessel shape and decor) and producers (*chaîne opératoire*), and two, the study of social groups through the *chaîne opératoire* approach and the evolution of technical traditions requiring the acquisition of motor habits from a tutor showing a pertinent signal of cultural transmission—form the foundation for the following discussion about rigidity and flexibility in relation to the social groups within which learning, teaching, and transmission occurred and where borrowing was absent or limited.

Transmission Showing Flexibility, Rigidity, and Slippery Stability in Techniques and Methods

Ancestral Technical Traditions Facing the New Consumer Needs

This European Bronze Age context offers a wide variability of techniques (roughing-out, preforming, finishing, and surface treatment). I focus here on the Duffaits culture (Gomez de Soto 1995) in Charente (France) with the two roughing-out techniques mentioned above, coiling and molding, as well as some associated parameters and methods whose lineage is traceable in the tree. Here, I argue that new traditions involving both techniques reveal a cultural accumulation of knowledge fueled by new demands of consumers in a social group whose technical transmission is based on a deep branching process.

Coiling is a technique present in almost all *chaînes opératoires* of the Duffaits culture (57 out of 60; see figure 4.2, case 1), inducing a strong link between the Duffaits cultural identity and this technique. The phylogenetic tree confirms a rigid transmission anchored in the Atlantic Early Bronze Age. There was no transmission failure between the Early Bronze Age and the Middle Bronze Age and during the Middle Bronze Age—that is, coiling was never "reinvented." This is probably because part of the native population was maintained from the Early Bronze Age to the Middle Bronze Age (Gomez de Soto 1995; Manem 2020), always occupying the same caves in the karst of La Rochefoucauld (Gomez de Soto 1996; Manem 2008, 2012b), the most picturesque area of the Duffaits territory (Manem 2012a). This roughing-out technique crosses the centuries unchanged, regardless of the dramatic expansion in the variability of ways of doing things during the Middle Bronze Age.

However, we also see how an invention such as molding transformed into an innovation by inserting itself within an ancestral tradition based on coiling. As discussed above, cladistics builds nested hierarchical relationships between taxa only with shared derived characters. Once a novel character appears—an invention—it can diffuse by being adopted across a community and across generations, thus becoming an innovation (Roux 2020). In our present case study, coiling appears as a shared ancestral character (i.e., a similarity shared between two or more taxa inherited from a more distant and older common ancestor) for all of the *chaînes opératoires* under consideration and thus forms the outgroup (e.g., the Early Bronze

Age *chaîne opératoire*, here taxon 10; see figure 4.2). Therefore, coiling is not informative here as it merely indicates the starting point of the present case of evolutionary change. In contrast, molding appears as an innovation during the Middle Bronze Age as it is shared between several *chaînes opératoires* (7 to 60), which constitute a monophyletic group on the tree (figure 4.3, A). According to the stratigraphic context of the caves (Gomez de Soto 1996; Manem 2008), the molding technique was transmitted for several generations of potters, inducing "pottery stability in terms of the stability of traditional patterns of demand" (Nicklin 1971, 18). The new traditions involving molding are not done at the expense of the coiling technique, since both are used but for different phases (i.e., the fashioning of the different parts of the recipient). Both techniques were rigidly transmitted within their respective potters' lineages, even though the appearance of coiling preceded that of molding and was transmitted over a longer period.

The molding technique innovation thus seems to be the result of a flexible behavior—that is, offering a solution to new functional demands of their market. Indeed, this technique appears with two new ceramic shapes, pots and cups (figure 4.3, A), indicating a response by potters to new demands by consumers.[4] I suggest that it is the rigidity of the original coiling tradition, faced with a new functional demand, which brought about this flexibility. Indeed, the ancestral coiling technique was not replaced at the time, as coiling coexisted with the new molding technique, exhibiting instead a form of slippery stability with a ratchet effect.

Slippery Stability: Bridging Rigidity and Flexibility

If these results allow us to see that flexibility and rigidity are intertwined and respond to different sources—one the original technical tradition and the other emerging from a new functional need—it is important to understand *how* and not only *why* ancestral rigidity at once engenders flexibility during an invention process (i.e., at the individual level) to form a kind of symbiosis before becoming an innovation (i.e., at the collective level), and thus a new rigid behavior in the practice of potters. This is where using cladistic analysis that takes into account the techniques and their parameters, methods, and *chaînes opératoires* (as defined above) proves its worth in addressing these key issues.

Tracking the evolution of the parameter dedicated to the elementary volume on which the forces act shows the architecture of the bridge that leads from the coiling to the molding technique. The figure (figure 4.3, B) illustrating the character 268 dedicated to the lump of clay shows that it supports a fundamental clade in the evolution of the majority of *chaînes opératoires*: from this innovation, a high variability of *chaînes opératoires* appears (44 to 60) at various times, characterized by modeling, molding, or a clay mass prepared into a flattened disc of clay (figure 4.3, B). This transition between a "family" of roughing-out techniques related to assembled elements to another "family" related to the lump of clay promoted a diversity of new *chaînes opératoires*. Thus, ancestral pressure-related techniques (e.g., coiling) ended up being combined with percussion-related techniques for roughing-out (e.g., molding). However, when looking at the whole evolution of traditions during the Bronze Age, the parameters dedicated to the elementary volume on which the forces act (assembled elements and lump of clay) are much more rigid than the roughing-out techniques themselves. One of the possible reasons is that the clades supporting the characters of the concerned parameters express *robust systems* (Roux 2010, 228)—that is, large transmission networks including a large number of potters sharing the same parameter and thus contributing to increasing its stability.

Since the Early Bronze Age, and for many generations of Middle Bronze Age potters, the bases are shaped by assembled elements with coiling (figure 4.2, case 2), as are the body and the neck (16 out of 60). With the lump of clay innovation, the method involving base construction by coiling disappears in the communities of practice. The coiling technique and some methods involving coiling continue to be transmitted yet only for the shaping of the other parts of the vessels (i.e., body and neck or only the neck), as well as the sequencing of the three phases (i.e., base to neck), independently of the two techniques used. These express a resistance to change marked by a rigid behavior persisting from one generation to another.

Let us examine in closer detail the two aforementioned shapes—pots and cups—that were the result of changes in consumer needs. Molding is involved only in the pots' base (figure 4.3, A, character 142) and in the base and body for the cups (figure 4.3, A, character 141). Therefore, coiling is involved in forming the body and neck of pots and only for the neck of cups (figure 4.3, A). Pots and cups share the same roughing-out techniques but not the same methods.

Molding has not only been a technique expressing flexibility in responding to the demand for new shapes, but this flexibility has also played out in the methods used to distinguish these two shapes. In a sense, from an evolutionary point of view, the pots are closer to the previous traditions since two out of three parts are made by coiling. The cups, on the other hand, show a greater evolutionary distance since molding is used to shape both their base and body. The cups' necks are the only "fossil witness" of the old ways harkening back to the Early Bronze Age that were transmitted across generations as a rigid or conservative behavior generated largely by motor habits (Minar and Crown 2001; Roux 2019). While the available stratigraphic information cannot answer the question of whether the pots appeared before the cups, I hypothesize this order to be correct.

Techniques and methods are central elements of innovation and, in our present case study, contribute to the branching process of the technical lineages. But innovations in methods appear much more dynamic than innovations in techniques. They cement the bridge between rigid and flexible behavior, between tradition and innovation, as a *slippery stability* where a technique slides from one part of the vessel (i.e., base) to another (i.e., body). This slippery stability is characterized by the more recently adopted technique "advancing" from one phase to another (e.g., from molding the base to molding both the base and body), with the more traditional technique "retreating" (i.e., coiling body and neck to only the neck). Moreover, it seems that this process operates according to the so-called ratchet effect, where each modification of the original tradition "stays firmly in place in the group until further modifications are made" (Tomasello, Kruger, and Ratner 1993, 495). We see this slippery stability here with some roughing-out techniques, but the observation is similar for some preforming techniques (i.e., beating) and methods that are shown in this phylogenetic tree (details in Manem 2020). However, as mentioned above, this slippery stability is conditioned by a particularly rigid perception of the method as a sequencing of the three phases from the bottom to the top of the ceramic (for all 60 *chaînes opératoires* illustrated in the tree).

The two examples discussed above also show that potters juggled tradition and novelty, and they absorbed and solved this tension according to a preexisting mental pattern linked to the individual potter's own initial learning. If the previous discussions were mainly focused on the concretization of the project (i.e., the manufacturing of a ceramic for a given function), it is important to also examine the subject at the level of the *conceptual scheme*. By conceptual scheme, I refer to the intellectual scheme, which is itself implemented through a *chaîne opératoire* (Inizan et al. 1999, 15). It can be extended to the "mappa mundi" mentioned in van der Leeuw (1993, 241). The sequencing of the three phases from the bottom to the top reflects a rigid conceptual scheme in all potters, whether they are involved in an invention process or not. In parallel, phases related to techniques appear flexible.

Focusing only on the roughing-out stage here, before the adoption of the lump of clay, the conceptual scheme was "simple": a single technique (coiling) and the same elementary volume on which the forces act (assembled elements) were used for the different phases (base, body, neck) to realize a diversity of vessels and functions. Because coiling was used for all phases as an invariant, the phases were differentiated only morphologically (i.e., in terms of their shape). After the invention and innovation characterized by the mass of clay— visible on the tree by the clade supporting the character 268—the conceptual scheme became "complex," with different techniques now being "located" with specific morphological parts (or phases) of the ceramic. This does not mean, however, that one scheme replaced the other wholesale; instead, the technical traditions resulting from a simple or complex conceptual scheme coexisted. In any case, the more complex conceptual scheme required a diversity of know-hows as it involved two or more techniques depending on different choice of parameters. This means that the conceptual scheme induced a broader perception of the parameter dedicated to the elementary volume on which the forces act by the systematic combination of assembled elements and clay mass. Moreover, this new conceptual scheme would give a more important role or meaning to the phases because these were no longer perceived solely in terms of the parts' morphology but now also in terms of the technique used for each phase (e.g., base with technique X, body with technique Y, etc.).

The apparition of a complex conceptual scheme raises an important question: Does giving more technical specificity to the phases rather than to the whole lead to more evolutionary independence of these phases, opening the door for more potential flexibility in modifying the whole technical tradition?

The Potters' Adaptation to the Consumers' Needs versus the Order of Development of Techniques

Looking for the Whom

We have seen *why* potters adopted molding—they were answering new consumer needs— and *how* they adapted to those demands. We still need to understand the *whom* at the origin of the invention. In this section, we will see that looking for the *whom* leads us to a better understanding of *how* the invention came about. I follow van der Leeuw's approach (1993, 241) by examining the choices made by the potters contraposed with some choices that were not made.

Techniques are not ineluctably linked to the diversity of vessel shapes or functions. There are many cultures where potters make ceramics that are different in shape and size using the same technique (e.g., Longacre 1991). This is especially true when the way of doing things offers a compromise in constraints encountered by the potter (van der Leeuw 1991, 34). For instance, one of the main difficulties for potters is to prevent the vessel from collapsing or being deformed during the roughing-out (van der Leeuw 1993, 243). If molding is perfectly adapted to produce an open shape with a very curved and regular profile (figure 4.3, A), other solutions exist. For instance, one could use a spiral coiling on a mold made of an old pot turned upside down, as observed in Uganda (O'Brien and Hastings 1933), or by using a mold made of an inverted basket, as has been reported for Zuni pottery in the American Southwest (Cushing 1886, 497). These solutions would even be the most "logical" for Duffaits potters since coiling was traditionally used for all phases. The flexibility here would have been to switch between molds of different sizes or shapes, as well as the quantity of mass of clay, to adapt to the demand and function. But Duffaits potters never adopted this way of doing things. It is therefore necessary to understand why if we are to understand the *whom*.

The parameter of the force in percussion is not only a characteristic of the molding traditions but is present in some *chaînes opératoires* involving beating, from a roughing-out by coiling. The major difference between molding and beating is that the percussion is limited in the latter case to preforming. This situation is reminiscent of a field experiment in Senegal with potters switching from a tradition based on modeling by drawing and involving hammering for preforming to another way of doing based on molding. In both traditions, percussion gestures are shared. There is no motor constraint impeding the switch from one tradition to the next (Gelbert 2002, 275–276). A conceivable hypothesis would thus be that Bronze Age potters practicing beating also made cups by molding. This would explain the ways in which many original behaviors are maintained in pots and cups (i.e., coiling for body and/or neck) and would reveal the flexibility of the force parameter in percussion. It would be an optimization of practices, in the sense that the percussion on assembled elements (and perhaps also some tools) would be "recycled" in another family of techniques, those on clay mass (i.e., molding). This situation can be found in ethnographic contexts where some potters juggle different mastered techniques in order to meet consumer demands (e.g., Southwest Ethiopia with some Gamo potters; see Arthur 2006). However, the tree shows that these two traditions—coiling/beating and molding/coiling[5]— are not in the same monophyletic group, and that the beating trait appears as a homoplasy in the tree (figure 4.3, B, character 2). That is, beating is a similarity that is not inherited from a common ancestor but is only shared by the *chaînes opératoires* concerned.

I suggest that the more likely explanation to the question of *whom* is that an expert potter (or group of potters) that mastered beating was able to transmit the percussion parameter to another group whose traditions were based on mass of clay and assembled elements. This form of horizontal transmission—between peers from two distinct learning networks but who share an older kinship—in turns suggests that a charismatic expert was able to bring the other group, composed of less skilled or nonexpert potters with different technical traditions, to change those traditions in one direction. The charismatic profile of a potter would be determining since it defines *when* the change took place. According to this hypothesis, a charismatic profile will contribute to a process of cultural selection: the percussion parameter from the beating tradition would have diffused through a "biased" transmission process (Boyd and Richerson 1985; Henrich and McElreath 2003; Shennan 2011).

In addition to the fact that molding is perfectly adapted to produce very open forms such as cups, this hypothesis—that the percussion parameter was transmitted from an expert to a less skilled or nonexpert group—is supported by the advantages of adopting the molding technique. As van der Leeuw (1993, 243) pointed out, one of the main difficulties for a potter is to control the shape of the vessel. The use of a mold offers a firm guide so that the potter is thus discharged from having to control the shape (Fewkes 1937), which in turn compensates for a lack of flexibility due to motor habits not initially related to percussion. Therefore, learning this technique is less difficult and requires less time than its alternative (Arnold 1985). For instance, by simply changing the type of mold used, modeling allows any potter "to make a wide range of different shapes or sizes, with minimal adaptation of the technique" (van der Leeuw 1993, 244). Thus, potters with less expertise can modify from the beginning their way of doing to successfully obtain the new desired shape. To summarize, molding is probably in what could be called the gray area of invention. Molding is an invention as an individual act of creation by a potter working on a mass of clay. However, the percussion parameter was probably learned from a more expert potter mastering the beating technique.

Beyond Social Motivation: The Order of Development

The process of invention has been so far characterized from a synchronic perspective by examining the why, how, and whom of the specific invention. However, invention is always located in a cultural context that precedes it. This calls for a diachronic analysis that, I argue, allows us to identify an "internal dynamic" to the evolutionary direction of technical behavior—a dynamic that the relevant generation of potters is not necessarily aware of being embedded within.

Robert Cresswell (1996, 21) indicates that techniques naturally tend to develop without the need for a social motivation, with traditions in cultures exhibiting different "orders of development." Cresswell argues that technology and society have a cyclical relationship, oscillating between interdependence and independence. The internal dynamic of interest here operates by evolutionary stages that do not only concern the specific case study of the invention of the pots and cups and their *chaînes opératoires*. It relates more generally to a large number of *chaînes opératoires*, starting with the innovation of mass of clay for the base (parameter: elementary volume on which the forces act), which is perfectly illustrated by the vast monophyletic group reflecting the majority of the *chaînes opératoires* (figure 4.3, B, black clades and character 268).

In the Bronze Age context, the order of development of the molding could appear complex. The cladistic approach presented here charts the process in detail. However, the order of development from parameters, techniques, and methods can be summed up in three main evolutionary stages starting from a coiling tradition: (1) mass of clay, (2) molding for base, and (3) molding for base and body. Thus, the slippery stability and the ratchet effect appear as actors in the order of development.

Molding is an evolutionary step of the mass of clay parameter of innovation and not of the assembled elements or the percussion. This may seem obvious, but coiling was not the bridge that led to molding. Otherwise, the Duffaits potters would probably have invented intermediate ways of doing things. Examples are (1) rolling a coil in the palm of the hand and then placing it against a mold to draw the coils against the mold, as in northern Cameroon with the Mofu potters (Barreteau and Delneuf 1990); (2) fixing coils directly against a mold, as described above with the case in Uganda (O'Brien and Hastings 1933); or (3) molding a base from a lump of clay and then applying a coil against the mold to extend the base, as in the Southern Diegueño Yuman ceramic tradition (Rogers 1936, 8–9).

Unlike the coiling technique, molding allows potters to achieve at the same time rough-out and preform for base and body while avoiding traces of joint. This technique also reduces the drying time between each step of the *chaîne opératoire* since two parts (base and body) are dried at the same time before the potter continues shaping the neck without the lower part sagging (Arnold 1985, 202). In other words, these evolutionary steps—from assembled elements to mass of clay and then molding—have led to a saving of energy and time, at least for the molded part of the ceramic, with this benefit being, of course, different according to the potters and their dexterity (Guthe 1925, 40).

The integration of percussion could also prove beneficial from another evolutionary trajectory (i.e., another monophyletic group in this tree, not involving molding), leading to a more complex *chaîne opératoire* (e.g., roughing-out base, body, and neck by coiling and preforming body by beating), which is itself the evolution of a less complex tradition (i.e., without beating) and similar to the Early Bronze Age tradition. Percussion-related behaviors could thus appear to be particularly flexible, moving between transmission networks that do not share a direct common ancestor. In summary, molding would be the result of the encounter between a new demand, on the one hand, and a change of perception of the parameter of the elementary volume on which the forces act as well as the parameter of the forces, on the other. The whole evolutionary context for inventing molding was present; all that was needed was a nudge to bring about its invention. The percussion parameter could have been the spark that produced the flame.

From a broad perspective in line with Cresswell (1996), the invention of molding shows an interdependence between producers and consumers. This is unlike the invention of mass of clay, which may be independent of consumers, follows another temporality different from the techniques themselves, and depends solely on producers' lineages.

The Evolvability of Technical Traditions

The preceding discussions lead to a simple question: Why is the molding technique not gradually adopted in the other technical traditions of this culture? This question therefore implies a context where a technique would no longer be assigned to a particular type of vessel. But this has never happened in the cultural context studied. Beyond this monophyletic group involving molding, the ratchet effect is not necessarily observed for other roughing-out techniques because they do not evolve like some traditions involving the coiling for base, body, and the neck. Alex Mesoudi (2011, 236) points out a nuance in Michael Tomasello's approach: "human culture *can* accumulate modifications." A way of doing things can show both accumulation and the absence of accumulation. This emphasis on "can" is perceptible within a learning network when the way of doing is approached with the *chaîne opératoire*.

Potters' lineages from the Duffaits culture indeed show different evolutionary trajectories in terms of roughing-out techniques. While some behaviors result from an accumulation of modifications exhibiting a ratchet effect, others remain stable for centuries. Some others do change but in other steps of the *chaîne opératoire*, such as preforming or finishing. Thus, each technical tradition within this culture follows its own evolutionary trajectory.

We will see how the evolutionary trajectory related to molding leads to its double "isolation." This will bring us to the notions of concretization, evolvability, and modularity, by insisting on the importance of the phases—namely, the connection between the base and the body on one side and the independence of the neck on the other side.

The molding technique likely facilitates the ratchet effect in the sense that this technique is not very complex and, accordingly, is easy to acquire (Arnold 1985, 205). This ease could have encouraged potters to abandon their original technique (Gosselain 2002, 133) or at least, as in Duffaits culture, limit it to the base and body for particular vessel shapes. It is even plausible that the evolution of the technical tradition tended toward a simplification of a part of the *chaîne opératoire*, as can be observed in certain ethnographic contexts (Pétrequin and Pétrequin 1999), since this technique offers a better control over the shape than coiling.

Contrary to the interdependence discussed above between base and body, the neck appears to be more independent from the other phases. Change in techniques and methods for base and body follows an interdependent evolutionary trajectory because of the search for efficiency (molding). However, by widening the scale of observation to the entire fashioning *chaîne opératoire*, we see that the change in base and body were made without sacrificing the traditional way of making a neck (coiling). This suggests that the use of the coiling technique for the neck depended more on cultural factors—hence the importance of its rigid transmission—than on a search for efficiency. Potters could perfectly see that the same roughing-out technique could be used for all three phases (e.g., coiling), yet this was never generalized to molding. Indeed, coiling is systematically maintained for the neck even though it extends the manufacturing time (e.g., Barreteau and Delneuf 1990, 127).

Coiling for the neck and the sequencing of the three phases from the bottom to the top appear to be the "invariant 'backbones'" (van der Leeuw 1993, 240) and probably the heart of the technical traditions, whether or not the potters were aware of this aspect of their traditions (van der Leeuw 1994, 314). The idea of reversing the sequencing or changing the technique for the neck seems to have been completely unknown.

The three parts of the ceramics did not all have an interdependent evolutionary trajectory, as demonstrated by the resistance to change observed with the neck made with coils. This is in line with Olivier Gosselain's (2002, 190) observation that "changes may be made at almost any stage of the *chaîne opératoire* without jeopardizing the whole system" (see also van der Leeuw 1993, 240).

These patterns of (a lack of) interdependence lead us to consider the issue of the evolvability of technical traditions. By evolvability, I refer to "the capacity of a lineage to evolve"—that is, "the capacity to generate heritable, selectable phenotypic variation" (Kirschner and Gerhart 1998, 8420). In a sense, the way of doing things was modified as much as possible without being "lethal"—that is, the disappearance of the *chaîne opératoire* and its replacement by another one that no longer contains any original technique or method—thus permanently maintaining the transmission signal from one generation of potters to the next. Flexibility was thus compartmentalized, in the sense of being restricted to certain steps of the *chaîne opératoire*.

In the present case, the walls of the compartment were fixed by what is most robust in the tradition or its core: the neck roughing-out in this Bronze Age cultural context. Stated otherwise, the evolutionary process "will never" touch the neck roughing-out, thus bridling any change beyond the base and the body—to address only the phases and the roughing-out techniques.

It is therefore only through a diachronic look at the evolution of the *chaînes opératoires* that we can perceive the limits of flexibility and invention, in the sense of evolution generating inheritable variations. Regarding the base and the body, the evolvability was doubly limited, if not null, as soon as the base and the body were made by molding.

This interpretation takes into account the process of concretization developed by Gilbert Simondon (1958) and exploited by Eric Boëda (2013) in prehistory and for lithic technology. By concretization, I refer to a technical object (the abstract technical object), which evolves by adaptation, unifying itself internally, into the concrete technical object (Simondon 1958, 20). An abstract technical object is characterized by elements (e.g., pieces of an engine) that intervene at a certain time in the functioning cycle, each in turn. Conversely, a concrete technical object implies a reciprocal exchange of energy between the pieces of the engine so that each piece cannot function without the other, since each piece plays several roles (Simondon 1958, 20–26).

The original ceramic tradition of using coiling for the three phases (base, body, neck) can be considered as an abstract object: the three phases are roughed out successively without any exchange of energy. Then, the potter proceeds to the preforming (if the potter does not do the preforming at the end of roughing-out of each part). On the other hand, the molding of the base and the body is done at the same time, making it impossible to distinguish a succession of operations: it is the same mass of clay. More than that, molding is both roughing-out and preforming technique: these two stages of the method cannot be distinguished. The lower part of the ceramic (base and body) can be considered as a concrete object. Therefore, the evolvability of the way of doing things seems particularly limited since base, body, roughing-out, and preforming are mutually linked, even cemented. The potter cannot change one without changing the rest. The upper part (neck by coiling) is still an abstract object but cannot be modified either by invention or borrowing because, as we have seen previously, it is the invariant cultural part bringing technical tradition and social group together. Thus, the solution brought by the potters—themselves embedded in an underlying dynamic linked to the order of development of techniques, and their parameters and methods—to answer some new needs of the consumers led to an evolutionary dead end, making an exclusive relation between forms (pots and cups), functions, and way of doing things, which could then explain why the molding was never introduced in the other *chaînes opératoires* by a borrowing process between learning networks and the potters' lineages. Furthermore, this monophyletic group involving this adequacy between molding and forms is doomed to disappear if consumer needs change. This situation thus reinforces the branching process observed for the European Bronze Age (Manem 2020).

Beyond this particular *chaîne opératoire*, the evolvability of a way of doing things seems much more important when it can be considered as an abstract technical object because of the relative independence of each phase and stage—that is, roughing-out and preforming techniques for base, body, and neck. The tree shows this very well: the roughing-out of the three phases by coiling generated a strong diversity of *chaînes opératoires*.

This leads us to complete this triptych by briefly extending the discussion of evolvability and concretization to some debates in cultural evolution concerning cultural recombination and modularity under the scope of the *chaîne opératoire* approach. The recombination process (Mesoudi and O'Brien 2008; Basalla 1988; Lewis and Laland 2012) can be defined as "the bringing together of existing cultural traditions or of existing cultural traditions' subcomponents into novel, complex cultural composites" (Charbonneau 2016, 374). This implies that the cultural traits concerned are hierarchically structured (e.g., Mesoudi and O'Brien 2008). Adapted to the *chaîne opératoire* concept, the process of recombination would involve decomposing some stages of existing *chaînes opératoires* (techniques and methods) and recomposing them into a new *chaîne opératoire*. Cultural recombination is based on cultural traditions with a modular structure, with the modules extracted for recombination being relatively independent of the original traditions (the questioning of this relative independence and thus our ability to predict or detect potential modules is highlighted in Charbonneau 2016).

The molding technique was an invention in our Bronze Age context. This technique is therefore not the result of a recombination. Moreover, molding "will never" be the decomposed stage of a *chaîne opératoire* that becomes recomposed into a new *chaîne opératoire* in this cultural context because of the concretization process discussed above, making an exclusive relation between technique and method, forms (pots and cups), and functions. Therefore, detecting Simondon's process of concretization in cultural traditions could predict (1) what cannot be a module in a cultural tradition—even though this stage of the *chaîne opératoire* is hierarchically structured and involved in learning and transmission—and (2) what can significantly curb modularity and thus recombination, since the nonmodular part of a cultural tradition comes to restrict its modular part and thus curb the evolvability of the involved traditions. However, a detailed perception of hierarchically organized technical behaviors (Charbonneau 2016; Simon 1962)—those involving techniques and their parameters—opens up a more nuanced situation. The two parameters related to the molding technique—percussion (parameter: force) and clay mass (parameter: elementary volume on which the forces act)—could be (theoretically) modules under the condition that the actors concerned by the recombination have acquired the motor habits involved. In our Bronze Age context, these parameters are indeed already "embedded" in some existing *chaînes opératoires* (Manem 2020). The evolvability of these two parameters is consequently different from that of the molding.

Conclusion

We have seen in the experiment conducted at the individual scale (Roux et al., this volume) that the high-level experts changed their way of doing from the beginning of the shaping process to provide a relevant solution, while the low-level experts only changed their course of action. The molding technique adopted for base and body was, for the Duffaits potters, a change in the way of doing from the beginning. However, this technique is guided, meaning that the mold facilitates the construction of the final shape and therefore makes the technique suitable for a low-level expert potter under certain conditions.

Consequently, I support the hypothesis that, under these conditions, rigidity and flexibility could operate in symbiosis for low-level experts helped by a potter mastering the percussion: rigidity invents flexibility to preserve some stability in original technical tradition when the low-level expert, already embedded in a community of practice mastering the mass of clay parameter, is faced with a new consumer demand. The choice adopted—molding—constitutes another example (Gosselain and Livingstone Smith 2005, 44; Birmingham 1975, 371) where the idea of choice does not necessarily mean that it is random but instead corresponds to a unique combination of factors of different natures described above, perceptible in the synchronic and diachronic and well-illustrated by the Swiss Cheese Model (Reason 2000).

The exploitation of data associated with the techniques, and especially with the parameters of the techniques and the method, shows how they generated cumulative culture and how each component plays a different role in the evolutionary origins of cumulative cultures (Charbonneau 2015). The tree shows that it is above all certain technical parameters and the method—not the techniques themselves—that are the real bridge leading from coiling tradition to molding by gradual changes, along a pattern of slippery flexibility. It constitutes another example of "descent with modification" (Lycett 2015). The ancestor-descendant relationship study thus allows "identifying a transmission history and characterizing the forces affecting it" (Shennan 2011, 1072).

Acknowledgments

Thanks to Mathieu Charbonneau and Dan Sperber for the initiative of this volume as well as their invitation. I am grateful to Mathieu Charbonneau, Stephen Shennan, and Felix Riede for useful comments on a previous draft. The results presented here are based on data from my former grants: Fyssen Foundation, hosted by University of Exeter, Department of Archaeology and with Anthony Harding; and Marie Curie Action ID 274395, FP7-People-2010-IEF-Bronze Age, hosted by University College London, Institute of Archaeology with Stephen Shennan.

Notes

1. A taxon is group of organisms, a species, or genus. In cultural evolution, a taxon can be other a priori defined analytical operational units. It is shown in a matrix as a list of characters and character states. Characters are observed traits.

2. "Cultural similarities and differences among human populations are primarily the result of a combination of within-group information transmission and population fissioning." (Collard, Shennan, and Tehrani 2006, 171).

3. "Relationships among human populations [can be perceived as] a braided stream, with different channels flowing into one another, then splitting again."

4. New demands may have multiple natures (e.g., Arnold 1985, 127–167) that will not be discussed here.

5. In figure 4.3, B, the monophyletic group characterizing the tradition involving coiling (roughing-out) and beating (preforming) is plotted on the tree (clade supported by character C2 and involving *chaînes opératoires* or taxon 13–18). The monophyletic group characterizing the tradition involving molding is plotted on the tree (clade supported by character C3 and involving *chaînes opératoires* or taxon 19–25).

References

Arnold, D. E. 1985. *Ceramic Theory and Cultural Process*. Cambridge: Cambridge University Press.

Arthur, J. W. 2006. *Living with Pottery: Ethnoarchaeology among the Gamo of Southwest Ethiopia*. Salt Lake City: University of Utah Press.

Barreteau, D., and M. Delneuf. 1990. "La Céramique Traditionnelle Giziga et Mofu (Nord-Cameroun): Étude Comparée Des Techniques, Des Formes et Du Vocabulaire." In *Relations Interethniques et Culture Matérielle Dans Le Bassin Du Lac Tchad*, edited by D. Barreteau and H. Tourneux, 121–155. Paris: Editions de l'ORSTOM.

Basalla, G. 1988. *The Evolution of Technology*. Cambridge: Cambridge University Press.

Birmingham, J. 1975. "Traditional Potters of the Kathmandu Valley: An Ethnoarchaeological Study." *Man* 10 (3): 370–386.

Boëda, E. 2013. *Techno-Logique & Technologie. Une Paléo-Histoire Des Objets Lithiques Tranchants*. Prigonrieux: Archéo-éditions.

Boyd, R., and P. J. Richerson. 1985. *Culture and the Evolutionary Process*. Chicago: University of Chicago Press.

Charbonneau, M. 2015. "All Innovations Are Equal, but Some More than Others: (Re)Integrating Modification Processes to the Origins of Cumulative Culture." *Biological Theory* 10 (4): 322–335.

Charbonneau, M. 2016. "Modularity and Recombination in Technological Evolution." *Philosophy & Technology* 29 (4): 373–392.

Cochrane, E. E. 2009. *The Evolutionary Archaeology of Ceramic Diversity in Ancient Fiji*. Oxford: BAR.

Collard, M., S. Shennan, and J. Tehrani. 2006. "Branching, Blending, and the Evolution of Cultural Similarities and Differences among Human Populations." *Evolution and Human Behavior* 27 (3): 169–184.

Cresswell, R. 1976. "Techniques et Culture. Les Bases d'un Programme de Travail." *Techniques & Culture* 1:7-59.

Cresswell, R. 1996. *Prométhée ou Pandore? Propos de Technologie Culturelle*. Paris: Éditions Kimé.

Cushing, F. H. 1886. "A Study of Pueblo Pottery as Illustrative of Zuni Culture Growth." *Washington Bureau of American Ethnology Annual Report* 4:467–521.

Darlu, P., P. Tassy, C. d'Haese, and R. Zaragüeta i Bagils. 2019. *La Reconstruction Phylogénétique: Concepts et Méthodes*. Paris: Editions Matériologiques.

Fewkes, V. J. 1937. "Aboriginal Potsherds from Red River, Manitoba." *American Antiquity* 3 (2): 143–155.

Fewkes, V. J. 1940. "Methods of Pottery Manufacture." *American Antiquity* 6 (2): 172–173.

Fewkes, V. J. 1944. "Catawba Pottery-Making, with Notes on Pamunkey Pottery-Making, Cherokee Pottery-Making, and Coiling." *Proceedings of the American Philosophical Society* 88 (2): 69–124.

Gelbert, A. 2002. "Emprunt Technique et Changement Gestuel: Mesure Des Contraintes Motrices En Jeu Dans Les Emprunts Céramiques de La Vallée Du Sénégal." In *Le Geste Technique. Réflexions Méthodologiques et Anthropologiques*, edited by B. Bril and V. Roux, 261–281. Ramonville Saint-Agne: Éditions Érès.

Gomez de Soto, J. 1995. *Le Bronze Moyen En Occident: La Culture Des Duffaits et La Civilisation Des Tumulus*. Paris: Picard.

Gomez de Soto, J. 1996. *Grotte Des Perrats à Agris (Charente)—1981–1994. Étude Préliminaire*. Chauvigny: Editions APC.

Gosselain, O. 2002. *Poteries du Cameroun méridional: Styles techniques et rapports à l'identité*. Paris: CNRS Editions.

Gosselain, O., and A. Livingstone Smith. 2005. "The Source Clay Selection and Processing Practices in Sub-Saharan Africa." In *Pottery Manufacturing Processes: Reconstitution and Interpretation*, edited by A. Livingstone Smith, D. Bosquet, and R. Martineau, 33–44. Oxford: BAR.

Guthe, C. E. 1925. *Pueblo Pottery Making: A Study at the Village of San Ildefonso*. New Haven, CT: Yale University Press.

Henrich, J., and R. McElreath. 2003. "The Evolution of Cultural Evolution." *Evolutionary Anthropology* 12 (3): 123–135.

Inizan, M.-L., M. Reduron-Ballinger, H. Roche, and J. Tixier. 1999. *Technology and Terminology of Knapped Stone*. Nanterre: CREP.

Kirschner, M., and J. Gerhart. 1998. "Evolvability." *Proceedings of the National Academy of Sciences* 95 (15): 8420–8427.

Krause, R. A. 1985. *The Clay Sleeps: An Ethnoarchaeological Study of Three African Potters*. Tuscaloosa: University of Alabama Press.

Lachenal, T., C. Mordant, T. Nicolas, and C. Véber, eds. 2017. *Le Bronze Moyen et l'origine Du Bronze Final En Europe Occidentalre (XVIIe-XIIIe Siècle Av. J.-C.)*. Strasbourg: Mémoires d'Archéologie du Grand-Est.

Lewis, H. M., and K. N. Laland. 2012. "Transmission Fidelity Is the Key to the Build-up of Cumulative Culture." *Philosophical Transactions: Biological Sciences* 367 (1599): 2171–2180.

Longacre, W. A. 1991. "Sources of Ceramic Variability among the Kalinga of Northern Luzon." In *Ceramic Ethnoarchaeology*, edited by W. A. Longacre, 95–111. Tucson: University of Arizona Press.

Lycett, S. J. 2015. "Cultural Evolutionary Approaches to Artifact Variation over Time and Space: Basis, Progress, and Prospects." *Journal of Archaeological Science* 56:21–31.

Manem, S. 2008. "Les Fondements Technologiques de La Culture Des Duffaits (Âge Du Bronze Moyen)." PhD diss., Université Paris X-Nanterre.

Manem, S. 2012a. "Les Lieux Naturels Atypiques, Sources Du Paysage Rituel: Le Karst de La Rochefoucauld et La Culture Des Duffaits (Charente, France)." In *Paysages Funéraires de l'âge Du Bronze*, edited by D. Béranger, J. Bourgeois, M. Talon, and S. Wirth, 573–594. Darmstadt: Verlag Philipp von Zabern.

Manem, S. 2012b. "The Bronze Age Use of Caves in France: Reinterpreting Their Function and the Spatial Logic of Their Deposits through the *Chaîne Opératoire* Concept." In *Caves in Context: The Cultural Significance of Caves and Rockshelters in Europe*, edited by K. A. Bergsvik and R. Skeates, 138–152. Oxford: Oxbow Books.

Manem, S. 2017. "Bronze Age Ceramic Traditions and the Impact of the Natural Barrier: Complex Links between Decoration, Technique and Social Groups around the Channel." In *Movement, Exchange and Identity in Europe in the 2nd and 1st Millennia BC: Beyond Frontiers*, edited by A. Lehoërff and M. Talon, 227–240. Oxford: Oxbow Books.

Manem, S. 2020. "Modeling the Evolution of Ceramic Traditions through a Phylogenetic Analysis of the *Chaînes Opératoires*: The European Bronze Age as a Case Study." *Journal of Archaeological Method and Theory* 27 (4): 992–1039.

Manem, S., C. Marcigny, and M. Talon. 2013. "Vivre, Produire et Transmettre Autour de La Manche. Regards Sur Les Comportements Des Hommes Entre Deverel Rimbury et Post Deverel Rimbury En Normandie et Dans Le Sud de l'Angleterre." In *Échanges de Bons Procédés. La Céramique Du Bronze Final Dans Le Nord-Ouest de l'Europe*, edited by W. Leclercq and E. Warmenbol, 245–265. Bruxelles: CReA-Patrimoine.

Marcigny, C., J. Bourgeois, and M. Talon. 2017. "Rythmes et Contours de La Géographie Culturelle Sur Le Littoral de La Manche Entre Le IIIe et Le Début Du Ier Millénaire." In *Movement, Exchange and Identity in Europe in the 2nd and 1st Millennia BC: Beyond Frontiers*, edited by A. Lehoërff and M. Talon, 63–78. Oxford: Oxbow Books.

Mesoudi, A. 2011. *Cultural Evolution: How Darwinian Theory Can Explain Human Culture and Synthesize the Social Sciences*. Chicago: University of Chicago Press.

Mesoudi, A. 2016. "Cultural Evolution: A Review of Theory, Findings and Controversies." *Evolutionary Biology* 43 (4): 481–497.

Mesoudi, A., and M. J. O'Brien. 2008. "The Learning and Transmission of Hierarchical Cultural Recipes." *Biological Theory* 3 (1): 63–72.

Minar, C. J., and P. L. Crown. 2001. "Learning and Craft Production: An Introduction." *Journal of Anthropological Research* 57 (4): 369–380.

Nicklin, K. 1971. "Stability and Innovation in Pottery Manufacture." *World Archaeology* 3 (1): 13–48.

O'Brien, M. J., J. Darwent, and R. L. Lyman. 2001. "Cladistics Is Useful for Reconstructing Archaeological Phylogenies: Palaeoindian Points from the Southeastern United States." *Journal of Archaeological Science* 28 (10): 1115–1136.

O'Brien, M. J., and R. L. Lyman. 2003. *Cladistics and Archaeology*. Salt Lake City: University of Utah Press.

O'Brien, T. P., and S. Hastings. 1933. "Pottery Making among the Bakonjo." *Man* 33:189–191.

Pardo-Gordó, S., D. G. Rivero, and J. B. Aubán. 2019. "Evidences of Branching and Blending Phenomena in the Pottery Decoration during the Dispersal of the Early Neolithic across Western Europe." *Journal of Archaeological Science: Reports* 23 (February): 252–264.

Pétrequin, A.-M., and P. Pétrequin. 1999. "La Poterie En Nouvelle-Guinée: Savoir-Faire et Transmission Des Techniques." *Journal de La Société Des Océanistes* 108 (1): 71–101.

Pujos, F. 2006. "Megatherium Celendinense SP. Nov. from the Pleistocene of the Peruvian Andes and the Phylogenetic Relationships of Megatheriines." *Palaeontology* 49 (2): 285–306.

Reason, J. 2000. "Human Errors: Models and Management." *British Medical Journal* 320 (7237): 768–770.

Rice, P. M. 2015. *Pottery Analysis: A Sourcebook, Second Edition*. Chicago: University of Chicago Press.

Riede, F. 2005. "Darwin vs. Bourdieu: Celebrity Deathmatch or Postprocessual Myth? Prolegomenon for the Reconciliation of Agentive-Interpretative and Ecological-Evolutionary Archaeology." In *Investigating Prehistoric Hunter-Gatherer Identities: Case Studies from Palaeolithic and Mesolithic Europe*, edited by H. Cobb, S. Price, F. Coward, and L. Grimshaw, 45–64. Oxford: BAR.

Riede, F. 2006. "*Chaîne Opératoire, Chaîne Evolutionaire?* Putting Technological Sequences into an Evolutionary Perspective." *Archaeological Review from Cambridge* 21 (1): 50–75.

Riede, F. 2008. "Maglemosian Memes: Technological Ontology, Craft Traditions and the Evolution of Northern European Barbed Points." In *Cultural Transmission and Archaeology: Issues and Case Studies*, edited by M. J. O'Brien, 178–189. Washington, DC: Society for American Archaeology Press.

Rogers, M. J. 1936. *Yuman Pottery Making*. San Diego Museum Papers, No. 2.

Roux, V. 2010. "Technological Innovations and Developmental Trajectories: Social Factors as Evolutionary Forces." In *Innovation in Cultural Systems. Contributions from Evolutionary Anthropology*, edited by M. J. O'Brien and S. Shennan, 217–233. Cambridge, MA: MIT Press.

Roux, V. 2019. *Ceramics and Society: A Technological Approach to Archaeological Assemblages*. Cham, Switzerland: Springer International Publishing.

Roux, V. 2020. "Apprentissage et Inventions: Des Individus Qui Font l'histoire." In *Apprendre: Archéologie de La Transmission Des Savoirs*, edited by P. Pion and N. Schlanger, 211–220. Paris: La Découverte.

Shennan, S. 2008. "Evolution in Archaeology." *Annual Review of Anthropology* 37 (1): 75–91.

Shennan, S. 2011. "Descent with Modification and the Archaeological Record." *Philosophical Transactions of the Royal Society B: Biological Sciences* 366 (1567): 1070–1079.

Shepard, A. O. 1956. *Ceramics for the Archaeologist*. Washington, DC: Carnegie Institution of Washington.

Simon, H. A. 1962. "The Architecture of Complexity." *Proceedings of the American Philosophical Society* 106 (6): 467–482.

Simondon, G. 1958. *Du Mode d'existence Des Objets Techniques*. Paris: Aubier.

Tixier, J. 1967. "Procédés d'analyse et Questions de Terminologie Dans l'étude Des Ensembles Industriels Du Paléolithique Récent et de l'épipaléolithique En Afrique Du Nord-Ouest." In *Background to Evolution in Africa*, 771–820. Chicago: University of Chicago Press.

Tomasello, M., A. C. Kruger, and H. H. Ratner. 1993. "Cultural Learning." *Behavioral and Brain Sciences* 16 (3): 495–511.

van der Leeuw, S. E. 1991. "Variation, Variability, and Explanation in Pottery Studies." In *Ceramic Ethnoarchaeology*, edited by W. A. Longacre, 11–39. Tucson: University of Arizona Press.

van der Leeuw, S. E. 1993. "Giving the Potter a Choice: Conceptual Aspects of Pottery Techniques." In *Technological Choices: Transformation in Material Cultures since the Neolithic*, edited by P. Lemonnier, 238–288. London: Routledge.

van der Leeuw, S. E. 1994. "Innovation et Tradition Chez Les Potiers Mexicains Ou Comment Les Gestes Techniques Traduisent Les Dynamiques d'une Société." In *De La Préhistoire Aux Missiles Balistiques. L'intelligence Sociale Des Techniques*, by B. Latour and P. Lemonnier, 310–328. Paris: Éditions La Découverte.

Wiley, E. O., and B. S. Lieberman. 2011. *Phylogenetics: Theory and Practice of Phylogenetic Systematics, Second Edition*. Hoboken, NJ: Wiley-Blackwell.

II FROM RIGID COPYING TO FLEXIBLE RECONSTRUCTION

5 Relevance-Based Emulation as a Prerequisite for Technical Innovation

György Gergely and Ildikó Király

Introduction

To understand the cumulative nature of human culture and its transmission across generations, it is important to shed light on the evolutionary origins and multiple determinants of humans' unique capacity to produce new, efficient, and innovative technological solutions to relevant problems. As a result of their increased efficiency, new innovative techniques are more likely to be selected, to be transmitted, and to stabilize in a culture, thereby extending the scope of the existing repertoire of cultural skills. The emergence of new and innovative technologies, therefore, contributes significantly to the cumulative nature of human cultural knowledge. The increasing scope and complexity of the repertoire of technological skills that are maintained and transmitted to new generations imply two different challenges that evolved mechanisms for transmitting cultural knowledge must solve. On the one hand, the transmission process must lead to the production of sufficient variability among the reproduced variants of the original technological skills if it is to enable the emergence of new and more efficient innovative technological solutions. On the other hand, the transmission mechanisms must also be able to reproduce cultural skills sufficiently faithfully for these to be successfully transmitted and maintained in a cumulative cultural repertoire.

The adaptive solution satisfying these two requirements may lie in the special design properties of evolved cultural transmission mechanisms. In order to account for transmission processes that produce both variability in alternative form variants and learning strategies generating rigid and conservative motor replicas of acquired technical skills, recent theoretical accounts postulate a strict teleological dichotomy between instrumental actions and ritual-like action kinds. In contrast, in this chapter we propose and argue for the existence of a *relevance-based emulative learning mechanism* by focusing on the central role played by *ostensive communication* in technical knowledge transmission. Our theoretical account differs in significant respects from the dichotomy-based approach because it focuses on the central role played by pedagogical ostensive communicative demonstrations—demonstrations that alternative accounts largely overlook.

Our approach is built on the cognitive foundations of human ostensive communication (Csibra and Gergely 2009, 2011; Heintz and Scott-Phillips 2022; Sperber and Wilson 2002). Our main claim is that when someone is communicatively addressed by knowledgeable others, it induces an expectation of relevance in naive learners, which is sufficient to account for both flexible (and potentially innovative) as well as rigid, high-fidelity aspects of cultural transmission. Communicative demonstrations function to bring into focus those parts of an action that are manifested as relevant by ostensive communication and, as such, are highlighted to be reenacted and learned. Notably, ostensive communicative cues identify the context and conditions in which an action sequence is adequate for use.

For instance, imagine someone waving their hand in the air. This behavior could be interpreted differently in particular contexts. If one discovers that a fireplace is close to the person, one might think that the goal of the behavior is to avert the smoke, in which case this behavior would qualify as a transparent instrumental action. However, if one observes the same action—waving the hand—without detecting any justificatory physical aspect in the situation, then the goal of the action (e.g., possibly serving as a gesture to greet a social partner or to express respect) remains teleologically opaque to the juvenile. In this example, the same instrumental action could be interpreted and represented as being transparent or opaque simply as a function of the specific context in which it is performed.

Given that the physical-causal relations between the observed or demonstrated actions and their subsequent outcomes often appear opaque to naive learners, relying on trustworthy and knowledgeable social partners' relevance-guided communicative demonstrations is an efficient strategy that novices can exploit to figure out what is worth acquiring as relevant in particular contexts. In other words, when the goal and goal structure of a behavior are opaque, the ostensive demonstrative behavioral cues accompanying the performance of the opaque action can be used to inform the naive learner that, despite its apparent teleological opacity, the ostensive highlighting itself means that the action is relevant for the novice to acquire and reenact in the kind of contexts where it was demonstrated.

Cognitive Mechanisms for Knowledge Transmission

Humans are certainly unique among other social and cultural species in their remarkable ability to create, maintain, and transmit across generations an ever-growing body of knowledge. This knowledge includes instrumental skills, know-hows, innovative techniques, and technical inventions, as well as traditional, conventional, normative, and rule-governed forms of social activities. These various kinds of culturally transmitted action forms allow humans to pursue a variety of instrumental and social goals that serve different adaptive functions in their cultural community. Achieving even a partial understanding of the origins, evolution, and multiple determinants of the processes involved in human cumulative culture (e.g., generating innovations and successfully transmitting and stabilizing them across generations) is a huge scientific enterprise requiring a massively multidisciplinary approach and interdisciplinary cooperation.

The classical view shared by psychologists, cultural anthropologists, ethnographers, and (until recently) archaeologists held that the obvious adaptive mechanism selected for the human transmission of novel cultural skills is the capacity for behavioral imitation, which allows for the faithful motor reproduction of novel action forms that naive human learners observe their culturally knowledgeable social partners performing (Tennie, Call, and Tomasello 2009; Legare et al. 2015; Boyd and Richerson 1996).

Similarly, many ethnographers and cultural anthropologists have claimed that novel cultural skills are learned in a straightforward manner by young children through passive participant observation, imitative copying, and practicing the imitatively reproduced behavioral forms with peers during joint play (Paradise and Rogoff 2009; Lancy, Gaskins, and Bock 2009; Lancy 1996). They argue that parental investment in the costly forms of active teaching, communicative demonstration, ostensive attention guidance to relevant properties, correction, feedback, or explanation are not cultural universals but culture-specific inventions related to the emergence of formal schooling (Lancy 1996) and practiced only in so-called WEIRD (Western, educated, industrialized, rich, and democratic) societies (Henrich, Heine, and Norenzayan 2010; for a contrary view, see Hewlett and Roulette 2016). Thus, the standard and still widespread view holds that humans' special adaptation for imitative learning by behavioral copying serves as the central and unique psychological mechanism of human cultural transmission, enabling us to reproduce and acquire novel skills.

This view has also received support from intriguing research on the striking phenomenon called "over-imitation" (Lyons, Young, and Keil 2007) in children from about four years onward and that remains present even in adults (Hoehl et al. 2019). The numerous studies on over-imitation appear to provide evidence that naive human cultural learners evolved a strong and spontaneous inclination to reenact faithful behavioral replicas of novel action sequences performed by adult models. In these experiments, children are presented with a transparent box and provided demonstrations of a sequence of novel actions used to obtain a goal object from the box, such as a sticker. These experiments intentionally vary an essential feature of human behavior: its causal transparency.[1] These studies revealed that even when the initial actions in a sequence are obviously causally unnecessary or irrelevant to the task (such as performing "magical" circular movements with a feather without making contact with the transparent box) and only the final action in the sequence is causally necessary and instrumentally relevant, children still readily and faithfully reproduce the whole sequence. Such a spontaneous inclination for over-imitation has been demonstrated by cross-cultural researchers in WEIRD societies and in a variety of non-WEIRD societies as well (Nielsen and Tomaselli 2010; Berl and Hewlett 2015). Even more surprising is that, upon being subsequently questioned about which of the imitated actions were necessary to retrieve the goal object and which were "silly," the children readily identified the causally irrelevant and unnecessary actions as the "silly" ones that they, nevertheless, faithfully reenacted (Lyons, Young, and Keil 2007).[2]

In sum, a variety of studies of imitative learning of novel actions converge in diagnosing humans' unique cognitive adaptation as a specialized behavior-copying imitation mechanism that naive cultural learners rely on to reproduce faithful motor replicas of novel and often causally opaque actions performed by knowledgeable cultural models. By enabling the faithful reproduction of actions and skills, this uniquely human mechanism has been, therefore, assumed to play a crucial role in the intergenerational cumulative growth and maintenance of a shared repertoire of skills within human social groups (Tomasello 1999).

At the same time, however, a growing body of evidence from developmental and comparative studies of imitative learning in humans and primates has led to the increasing recognition that the processes of imitative reenactment are often characterized by a significant degree of flexibility and variability. Indeed, while reproducing the same goal outcome as the observed novel action, learners frequently introduce alternative (and often less costly) action variants of varying efficiency. This difference was first explicitly recognized by Michael Tomasello (1996, 1998), who proposed a distinction between two types of reenactment mechanisms for reproducing novel action skills during cultural transmission (Gergely and Csibra 2006).

Tomasello (1998) termed the *goal-* or *outcome-emulation mechanism* the more flexible kind of reenactment mechanism that can generate variable forms of actions reproducing the same goal outcome as the originally observed action. He situated this mechanism in contrast to the more rigid *imitative behavior-copying mechanism* that produces faithful motor replicas of observed means actions. In goal emulation, the learner's attention is focused primarily on reproducing the goal outcome of the modeled behavior, while the specific means actions performed by the model can be flexibly substituted by alternative action variants that realize the same goal state. In fact, some of these variants can prove to be more efficient than the original observed action and be retained in the social group. It was therefore hypothesized that the relative flexibility of processes of emulation and the consequent degree of variability in modified action forms may play an important role in the generation of innovative and increasingly efficient action routines—thereby cumulatively and adaptively enriching the cultural repertoires transmitted to future generations (Legare and Nielsen 2015; Legare et al. 2015).

Determinants and Functions of Rigidity versus Flexibility of Cultural Transmission Mechanisms

A theoretical proposal (Clegg and Legare 2016; Legare and Nielsen 2020) postulates two basic dichotomies that human cultural learners are assumed to possess. First, it is proposed that humans are cognitively adapted to distinguish and recognize two ontologically distinct types of actions: (a) *causally transparent*, efficient goal-directed actions serving instrumental functions (such as tool use), and (b) *causally opaque* conventional or traditional types of actions that are typically jointly and publicly performed practices serving social rather than instrumental functions (such as rituals or traditional joint activities; see Legare and Nielsen 2020).

The second dichotomy distinguishes between two alternative learning mechanisms that human cultural learners are hypothesized to possess (Clegg and Legare 2016; Legare and Nielsen 2020). The first is a *more flexible emulative action reproduction mechanism* that generates more variability and modifications in the reproduced action variants. This mechanism is hypothesized to be selectively triggered by the learner's recognition of the causal transparency of a novel goal-directed action that serves instrumental functions. The second kind of imitative learning capacity is the *more rigid behavior-copying mechanism that produces faithful motor replicas* of an observed novel action. This more conservative transmission mechanism is assumed to be selectively induced by the learner's recognition of the causal opacity of conventional or traditional noninstrumental action forms. This more rigid imitative motor action copying mechanism would generate significantly less variability in the novel cultural action forms that it produces (Legare 2019; Legare and Nielsen 2015; Legare et al. 2015; Tennie and van Schaik 2020).

This theoretical account attempts to explain the differential distributions of rigidity versus flexibility in the reproduction processes by linking the different kinds of actions (causally transparent and opaque actions) to the two different types of social learning mechanisms (emulation and imitation). The emulation-based reenactment mechanism activated by the causal transparency of novel instrumental actions is hypothesized to generate more variability and alternative functional variants to solve the same goal, thereby supporting the emergence of "innovations."[3] These potentially more adaptive and novel variants can then be retained and stabilized within the population, fueling cumulative change in the action repertoire transmitted to subsequent generations (Clegg and Legare 2016; Legare and Nielsen 2015). In contrast, the apparent causal opacity of conventional or traditional action forms that are novel to the learner induces the more rigid imitative action copying mechanism, resulting in faithful motor action replicas with little variability. This conservative strategy is hypothesized to play an adaptive role by inhibiting the generation of disruptive variability of causally opaque action forms, thereby promoting successful cultural transmission and the stable maintenance of repertoires of conventional traditions. The suggestion here is that if such causally and teleologically opaque ritual acts induced the more flexible and emulative strategy, which would generate variability in the ritual behavior, the conventional behavior's recognition later would be endangered, thereby reducing the likelihood of the conventional action's successful maintenance and stability across generations (Kapitány and Nielsen 2019).

One of the central problems faced by this dichotomizing approach is that it largely overlooks the important fact that the repertoire of novel cultural actions to which naive social learners are exposed includes many goal-directed actions that serve primarily instrumental functions but nevertheless have constituent parts (lower-level actions or subgoals) that appear causally opaque to the learners (e.g., by violating the causal efficiency requirement of goal-directed instrumental actions). Equally importantly, partially causally opaque instrumental actions often serve traditional or ritualistic functions as well. For example, while eating food with one's hands is a highly efficient and causally transparent instrumental act, adults in various cultural traditions present the novice with less efficient normative manners and styles of consuming food (e.g., eating with knives and forks or using chopsticks). These alternative culture-specific variants of instrumental actions are more complex and contain causally and teleologically opaque parts, which include performing more costly and less instrumentally efficient actions than the alternative of eating food with one's hands does. Yet, their primary function is still obviously instrumental, while also serving social and traditional display functions as well.

Such transitive actions with "mixed" functions involve culturally normative, subefficient manners of action execution that are not causally transparent to the learner and are in fact more costly to perform than other equally available, more efficient, and often more familiar alternatives. The specific causal and functional properties of these actions must thus appear causally opaque for the juvenile cultural learner.[4] However, the dichotomy-based approach holds that observing the apparent causal opacity of a novel action form induces in the naive cultural learner a high-fidelity, rigid copying mechanism that produces exact motor replicas of the opaque target action, thereby generating little or no variability in the reproduced action forms. This prediction holds whether the causal opacity is detected in the action serving noninstrumental social functions or at the level of the causally opaque subactions or subgoals of a clearly goal-directed, larger action that serves both an instrumental and a social function.

Relevance-Based Emulation as the Mechanism of Cultural Learning

We propose an alternative view according to which relevance-based goal emulation serves as the dedicated adaptation for social learning, selected for acquiring and transmitting the repertoire of culturally shared action skills. This mechanism is induced by ostensive communicative manifestations of culturally relevant intentional actions irrespective of their degree of causal transparency or opacity.

According to our view, children are adapted to recognize the communicative gestures of knowledgeable adults as indicating pedagogical (or demonstrative) contexts in which new and relevant cultural information is made manifest. Ostensive communicative signals induce their addressee to segment the demonstrated action into constituent parts such as its component means actions, the subgoals that these means actions bring about, the final goal to which the means actions ultimately lead, and the specific context in which the action is performed overall. The pedagogical signals highlight for the naive learner the novel and relevant aspects of the segmented action sequence that should be learned and reproduced in a given context. This pedagogical guidance allows the learner to identify both the overall goal and, when highlighted, the relevant subgoals leading to it—whether or not these behaviors are instrumental. In what follows, we will discuss the available empirical evidence as well as new experimental data supporting the selective and inferential nature of the relevance-guided emulation mechanism.

Our studies focus on the transmission of a goal-directed instrumental action that involves a novel, causally "opaque," and subefficient subgoal. These studies demonstrate the fast-learning, long-term retention, and flexible production of alternative action variants of the causally opaque and subefficient elements of novel instrumental actions. We take these results as evidence in support of our claim that the dedicated psychological mechanism serving cultural transmission in humans is one of *selective, inferential, and relevance-guided emulation*. This mechanism involves the ability to *flexibly choose alternative means actions* to reproduce relevant subgoals manifested by the pedagogical demonstration of knowledgeable partners. In this sense, we argue that the transmission of novel, ostensively demonstrated goal-directed actions supports the production of alternative behaviors to achieve goals manifested as relevant, thus providing the basis of the capacity for innovative use of accumulated knowledge.

As such, our alternative proposal equally supports the claim that a flexible cultural (emulative) learning mechanism may provide variation in the reproduced actions, fueling the emergence and selection of more efficient versions of the original action. As these more efficient action variants stabilize within a population, they can lead to the cumulative enrichment of cultural action repertoires. Therefore, variability of reproduction is, indeed, an important source of the ratchet effect in cultural transmission, contributing to the cumulative nature of human culture that characterizes its transformations through cycles of intergenerational transmission (Boyd and Richerson 1996; Tomasello 1999).

Context-Sensitive Selective Imitation of Causally Opaque Instrumental Actions

In this section, we present some studies that aim to test the above hypotheses by using new versions of the "head touch paradigm," originally designed to investigate the imitative reenactment of novel and causally opaque actions by infants. This paradigm (first presented in Gergely, Bekkering, and Király 2002) demonstrates that imitation is a selective, inferential, and context-sensitive learning mechanism.

In the original study, 14-month-old infants watched an adult sitting in front of a table with a touch-sensitive lamp on it. The experimenter first placed her hands on the table next to each side of the lamp and then performed an unusual and subefficient means action to illuminate the lamp: she bent over the lamp to press its touch-sensitive surface with her forehead (the "hands free" condition). A separate group of infants were tested in an alternative context condition where the model first pretended to be freezing, telling the infant that she was really cold, and so she put a blanket around her shoulders and held onto it tightly with both hands (the "hands occupied" condition). She then went on to demonstrate the very same unusual and subefficient head touch action to light up the lamp, bending over it and activating it with her forehead.

In the test phase, the infants were given the touch-sensitive lamp and were encouraged to play with it on their own. In the hands-free condition, most of them (69%) used their head to activate the lamp (cf. Meltzoff 1988). However, in the hands-occupied condition, only a small proportion of infants (21%) performed the head action to light the lamp; most of them just used the more efficient (but undemonstrated) method of pressing the light box with their free hands.

These patterns of selective imitation of the demonstrated head touch action indicated that infants were sensitive to the context in which the model presented the unusual action. In the hands-free condition, the model's subefficient and unusual head touch action must have appeared causally opaque to the infants, given that the demonstrator's hands were resting freely on the table and thus could have been used to press the lamp (a more efficient and familiar alternative means action to light up a lamp). Nevertheless, the demonstrator opted not to use her free hands, instead presenting to the infants the causally opaque (subefficient and more costly) head touch action. In contrast, in the hands-occupied condition, where the model's hands were not free (being occupied with holding the blanket around her shoulders), the demonstrator's head touch action must have appeared causally transparent to the infants, since using the head to activate the lamp was contextually justified as a causally efficient alternative action, given the constraints of her hands being occupied.

These findings have been interpreted as evidence that imitative reenactment is a selective, inferential, context-sensitive learning mechanism that relies on evaluating the relative efficiency of the target action observed as a function of the situational constraints on possible actions (cf. Gergely and Csibra 2003). In other words, when deciding whether or not to reenact a causally opaque behavioral component or subgoal of the novel action, the naive learner took into account the action's relative efficiency in obtaining the specific goal outcome as a function of the constraints imposed by the particular context (Gergely, Bekkering, and Király 2002; Gergely and Jacob 2012; Király, Csibra, and Gergely 2013).

Advocates of the dichotomy-based approach to cultural transmission (discussed above) could argue that it can account for these findings of selective reenactment insofar as it appears to support the central assumption of their theory—namely, that the causal opacity of an observed action induces high-fidelity, rigid behavior-copying in the cultural learner (Clegg and Legare 2016; Legare and Nielsen 2015). Additionally, these results appear to be in line with the idea that the faithful copying of novel instrumental actions is the result of a copy-when-uncertain strategy in social learning (Rendell et al. 2011; Toelch, Bach, and Dolan

2014). It has also been proposed that high-fidelity imitation may be so useful when learning novel but opaque behaviors that its benefits outweigh potential efficiency costs in the transmission process (McGuigan et al. 2007).

In contrast in the hands-occupied context, the demonstrated head touch action must have appeared *causally transparent* to the infants. Yet, despite its apparent causal transparency, the head touch action failed to be imitated. Instead, infants—whose own hands were unoccupied—chose far more frequently to use their free hands to produce the more efficient "hand touch" action to activate the lamp. Note that this finding appears hard to accommodate within the framework of the dichotomy-based approach (Legare et al. 2015; Legare and Nielsen 2015).

The challenges that need to be handled by a dedicated adaptive cultural learning mechanism are twofold. First, such mechanisms should support the fast-learning, long-term retention, and functionally adequate delayed reenactments of novel means actions. These aspects are the main signs of the reliable and long-lasting acquisition of a new action. Second, it should at the same time allow for the flexible and functionally appropriate generalization and selective reproduction of the newly acquired motor skill across a variety of relevant and novel contexts. Learning a new skill is only adequate if it is applicable and useful in future situations as well, and this requires the identification of new contexts where it is relevant.

These criteria represent a challenge for the dichotomy-based approach. In response to the first challenge, this view does not offer a means to distinguish instrumental transitive actions that contain *causally opaque* subcomponents while serving a primarily transparent instrumental function (while also serving social functions) from conventional and traditional action routines, which also involve causally opaque actions as well as actions allegedly serving only social and affiliative functions. Thus, the dichotomy-based approach faces a challenge in providing the criteria necessary for the adequate and flexible selection of adaptive means.

With regard to the second challenge, the cultural learning model proposed by the dichotomy-based approach does not adequately support generalization across functionally relevant new contexts. The dichotomy-based view proposes that based on its apparent causal opacity, naive cultural learners recognize and categorize the observed novel action as belonging to the domain of conventional, traditional actions that serve noninstrumental social functions. Because of their causal opacity, such actions induce and are acquired by conservative imitative behavior-copying, and their rigidly reproduced motor replicas are probably stored separately as conventional, traditional actions serving primarily social functions. As a result, instrumental transitive actions that contain causally opaque components could become miscategorized as actions serving only social and conventional functions. However, in order to provide advantageous solutions for similar problems in other instrumental contexts, the acquired novel action routines should be applicable in functionally relevant new situations as well. Yet it is unclear how the strategy of rigid imitative copying of causally opaque action skills could promote generalization across such contexts.

Our alternative approach, however, holds that there are in fact no distinct imitative mechanisms to support the acquisition of conventional as opposed to instrumental actions. Rather, the same learning mechanism is used to learn both opaque and transparent novel actions. In the case of the causally opaque behavioral components that are ostensively manifested by the demonstrator as culturally relevant, more elements of these causally opaque but relevant action components are likely to be reproduced, allowing better identification of appropriate contexts and, consequently, flexible generalization of behavior.

Let us investigate what strategies infants pursue when they attend closely to ostensively modeled behavioral elements in relation to the relevant context within which they are demonstrated. We propose that ostensive attention guidance to the relevant (though possibly opaque) action components highlighted by the communicative action manifestation, together with the experience of alternative variants and their repeated application in variable contexts, should provide the relevant informational basis to identify instrumental actions as separate from conventional ones in such contexts.

The Illusion of Imitation: Is There Imitative Form-Copying?

With the aim of investigating more closely the underlying mechanism behind the selective learning of causally opaque instrumental means actions, we (Király, Csibra, and Gergely 2013) ran follow-up studies using the head touch paradigm. They found that 14-month-olds selectively reenacted the novel, apparently arbitrary, and subefficient means action (lighting the touch lamp by contacting it with one's forehead) in the hands-free context condition only when the subefficient and opaque "head touch" action was demonstrated by an adult model addressing them in an ostensive communicative manner. This ostensive manifestation was interpreted by the naive learner as reflecting the adult's pedagogical intention (Csibra and Gergely 2011) to communicate that the causally subefficient and opaque means action was nevertheless culturally relevant and, as such, should be acquired by the naive learner. However, when the same action was performed but observed from a third-person perspective without being accompanied by ostensive cues of communication by the adult, infants did not reenact the causally opaque subefficient head touch action. Instead, they achieved the same goal more efficiently by using their hand to operate the touch-sensitive light box.

Király, Csibra, and Gergely (2013) proposed that the selective reenactment of the novel behavior observed in the communicative context—specifically, the imitation of the novel and arbitrary head touch means action in the hands-free context—demonstrates that young children are prepared to reenact and acquire novel actions even when their subefficient execution appears causally opaque. In the absence of the possibility of exploiting their individual learning strategies, they rely on the communicative signals of experienced others. Infants interpret the ostensive action demonstrations as pedagogically intended communicative manifestations of novel and (in spite of their causal subefficiency) culturally relevant means actions to be acquired. Because of the ostensive communicative signals that accompany the action demonstration, infants construe it as conveying new and relevant information that the demonstrator intends to communicate and not as a purely instrumental action (Csibra and Gergely 2011; Gergely and Jacob 2012; Altinok, Király, and Gergely 2022). The infants would thus interpret the situation as a teaching context where the demonstrated action manifested to them with ostensive communicative cues is intended to guide them in learning a relevant (if causally opaque) novel instrumental action.

In the hands-occupied condition, the obvious constraint of having the hands busy with another goal is sufficient for young children to form a coherent interpretation of the demonstrator's choice to use her head to light up the lamp. However, this is not the case in the hands-free condition. Instead, in this context, the child observer must search for an alternative explanation to understand the use of the head action. The communicative demonstration context induces in them the presumption of relevance (Sperber and Wilson 2002), leading them to interpret the opaque and subefficient means action manifested as relevant (from some unspecified point of view) and, as such, a behavior to be acquired because it is culturally significant. As a result, in addition to interpreting the final instrumental goal as "lighting up the lamp," infants will also construe as relevant the apparently causally opaque subgoal (i.e., to achieve the final goal by making contact between the lamp and their forehead). This approach thus proposes that the ostensive communicative demonstration context can enrich the encoding of the overall goal-directed action by signaling that the specific manner of performing the means action (contacting the lamp with the head) is, in spite of its apparent subefficiency, the culturally relevant way of attaining the final goal. But is this interpretive process served by a rigid imitative copying of the demonstrated, causally opaque action?

A closer look at the performance of children allows us to answer this question. According to the dichotomy-based approach, when children turn to behavior-copying after observing a new, arbitrary, and causally opaque action used to attain a goal, their strategy would be to merely copy (or imitate) the performed behavior. With respect to the head touch action, this strategy implies that children would construe the head action as a causally opaque but successful way of bringing about an effect and would not use any other means to achieve the goal.

To pursue this question further, we first analyzed the reenactment behavior of children in relation to the goal object. We found that those children who performed the head action not only performed this novel means action, but they also performed the simpler (but undemonstrated) hand action to attain the goal—and did so without exception. In fact, the hand action in all cases preceded the head action and was successful in bringing about the effect (Király, Csibra, and Gergely 2013). This means that children encoded the overall goal of the situation; more importantly, they were inclined to try out alternative means that turned out to be more effective in attaining the same outcome.

Second, we coded the specific forms of head touch actions with the aim of assessing the potential variability in the particular manner of reenacting the modeled action. The level of variability could reflect the degree of fidelity of imitative behavioral responses and as such could be used as an indicator of the underlying mechanism of social learning. The results showed that children performed the head action with high variability. They used different parts of their head to contact (or approach) the lamp, including their cheeks, faces, mouth, eyes, and even their ears. Thirty-six percent of imitators performed two or even three different forms of head actions during the testing session, and only 11 percent produced a faithful version of touching the lamp with the forehead—however, they did so while also using other parts of their head to light up the lamp, contacting it with an ear or with the face (see figure 5.1). In addition, in some cases (25%), they only approached the lamp and never made the contact between their head and the lamp (Király, Csibra, and Gergely 2013).

Furthermore, children also alternated the way that they managed to bring the lamp into contact with or close to their head: they either followed the demonstrated version, leaning toward the lamp without moving it, or most interestingly, they grabbed the lamp by hand and lifted it up to touch their head. This intriguing variant of an emulative response was at first only a chance observation (because with overuse, the adhesive putty fixing the lamp to the table became loose, allowing the infants to grab and raise the lamp by hand). Based on this observation, we developed a further study (Chen, Király, and Gergely 2012).

Figure 5.1
The different ways of using the head to act on the lamp. *Source*: Photo by Ildikó Király.

Following the demonstration with a lamp fixed to the table, for the test phase different (though similar-looking) touch lamps were put on the table in front of the children, each being easy to lift. In this version, we could directly observe the variability that children revealed in choosing their specific version of means action during reenactment. We found that after the ostensive demonstration of the target act of leaning forward and contacting the lamp with the forehead, 85 percent of the imitators (i.e., 58% of the participants) lifted the lamp to their head. Moreover, in half of these cases, children performed both the lifting and the leaning-forward variants of the head action (Chen, Király, and Gergely 2012).

From the perspective of the dichotomy-based view, opaque action demonstrations should trigger imitative form-copying. According to this account, in the head touch paradigm, the arbitrary and opaque application of the head being bent down to touch the lamp should specifically and solely induce high-fidelity copying of the action sequence demonstrated. However, as described above, we found a great deal of variability, which allows us to suggest that children interpret the demonstration as a communicative action in which the demonstrator informs them about the instrumental goal and also about the subgoal relevant to achieving the final goal within the context (i.e., to activate the lamp by bending forward and touching it with the forehead).

These findings clarify the role of inferential processes involved in action analysis and reveal the important role of ostensive communication in enabling infants to represent the goal structure—that is, the overall goal and the specific means as the relevant subgoal—of novel actions even when the causal relations between the means and the end-states are causally opaque. We propose that the presumption of relevance induced by the ostensive cueing guides infants' interpretations of the relevant subgoal/final goal structure of the demonstrated action sequence. Indeed, it seems that while they reproduce the ostensively interpreted relevant goal structure of the manifested action sequence, infants continue to monitor the efficiency of actions when choosing, changing, or disregarding certain action elements, so long as these modifications leave constant—or make more efficient—the realization of the relevant subgoal/final goal structure. We further suggest that after the observation of a novel goal-directed action, infants can identify the goal of the action that they encoded. When they are invited to reenact the action, they recall the encoded goal and (re)enact an action variant to attain the same goal. Most importantly, they encode the novel means as a subgoal when it is signaled as novel and relevant by the ostensive communicative context.

We take these results as evidence against the behavior-copying hypothesis of the dichotomy-based approach and argue instead that relevance-based emulation serves as a central mechanism driving reenactment. In this process, the subgoal is accentuated as relevant by the ostensive communication, keeping open the possibility that either (a) the subgoal is causally linked to the overall goal and serves as a placeholder for potential technical variation (which remains opaque to the infant), or (b) the subgoal is a social goal representing a socially accepted alternative manner of attaining the final goal.

In essence, we posit that there is only one form of cultural learning that infants use to learn both causally opaque and transparent behaviors: the emulation of encoded goals or goal structures. Yet, the richness and detailed nature of this encoding of goals is modulated and guided ostensively during its communicative demonstration. The presumption of relevance induced by the communicative situation highlights novel information in relation to the overall goal, but it does not necessarily disambiguate initially the exact sense in which the ostensively highlighted aspect of the action is novel and relevant. Indeed, further communicative exchanges allow for disambiguating in what sense the behavior is relevant (e.g., understanding whether the subgoal has instrumental merits or is rather a convention serving social purposes). This later disambiguation, however, requires some grounding points or bases for further elaboration. How then do children differentiate instrumental from conventional behavior during observation?

It is widely accepted that behavioral reenactment contributes to the transmission of con-ventions and can serve social functions itself (Legare et al. 2015; Over and Carpenter 2012, 2013; Watson-Jones and Legare 2016; Wen, Herrmann, and Legare 2016). From a young age, children attend to a variety of social and contextual cues to determine the goal of behav-iors (Buchsbaum et al. 2011; Carpenter, Call, and Tomasello 2005). From this angle, the learning situation should highlight those contextual features that allow children to map when conventional manners are demonstrated and when there is more space for refinement of an instrumental action.

It is also a possibility that ostensive communication helps children to differentiate instru-mental goals from conventional ones. In the following section, we present novel studies that directly investigate whether the communicative demonstration of a novel action can provide disambiguating cues by differentiating the relevant action contexts. We argue that these cues can help infants infer whether and how to reenact demonstrated novel actions and guide them toward evaluating the conventional versus instrumental functions served by the same action.

Further Studies to Provide Evidence in Support of Relevance-Driven Goal Emulation

We argue that in order to learn novel actions, infants need to be able to identify the relevant target actions. Moreover, for this purpose, infants need to rely on the active inferential guidance provided by the demonstrator's ostensive communicative gestures and manner of manifesting the action (cf. Gergely 2007; Gergely and Csibra 2006; Csibra and Gergely 2011; Király, Csibra, and Gergely 2013). In the following experiments, we introduce modifications to previous experimental setups—namely, the "hands up" and "balls" conditions (Paulus et al. 2011)—in order to test whether the presumption of relevance induced by ostensive communication guides infants to infer and identify the relevant information that is applicable and generalizable for later use.

The Role of the Demonstrated Relevant Context in Interpreting the Manifested Opaque Means Action

Our first objective in this series of studies was to test the proposal that observing a causally opaque action leads to imitative behavior-copying (possibly as a result of the induced motor resonance of the infant's corresponding motor programs; see Paulus et al. 2011). Our second aim was to test the alternative hypothesis of our inference-based selective emulation model, which holds that ostensive demonstration of an opaque means action is interpreted by infants through context-sensitive inferences constrained and informed by the relevant aspects of the context in which the action takes place.

In proposing our alternative relevance-based emulation model, we predicted that the ostensive manifestation of the opaque head touch action would induce infants to attend to the relevant action context in which the head touch action is demonstrated. As a result, infants would interpret the demonstrated opaque head touch action of the model by evaluating it in relation to the relevant action context. In particular, we predicted that by varying relevant aspects of the demonstration context in which the very same opaque head touch action is observed, we could induce differential and selective reenactments of the opaque action by the infants. To achieve these aims, we borrowed and modified the experimental conditions that were initially designed to provide evidence for the motor resonance-based, automatic behavior-copying theory of imitative learning (see Paulus et al. 2011).

Study 1. Imitation of the Opaque Head Touch Action in the "Palms in Air" Demonstration Context

Markus Paulus and colleagues (2011) criticized the inference-based selective imitation account of the original "head touch" study (Gergely, Bekkering, and Király 2002) by suggesting that the reason infants failed to imitate the demonstrated "head action" in the "hands occupied" condition could be that 14-month-olds simply cannot bend over to touch the lamp with their head without supporting their body by putting their hands on the table. In fact, Paulus and colleagues suggested that infants imitated the "head touch" action in the "hands free" condition precisely because they observed that the experimenter put her hands on the table to support her body by leaning on them when bending forward to touch the light box with her head—that is, precisely the way that the infants were themselves constrained to perform the action. According to Paulus and colleagues, observing this configuration of the model's actions induced motor resonance in the infants' corresponding body parts and activated the motor imitation of both the model's body supporting hand actions and the bending over to touch the lamp with their forehead. So, contrary to the account provided by the inference-based rational imitation theory, infants did not infer that the model's hands were free. Instead, they observed that the hands were occupied by supporting the model's body during the head action.

Our version of the "hands up" condition described by Paulus and colleagues (2011) is the "palms in air" demonstration. In our version, the context differed from Gergely, Bekkering, and Király's (2002) original "hands free" condition. While in both conditions the model's hands were free when the head action was performed, in the "palms in air" condition, hands were not placed on the table—thus, they could not have been used to support the model's body while bending forward. After sitting down in front of the table and the touch lamp, the model in the "palms in air" condition demonstrated two different salient actions separately in a sequence. The model first presented a hand action extending her two hands toward the light box on the table while turning her palms upward midair (see figure 5.2). This hand action corresponds to the kind of semi-conventionalized ostensive referential manual gesture that humans often use to "show" or "highlight" an object or event as relevant for another to attend to. In everyday communication, this demonstrative manual gesture is often accompanied by some verbal referential expression, such as "Here!" in English or "Voilá!" in French—something that we have adopted in our ostensive communicative demonstration condition.

After the model's hands had finished their referential gesture highlighting the lamp (accompanied by the referential vocal gesture "Voilá!"), a slight pause followed, and then the model proceeded to perform a second action with her head. She bent forward from the waist and lit up the lamp by touching it with her forehead. During the performance of the head touch action, her hands remained stationary in their previous position (held with palms up midair). Therefore, the "palms in air" demonstration context provided clear temporal and contextual segmentation cues to help infants interpret the model's demonstrated hand gesture as a separate referential action that established a relevant context for interpreting the subsequent head touch means action. (Note also that the hands' referential act was clearly completed, and so the hands in the air in front of the infants' eyes were clearly "free" to be used for a new action.)

We presented two groups of 14-month-olds with this action sequence in two demonstration conditions: in a second-person ostensive communicative context and in a third-person noncommunicative observation context. In the communicative second-person condition, apart from providing temporal action segmentation cues, the demonstrator also addressed the infant through ostensive referential gestures and presented the action demonstrations in an ostensive way (i.e., in a slightly exaggerated "motionese" manner). This provided infants with ostensive highlighting and temporal parsing cues to guide them to separately interpret the initial hand action demonstration as an ostensive referential manual gesture. We hypothesized that the presence of these ostensively provided informative cues—similar to Gergely, Bekkering, and Király's (2002) original hands-free condition—would direct the infant to parse and interpret the hand gesture as forming part of the relevant action demonstration context rather than being part of the demonstrated head touch target action itself. In contrast, in the third-person noncommunicative observation context of the "palms in air" study, the demonstrator presented the exact same action sequence to a different group of 14-month-olds without any ostensive communicative gestures.

For the "palms in air" condition, our relevance-guided inferential account predicts that, guided by the provided ostensive signals and temporal parsing cues, infants will be able to infer the new and relevant information manifested for them by the ostensive demonstration of the unusual head touch means action, and they will thus be able to learn and reenact it. For the third-person condition, our account predicts that without the presence of such ostensive communicative cues, there will be no imitative reenactment of the causally opaque and teleologically subefficient head touch action. In contrast, according to the dichotomy-based view, which claims that observing a causally opaque cultural action induces faithful behavior-copying, both conditions should induce a reenactment by the infant since both the communicative (second-person) and the noncommunicative (third-person) observation conditions present the exact same (partially opaque) action sequences to the infants.

Method

Participants. Twenty-six 14-month-old infants were recruited (two were excluded because of parental interference or fussiness). Twenty-four children were assigned to one of the two experimental conditions (12–12).

Test phase. The modeling phase in both conditions (the second-person communicative demonstration context and the third-person noncommunicative observation context) was immediately followed by the test phase, in which the infants received the light box. Infants were given 60 seconds to explore and play with the lamp.

Data Analysis and Scoring

The video records of the test phase were scored by two independent observers who were uninformed about which of the conditions the participants belonged to. The dependent measure was whether the infant attempted to perform the head-on-box action within a 60-second time window. An attempt was defined as either touching the lamp with the head or approaching the lamp with the head (e.g., leaning forward) within 10 centimeters or less (see Meltzoff 1988). We also coded for the direction of approach of the target action. The potential ways of approaching the lamp were either leaning forward or lifting up the lamp. The two coders' evaluations of the participants' performances were in 97 percent agreement (kappa = 0.94).

Results

The proportion of infants who performed the target action is presented in figure 5.2.

We compared the performance in the two conditions. The frequency of target action reenactment was lower in the third-person "noncommunicative observation context" condition than it was in the "communicative demonstration context" condition (Fisher exact p = 0.05). Calculating the odd ratio confirmed (OR = 5.431) that infants in the "communicative context" condition were more likely to reenact the head action in comparison to the group of infants in the "noncommunicative context" condition. As in previous studies (Gergely, Bekkering, and Király 2002; Paulus et al. 2011), at least one hand action preceded the head action in 92 percent of cases. The frequency of hand actions was 7.8 for one head touch within the first 60 seconds.

Interestingly, the head touches appeared in different forms than the modeled behavior. Most importantly, 30 percent of infants who performed the head action (three infants in the communicative demonstration context and one infant in the noncommunicative observation context) lifted the lamp up to their heads instead of leaning forward to touch it. Moreover, in 30 percent of cases, there was no contact between the approaching head and lamp (two

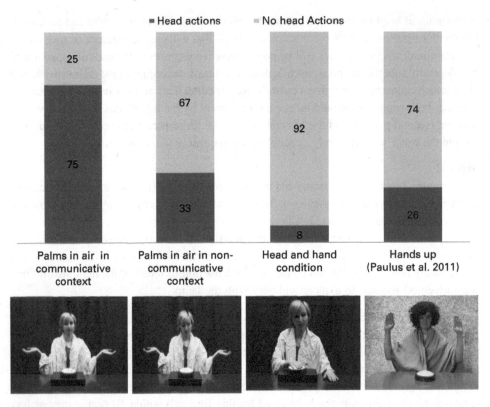

Figure 5.2
The proportion of imitators in each condition in study 1 and study 3. *Source*: Photo by Ildikó Király.

infants lifted the lamp up to the head but did not make physical contact with it, while two other infants bent forward but did not make contact with the lamp; all of these were in the communicative demonstration context).

These results confirm that infants performed voluntarily chosen variations of the originally observed behavior, rather than reenacting a matching motor replica of the observed action. The selective pattern of results in the different demonstration conditions provides further empirical evidence for the natural pedagogy account of learning (Gergely and Csibra 2006)—namely, that the presence of a demonstrator's ostensive and referential communicative signals addressing the infant is a critical factor that is necessary to induce the imitative reenactment of a novel and apparently subefficient means action.

Study 2. Situating the Goal in Context: The "Balls" Study

The main objective of this study was to demonstrate that ostension plays a crucial role in linking the overall goal of an action with its manifested subgoals, and that this integration is driven by the learner relying on the relevant aspects of the action context being manifested. Furthermore, we also aimed to show that the ostensive demonstration of opaque means actions of an instrumental transitive act could induce variability in the action, thereby leading to alternative actions when reproducing the opaque subgoal. We highlight the role of osten-

sive communicative and temporal parsing cues that could guide infants' interpretive inferences to identify the relevant aspects of the new information manifested to them given the
demonstrated action context.

In the "balls" condition of Paulus and colleagues (2011), there were two softballs lying
on the table next to the lamp. In the demonstration context preceding the manifestation of
the opaque means action, the experimenter took a seat and played with the two softballs
for approximately eight seconds. Then, keeping one ball in each hand, the experimenter
put her hands on the table next to the lamp. From then on, the procedure followed exactly
that of the hands-free condition in Gergely, Bekkering, and Király (2002), with the only
difference being that the experimenter was holding the two softballs in her hands on the
table while performing the opaque head action to activate the lamp.

In this condition, one can also argue that observing the hand action itself (i.e., putting the
hands with balls on the table) may not be sufficient for the infant to infer if the hands are
free or occupied. Such an inference must rely on and is constrained by the relevant aspects
of the context in which the hand actions were demonstrated. Here, the model was playing
with the balls and then stopped—a context in which infants could infer that her hands were
now free to act (they do not necessarily have to continue holding the balls).

Nevertheless, to clearly disambiguate the interpretation of the relevant action and help the
infants parse the manifested action sequence, we introduced two different versions of the
demonstration context. In the "hands free resting on balls" condition, we followed the procedure of the "balls" condition of Paulus et al. (2011), except that in our study (1) the two
balls were lying on two little plates on the table, and (2) after the model had put her hands
with the balls next to the light box, she lifted her hands up without the balls for two to three
seconds. The balls remained in the plates and could not roll away. Then the model put her
hands down again, grasping the balls on the plates as before. After this short event, the model
performed the head action with her hands resting on the balls. In this context, it was made
explicit during the demonstration that the hands were free to act because they were not
occupied with holding the balls so that they would not roll away.

In the "hands occupied with holding balls" condition, there were no plates next to the
light box, and the model performed the exact same action sequence performed in the "hands
free resting on balls" condition. Accordingly, when the model lifted her hands for a moment,
the balls started to roll away, so she had to quickly reach back and grasp them again to keep
them from moving farther. This situation unambiguously manifested that the hands were
occupied and were not free to engage in another action. In both situations, however, the
model's hands (with the balls in them) were placed on the table, so they could provide support
for her body when she bent forward to touch the lamp with her head.

According to the motor resonance theory and the dichotomy-based view, there should be
no difference in the number of imitators in the two conditions. However, the different situational constraints demonstrated relevant contextual information for the infant to infer
whether the hands were free or occupied during the performance of the head touch action.
This allowed the infants to interpret the head action as an efficient means to perform in the
condition in which the hands were occupied with holding the ball ("hands occupied with
holding balls" condition). In contrast, given the relevant contextual information demonstrating that the hands were free and could have been used to touch the lamp ("hands free resting

on balls" condition), infants could infer the demonstrated relevance manifested by performing the subefficient and causally opaque head touch action. This generates the prediction
that the number of imitators should differ in the two conditions.

Method

Participants. Thirty 14-month-old infants were recruited; three of them were excluded
from the final sample because of fussiness (n = 1), technical error (n = 1), and parental
interference (n = 1). Participants were randomly assigned to the two experimental conditions. As a result, 14 infants were tested in the "hands free resting balls" condition, and
13 infants were tested in the "hands occupied with holding balls" condition.

Test phase. The test phase immediately followed the modeling phase in both conditions.
The model pushed the lamp across the table in front of the infant and said, "It is your turn
now! You can try it!" She encouraged the infant to play with it and stayed in the room.
Infants were given 60 seconds to explore and play with the lamp.

Data Analysis and Scoring

The video records of the test phase were scored by two independent observers who were
uninformed about which of the conditions the participants belonged to. The dependent
measure was whether the infant attempted to perform the head action within a 60-second
time window (as in study 1). The two coders' evaluations of the participants' performance
was in 92 percent agreement (kappa = 0.85).

Results

The number and proportion of infants who performed the target action are presented in
figure 5.3.

When we compared the performance in the two conditions, it was revealed that the
frequency of target action tended to be lower in the "hands occupied with holding balls"
condition than it was in the "hands free resting on balls" condition (Fisher exact p = 0.054).
Odd ratio (OR = 5.177) examination revealed that the probability of performing a head

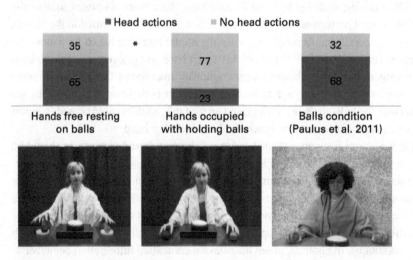

Figure 5.3
The proportion of imitators in each condition in study 2. *Source*: Photo by Ildikó Király.

touch is more likely in the "hands free resting on balls" than it is in the "hands occupied with holding balls" condition.

Hand touch actions preceded head touch action in 94 percent of cases. The frequency of hand touch actions was almost six for a head touch. Moreover, like in study 1, the head touch actions did not follow the modeled head touch with high fidelity. Intriguingly, in 50 percent of cases, infants lifted the lamp up to their heads instead of leaning forward to it (four infants in the "hands free resting on balls" condition and two infants in the "hands occupied with holding balls" performed the head touch action this way). Also, in 58 percent of cases (out of which 25%, or three infants, were lifters), there was no contact between the approaching head and the lamp.

Study 3. Control Condition: Ostensive Demonstration of Both the Head Touch and the Hand Touch Actions

Our main suggestion is that when children observe in a communicative context the ostensive demonstration of a novel subefficient goal-directed means action that cannot be justified as efficient, they will interpret it as a causally opaque subgoal that represents a socially relevant manner in which to attain the final goal in the demonstrated context. In the present experiment, we explore the further assumption that when children have already acquired an efficient (causally transparent) means action to operate an artifact (which is therefore not novel to them), they will not learn to reenact an alternative nonefficient (and so casually opaque) means action to achieve the same goal, even if it is presented in an ostensive communicative context (Pinkham and Jaswal 2011), since infants would infer from the ostensive demonstration that the hand action is a socially accepted and efficient means to attain the goal. However, here, the dichotomy-based approach would predict that children will imitate the head touch action, since in this view, seeing a causally opaque action automatically triggers this form of social learning.

Method

Participants. Fourteen 14-month-old infants were tested in the "head and hand" control condition.

Apparatus. The apparatus used was identical to that of study 1.

Modeling phase. Here the model demonstrated two actions (in ostensive-communicative context), while she also demonstrated that her hands were free (lying on the table next to the light box). The "head and hand" condition was thus exactly the same as the hands-free condition of Gergely, Bekkering, and Király (2002), with the only difference being that the model demonstrated the prepotent hand touch action as well, and both actions resulted in activating the light box. The head touch action was identical to the one employed in the previous experiments, while the hand action consisted of touching the top of the lamp and lighting it up by hand. Both actions were modeled twice, in alternating order, with the first action counterbalanced across participants.

Test phase. The test phase in this condition was similar to the test phase of the communicative context condition in study 1.

Results

Briefly, in the "head and hand" condition, only 8 percent of infants (one child) imitated the head touch (see figure 5.2). We compared the results of the present study to the results of the two conditions of study 1. The frequency of target action reenactment was lower in the "head and hand" condition than it was in the second-person "communicative demonstration context" condition of study 1 (Fisher exact p = 0.001 and OR = 48.1), and the frequency of target actions did not differ significantly between the third-person "noncommunicative observation context" condition of study 1 and the "head and hand" condition (Fisher exact p = 0.148).

Discussion

The results of the "head and hand" condition are in line with the assumption of inference-based learning: it was ostensively demonstrated to infants that the efficient hand action is an established, socially sanctioned, and efficient way to attain the goal. Therefore, they acquired this means action, which they judged as efficient, socially shared, and relevant. At the same time, they also saw an ostensive manifestation of an opaque alternative way to attain the same goal (the head touch action). Since it was demonstrated to infants that both means (hand touch action and head touch action) are equally relevant and socially sanctioned ways to attain the same goal, they were free to choose between them. The results show that in this case they chose to perform the more efficient means action and did not reenact the opaque (subefficient and thus more costly) alternative. Note that this finding is hard to reconcile with theories according to which "children imitate behavior that is causally opaque with higher fidelity than behavior that has a transparent physical causal mechanism" (Legare 2019, 130; also see Legare et al. 2015). Moreover, the lack of imitation of the opaque head action in the present study is hard to explain with the variety of theories of cultural transmission that consider opacity of cultural actions as automatically inducing high-fidelity behavior-copying (e.g., Tennie, Call, and Tomasello 2009).

Conclusion

In our view, to provide an adequate explanatory model of the role of emulative reenactment in human cultural learning, any viable theory must be able to account for two significant empirical properties of the way human infants acquire novel skills from observing them performed by others. First, one must account for our remarkable species-unique ability for social learning that allows even preverbal infants to learn quickly, retain over the long term, and delay functionally appropriate reenactments of novel means actions. Indeed, even in cases when a skill has been presented to them only on a single occasion and its reenactment takes place weeks or even months later—for example, as demonstrated in Meltzoff (1988, 1996)—infants are able to acquire culturally relevant skills with remarkable success.

Second, it is crucial to account for the adaptive ability of human infants—and, more widely, of "human cultural novices"—to flexibly but appropriately generalize and selectively reproduce newly acquired motor skills across a variety of functionally relevant and novel contexts. As demonstrated in our studies, the proposed relevance-based inferential account can provide solutions to these two problems since infants can encode a novel behavior after only a few demonstrations. Furthermore, our studies provide evidence that infants' functionally appropriate application of the head touch action is in response to the specific demands of the different situations. These properties of inference-based selective emulative learning, however, represent challenges for the rigid behavior-copying model of imitative learning that accompanies the dichotomy-based approach.

The evidence presented in this chapter contradicts the basic assumptions of the dichotomy-based approach of cultural transmission by showing that when infants and young children observe demonstrations of novel goal-directed instrumental actions that contain causally opaque components and subgoals, the naive learner does not rigidly produce faithful motor replicas of the observed target action. On the contrary, the demonstration of causally opaque instrumental actions induces goal-emulative variability in reproduction.

A closer look at the concrete form of reenacted target actions uncovered how infants reproduced the action means in a remarkably flexible manner, freely generating alternative action variants. Infants did not always bend forward to contact the lamp with their forehead; instead, they either lifted the lamp up by hand or bent forward to approach the lamp with their head. Moreover, in many cases (30% in study 1 and 58% in study 2), even for infants who did not actually bring about the desired outcome, their performed actions clearly indicated their intention to achieve the observed goal. Hence, the main findings of the presented studies support the view that action understanding and goal inference precede, rather than follow, action mirroring processes (Csibra 2007). We acknowledge, however, that the studies presented in support of this general conclusion are restricted in scope and test the effect of causal opacity on cultural transmission processes only in the domain of transitive goal-directed instrumental actions—where the presence of causally opaque action components and subgoals have received less attention in theories of cultural transmission.

According to our interpretation of relevance-based goal emulation, infants encode the goal of the action contextually, and they retrieve a behavior that is effective in its attainment. In addition, it is proposed that natural pedagogy modulates what is learned in the situation (Gergely and Csibra 2006, 2009; Király, Csibra, and Gergely 2013). Ostensive communicative demonstrations can enrich the encoding of the goal by manifesting the particular means (or features thereof) as a culturally relevant manner to attain the goal, thereby indicating to the infant that the manifested action variant is worth acquiring. The presumption of relevance also induces infants' attention to the contextual factors in which the action variant is manifested, leading children to generate emulative variants to discover whether or not there is space for behavioral refinements. Finally, the presumption of relevance also leads infants to differentiate the instrumental from the conventional functions that are served by an action.

The role of ostension is to highlight a behavior, or an aspect of behavior, that is not causally accessible to the observer in that it is not analyzable with the individual's cognitive toolkit. However, ostensive demonstration induces the reproduction of behavior while allowing for variability in its reproduction. Indeed, ostensive demonstration brings into focus aspects of a novel action sequence to be encoded as subgoals that would otherwise remain opaque or unattended to by the learner. These subgoals are highlighted within specific contexts that help ground their relevance. However, being goals themselves, these subgoals can also be attained through variable means actions. This results in the fact that some subefficient means actions do not overwrite or rule out alternatives but rather often coexist with more efficient variants that are also acceptable in everyday situations (Altinok et al. 2020).

Our head touch studies underline the fact that subefficient means actions are used together with efficient alternatives. In the context of teaching, ostensive demonstration not only induces the encoding of a relevant novel aspect of an action as a subgoal of that action, but it also allows infants to segment the action and analyze the situation of its enactment, thereby promoting learning about other specific contexts in which the novel action could be applied as a relevant means. Changes in contexts can be responsible for variability in behavior and, at the same time, contribute to the survival of both efficient and subefficient yet socially determined and conventional formats of action. Ostensive demonstrations inform the learner about how to adapt their actions in a context-dependent way.

How and why do less efficient cultural versions of means actions survive? First, as seen in our illustrative head touch studies, when children try to reach a goal, they vary the means employed in a trial-and-error manner. This process can result in more flexibility on the part of the learner, who can compare alternative versions of an action—including the versions tried out by a model—for attaining the subgoal itself (e.g., the varying way that infants touched the lamp with different parts of their head). This potential for monitoring the variants and their success during reenactment, and for relating them to different features of the context in which they tend to be produced, can also facilitate the emergence of new variants and combinations of behaviors.

This process could also result in a deeper and more detailed understanding of the different kinds of functional determinants that are involved in relating the subgoals to the final goal in particular contexts. In particular, by comparing and analyzing the alternative variants produced along the lines of efficiency and relating them to differences of the situational context in which they are more likely to appear, the learner could differentiate the relevant contributions of causal and social conventional functions in the use of alternative action variants to achieve a subgoal. Consequently, during the reproduction of the behavior in a context-dependent way, new variants could emerge. These new variants could result in new solutions and so could also fuel innovative processes (for a similar proposal, see Yu and Kushnir 2020).

If the primary mechanism of social learning were to copy blindly behaviors that appear causally opaque to the learner, there would be no room for instrumental refinement. Based on the empirical findings presented above, we suggest that the central assumption of the dichotomy-based approach—namely, that the apparent causal opacity of novel actions observed induces a rigid behavior-copying mechanism that produces faithful motor replicas and inhibits the production of variability in action alternatives—is fundamentally mistaken both when applied to the domain of noninstrumental ritual actions serving social functions and in the domain of transitive goal-directed actions associated with instrumental functions.

We propose that repeated ostensive demonstrations of causally opaque and subefficient actions aimed at attaining subgoals in certain types of social contexts function to maintain and stabilize the normative use of less efficient versions and to safeguard against their replacement in the cultural repertoire by new, successful, and more efficient alternatives. Ostensive manifestations highlight how the utilization of a subefficient action version is not accidental but instead serves a social function, which constitutes a culturally sanctioned alternative. This picture thus suggests that the ostensive demonstration of opaque action variants induces in the cultural learner a presumption of relevance and leads to relevance-based emulation of alternative variants. This source of reproductive variability in the domain of instrumental actions could foster the emergence of technological innovations by discovering more adaptive action variants or their combinations and generalizing the acquired functional skill over a broad range of contexts. Even in the case of subgoals perceived as conventional, relevance-based emulation can introduce variability and could therefore allow the emergence of more efficient alternatives.

Yet, as the studies discussed by Nicola Cutting (this volume) highlight, children are poor tool innovators. While children have an outstanding ability to quickly learn the use of a tool when following the demonstration of others, it is rare for them to solve problems by designing (even simple) new tools. Cutting emphasizes how innovation is likely to be more socially mediated and how previous studies lack attention to this social aspect. What is missing for children in such cases is the ability to activate the relevant knowledge that they have already learned and reorganize it in a novel way for functional and efficient goal attainment. A recent investigation brought into focus how in hunter-gatherer societies, some socialization practices can be observed that boost innovation in children (Lew-Levy et al. 2020). These socialization practices include the support of learning through autonomous exploration (see Boyette, this volume), the teaching of children by adults and peer-play, and sensitizing children for seeking novel forms of goal attainment (innovation seeking). Genuine innovation, consequently, builds on socially mediated, accumulated knowledge as well as the capacity to introduce variations. That is why innovative tool design seems to be a late-developing competence. In this sense, the flexibility of social learning is a prerequisite to the emergence of innovative capabilities.

To summarize, the view that we propose here is the following. On the one hand, relevance-based goal emulation is a social learning mechanism that promotes the discovery of variants in behavior through the understanding of goals and subgoals organized into hierarchies within an action. On the other hand, this mechanism contributes to the establishment of a robust and rich knowledge base by facilitating the production of alternative variants in different contexts. In our view, these two aspects actually represent different forms of learning carried out by the same mechanism—that is, relevance-based emulation—that are jointly necessary for achieving innovation.

Notes

1. When one observes a behavior, one employs mental causal models when interpreting it. A behavior is causally transparent when the available contextual information allows for interpretation of its causal structure using such model. For instance, when an agent makes a detour to reach a goal object, this behavior is causally transparent if one sees an obstacle to go around. In contrast, a behavior is causally opaque when the contextual information is insufficient to interpret the behavior's causal structure—for example, when an agent makes a detour to reach a goal object, this behavior is difficult to understand if there is no obstacle to go around (see Gergely, Bekkering, and Király 2002).

2. Note, however, that by identifying the causally unnecessary transitive actions as being "silly," one suggests that the children recognized them as being causally transparent, as indicated by their correct judgment that they were causally irrelevant for retrieving the goal object. Despite this understanding, these actions were faithfully copied and reproduced just as were the equally unnecessary but causally opaque intransitive "magic" gestures that were also part of the demonstrated series of actions.

3. The term "innovation" is defined by proponents of the dichotomy-based view of cultural transmission as "constructing new tools, or using old tools in new ways, to solve new problems" (see Legare and Nielsen 2015, 689).

4. Some forms of everyday behaviors appear causally opaque even for skilled adults, who habitually eat food in the normative manner of their own cultural tradition or do so selectively as a function of the particular social context, such as at a formal dinner with the president of a university.

References

Altınok, N., M. Hernik, I. Király, I., and G. Gergely. 2020. "Acquiring Sub-efficient and Efficient Variants of Novel Means by Integrating Information from Multiple Social Models in Preschoolers." *Journal of Experimental Child Psychology* 195:104847.

Altınok, N., I. Király, and G. Gergely. 2022. "The Propensity to Learn Shared Cultural Knowledge from Social Group Members: Selective Imitation in 18-Month-Olds." *Journal of Cognition and Development* 23 (2): 273–288.

Berl, R. E., and B. S. Hewlett. 2015. "Cultural Variation in the Use of Overimitation by the Aka and Ngandu of the Congo Basin." *PLOS One* 10 (3): e0120180.

Boyd, R., and P. J. Richerson. 1996. "Why Culture Is Common, but Cultural Evolution Is Rare." In *Evolution of Social Behaviour Patterns in Primates and Man*, edited by W. G. Runciman, J. M. Smith, and R. I. M. Dunbar, 77–93. Oxford: Oxford University Press.

Buchsbaum D., A. Gopnik, T. L. Griffiths, and P. Shafto. 2011. "Children's Imitation of Causal Action Sequences Is Influenced by Statistical and Pedagogical Evidence." *Cognition* 120 (3): 331–340.

Carpenter, M., J. Call, and M. Tomasello. 2005. "Twelve- and 18-Month-Olds Copy Actions in Terms of Goals." *Developmental Science* 8 (1): 13–20.

Chen, M., I. Király., and G. Gergely. 2012. *"Opacity and Relevance in Social Cultural Learning: Relevance-Guided Emulation in 14-Month-Olds."* Paper presented at Budapest CEU Conference on Cognitive Development, Budapest, January 12–14.

Clegg, J. H., and C. H. Legare. 2016. "Instrumental and Conventional Interpretations of Behavior Are Associated with Distinct Outcomes in Early Childhood." *Child Development* 87 (2): 527–542.

Csibra, G. 2007. "Action Mirroring and Action Understanding: An Alternative Account." In *Sensorimotor Foundations of Higher Cognition: Attention and Performance XXII*, edited by P. Haggard, Y. Rosetti, and M. Kawato, 435–458. Oxford: Oxford University Press.

Csibra, G., and G. Gergely. 2006. "Social Learning and Social Cognition: The Case for Pedagogy." In *Processes of Change in Brain and Cognitive Development: Attention and Performance XXI*, edited by Y. Munakata and M. H. Johnson, 249–274. Oxford: Oxford University Press.

Csibra, G., and G. Gergely. 2009. "Natural Pedagogy." *Trends in Cognitive Sciences* 13 (4): 148–153.

Csibra, G., and G. Gergely. 2011. "Natural Pedagogy as Evolutionary Adaptation." *Philosophical Transactions of the Royal Society B: Biological Sciences* 366 (1567): 1149–1157.

Gergely, G. 2007. "The Social Construction of the Subjective Self: The Role of Affect-Mirroring, Markedness, and Ostensive Communication in Self-Development." In *Developments in Psychoanalysis. Developmental Science and Psychoanalysis: Integration and Innovation*, edited by L. Mayes, P. Fonagy, and M. Target, 45–88. London: Karnac Books.

Gergely, G., H. Bekkering, and I. Király. 2002. "Rational Imitation in Preverbal Infants." *Nature* 415 (6873): 755–755.

Gergely, G., and G. Csibra. 2003. "Teleological Reasoning in Infancy: The Naïve Theory of Rational Action." *Trends in Cognitive Sciences* 7:287–292.

Gergely, G., and G. Csibra 2006. "Sylvia's Recipe: The Role of Imitation and Pedagogy in the Transmission of Human Culture." In *Roots of Human Sociality: Culture, Cognition, and Human Interaction*, edited by N. J. Enfield and S. C. Levinson, 229–255. Oxford: Berg Publishers.

Gergely, G., and P. Jacob. 2012. "Reasoning about Instrumental and Communicative Agency in Human Infancy." In *Rational Constructivism in Cognitive Development*, edited by F. Xu and T. Kushnir, 59–94. New York: Academic Press.

Gergely, G., Z. Nádasdy, G. Csibra, and S. Bíró. 1995. "Taking the Intentional Stance at 12 Months of Age." *Cognition* 56 (2): 165–193.

Heintz, C., and T. Scott-Phillips. 2023. "Expression Unleashed: The Evolutionary and Cognitive Foundations of Human Communication." *Behavioral and Brain Sciences* 46 (January): 1–46.

Henrich, J., S. J. Heine, and A. Norenzayan. 2010. "The WEIRDest People in the World?" *Behavioral and Brain Sciences* 33 (2–3): 61–83.

Hewlett, B. S., and C. J. Roulette. 2016. "Teaching in Hunter-Gatherer Infancy." *Royal Society Open Science* 3:150403.

Hoehl, S., S. Keupp, H. Schleihauf, N. McGuigan, D. Buttelmann, and A. Whiten. 2019. "'Over-Imitation': A Review and Appraisal of a Decade of Research." *Developmental Review* 51:90–108.

Kapitány, R., and M. Nielsen. 2019. "Ritualized Objects: How We Perceive and Respond to Causally Opaque and Goal Demoted Action." *Journal of Cognition and Culture* 19:170–194.

Király, I., G. Csibra, and G. Gergely. 2013. "Beyond Rational Imitation: Learning Arbitrary Means Actions from Communicative Demonstrations." *Journal of Experimental Child Psychology* 116 (2): 471–486.

Lancy, D. F. 1996. *Playing on the Mother-Ground: Cultural Routines for Children's Development*. New York: Guilford Press.

Lancy, D. F., S. Gaskins, and J. Bock, eds. 2009. *The Anthropology of Learning in Childhood*. Lanham, MD: Altamira Press.

Legare, C. H. 2019. "The Development of Cumulative Cultural Learning." *Annual Review of Developmental Psychology* 1 (1): 119–147.

Legare, C. H., and M. Nielsen. 2015. "Imitation and Innovation: The Dual Engines of Cultural Learning." *Trends in Cognitive Sciences* 19 (11): 688–699.

Legare, C. H., and M. Nielsen. 2020. "Ritual Explained: Interdisciplinary Answers to Tinbergen's Four Questions." *Philosophical Transactions of the Royal Society B: Biological Sciences* 375 (1805): 1–5.

Legare, C. H., N. J. Wen, P. A. Herrmann, and H. Whitehouse. 2015. "Imitative Flexibility and the Development of Cultural Learning." *Cognition* 142:351–361.

Lew-Levy, S., A. Milks, N. Lavi, S. M. Pope, and D. E. Friesem. 2020. "Where Innovations Flourish: An Ethnographic and Archaeological Overview of Hunter-Gatherer Learning Contexts." *Evolutionary Human Sciences* 2 (2020): e31.

Lyons, D. E., A. G. Young, and F. C. Keil. 2007. "The Hidden Structure of Overimitation." *Proceedings of the National Academy of Sciences* 104 (50): 19751–19756.

McGuigan, N., A. Whiten, E. Flynn, and V. Horner. 2007. "Imitation of Causally Opaque versus Causally Transparent Tool Use by 3- and 5-Year-Old Children." *Cognitive Development* 22 (3): 353–364.

Meltzoff, A. N. 1988. "Infant Imitation after a One-Week Delay: Long-Term Memory for Novel Acts and Multiple Stimuli." *Developmental Psychology* 24 (4): 470–476.

Meltzoff, A. N. 1996. "The Human Infant as Imitative Generalist: A 20-Year Progress Report on Infant Imitation with Implications for Comparative Psychology." In *Social Learning in Animals: The Roots of Culture*, edited by C. M. Heyes and B. G. Galef, 347–370. New York: Academic Press.

Nielsen, M., and K. Tomaselli. 2010. "Overimitation in Kalahari Bushman Children and the Origins of Human Cultural Cognition." *Psychological Science* 21 (5): 729–736.

Over, H., and M. Carpenter. 2012. "Putting the Social into Social Learning: Explaining Both Selectivity and Fidelity in Children's Copying Behavior." *Journal of Comparative Psychology* 126 (2): 182–192.

Over, H., and M. Carpenter. 2013. "The Social Side of Imitation." *Child Development Perspectives* 7:6–11.

Paradise, R., and B. Rogoff. 2009. "Side by Side: Learning by Observing and Pitching In." *Ethos* 37:102–138.

Paulus, M., S. Hunnius, M. Vissers, and H. Bekkering. 2011. "Imitation in Infancy: Rational or Motor Resonance?" *Child Development* 82:1047–1057.

Pinkham, A. M., and V. K. Jaswal. 2011. "Watch and Learn? Infants Privilege Efficiency over Pedagogy during Imitative Learning." *Infancy* 16:535–544.

Rendell, L., L. Fogarty, W. J. Hoppitt, T. J. Morgan, M. M. Webster, and K. N. Laland. 2011. "Cognitive Culture: Theoretical and Empirical Insights into Social Learning Strategies." *Trends in Cognitive Sciences* 15 (2): 68–76.

Sperber, D., and D. Wilson. 2002. "Pragmatics, Modularity and Mind-Reading." *Mind & Language* 17 (1–2): 3–23.

Tennie, C., J. Call, and M. Tomasello. 2009. "Ratcheting Up the Ratchet: On the Evolution of Cumulative Culture." *Philosophical Transactions of the Royal Society B: Biological Sciences* 364 (1528): 2405–2415.

Tennie, C., and C. P. van Schaik. 2020. "Spontaneous (Minimal) Ritual in Non-human Great Apes?" *Philosophical Transaction of the Royal Society B: Biological Sciences* 375:20190423.

Toelch, U., D. R. Bach, and R. J. Dolan. 2014. "The Neural Underpinnings of an Optimal Exploitation of Social Information under Uncertainty." *Social Cognitive and Affective Neuroscience* 9 (11): 1746–1753.

Tomasello, M. 1996. "Do Apes Ape?" In *Social Learning in Animals*, edited by C. M. Heyes and B. G. Galef, 319–346. New York: Academic Press.

Tomasello, M. 1998. "Emulation Learning and Cultural Learning." *Behavioral and Brain Sciences* 21 (5): 703–704.

Tomasello, M. 1999. *The Cultural Origins of Human Cognition*. Cambridge, MA: Harvard University Press.

Watson-Jones, R. E., and C. H. Legare. 2016. "The Social Functions of Group Rituals." *Current Directions in Psychological Science* 25 (1): 42–46.

Wen, N. J., P. A. Herrmann, and C. H. Legare. 2016. "Ritual Increases Children's Affiliation with In-Group Members." *Evolution and Human Behavior* 37 (1): 54–60.

Yu, Y., and T. Kushnir. 2020. "The Ontogeny of Cumulative Culture: Individual Toddlers Vary in Faithful Imitation and Goal Emulation." *Developmental Science* 23 (1): e12862.

6 Flexible Social Learning of Technical Skills: The Case of Action Coordination

James W. A. Strachan, Arianna Curioni, and Luke McEllin

Introduction

Technical expertise can encompass a wide variety of skills and knowledge, such as the functional, mechanical, or material properties of objects or tools with which one must work or causal relationships between objects and their environment. The type of technical knowledge that we focus on is the motor form: that is, how one moves one's own body in order to successfully perform some technique. In cases of action-oriented techniques such as sport or dance, the way one moves is clearly integral to the intended outcome, but learning how to move one's body is also fundamental to object-oriented techniques such as playing a musical instrument, throwing clay on a potter's wheel, sailing a boat, or even tying one's shoelaces. Expert performance of many techniques requires precise motor plans, sophisticated control of the timing and trajectory of one's movements, and careful monitoring and integration of sensory and proprioceptive feedback.

However, the specific cognitive mechanisms responsible for cultural transmission of technical motor skills remain an open question (Heyes 2016; Stout 2002). Part of the challenge in addressing this lies in the flexibility with which technical skill transmission occurs: someone looking to learn how to prepare a specific meal may learn it by reverse-engineering the preparation process from its taste or looks, or they may follow a recipe book or cooking program, or they may observe an experienced chef prepare the dish, or they may ask an experienced chef to teach them. These are all cases of social learning, and each will offer unique opportunities and challenges to learning.

Despite the wide variety of scenarios, experimental studies of cultural transmission of techniques do not tend to focus on the cognitive factors allowing for such flexible adaptation to different learning environments. Indeed, transmission chains are the most common design in the study of cultural transmission, where a simple skill (such as constructing a paper airplane or a spaghetti tower) is transmitted through iterated generations of learners who observe either the actions or the end-product of a previous generation (Mesoudi and Whiten 2008; Miton and Charbonneau 2018). In turn, the transmission of a technique is evaluated by comparison of the artifacts produced at each generation on some prespecified coding metric (e.g., the distance the paper plane flies or the height of the tower). The information passed to learners at each generation is therefore tightly constrained, and the grain of analysis of this kind of data coding is too coarse to allow for much flexibility in transmission (Charbonneau and Bourrat 2021). As a result, such studies offer little insight into the cognitive mechanisms at play in these transmission episodes, as they "black box" the cognition of social learning (Heyes 2016) in favor of measuring input-output relationships between artifacts produced by different generations (Charbonneau and Bourrat 2021).

So far, the specific cognitive mechanisms supporting cultural transmission of technical motor skills remain an open question. In our chapter we propose that, to address the flexibility of technical traditions, it is important to situate social learning within an interaction context. To this end, we show how current discourse on observational learning through imitative copying typically presupposes a unidirectional type of interaction that imposes rather than demonstrates rigidity on the learned behavior. We go on to differentiate between *know-how* and *information flow* and show that while the former is necessarily unidirectional in social learning episodes, the latter rarely is. We then show that by expanding the range of social learning interaction contexts to allow for bidirectional information flow, it is possible to draw clear parallels with the literature on action coordination. Finally, we show that considering social learning as a type of action coordination can help to explain both the flexibility and rigidity of technical traditions in a way that is coherent with the anthropological record on complex skill learning.

The Unidirectionality Problem

Social learning has been defined as "learning that is influenced by observation of, or interaction with, another animal . . . or its products" (Heyes 1994), or more precisely, "learning that is *facilitated* by observation of, or interaction with, another individual or its products" (Hoppitt and Laland 2013, emphasis added). It is interesting to note that under either definition, observation is only one way by which individuals can socially learn from conspecifics. Yet many discussions of social learning treat observation as the basic phenomenon of interest, and discussions of mechanisms involved in social learning tend to focus on those that characterize the relationship between a learner's observational input and their behavioral output— for example, as the result of imitation, emulation, local enhancement, and so on (Caldwell and Millen 2009; Henrich 2016; Henrich and McElreath 2003; Heyes 2001; Mesoudi 2011; Whiten et al. 2009).

This emphasis on learners and how they unpack or decode the learning input helps us to simplify complex and variable social learning interactions and find a common denominator; then we can explore what specific adaptations or mechanisms allow a learner to acquire input by reading out information encoded in the actions or behaviors of others. The assumption inherent in this approach is that the mechanisms the learner uses to decode input in one scenario can be generalized to other scenarios where they are learning different content with a different model in a different environment. Observational learning therefore serves as a minimal working example where the social input to the learner from the model can be treated as if it were any other stimulus in the environment. Minimal working examples are important for guiding theoretical and empirical work, particularly on such varied phenomena as social learning, where interactions are diverse and complex and individuals can learn using a range of methods such as observation, local enhancement, or even direct teaching (Kline 2015).

To say that the example is minimal is not to say that the question of how an individual learns through observation is trivial—far from it. But explaining this episode of social learning does not require knowing what is going on in the mind of the model, nor anything of the prior knowledge or experience of either agent, nor the local ecology or social environment in which the interaction is situated. One need only identify the learning processes going on in a single individual's mind under the assumption that these processes can then be generalized to more conceptually complex scenarios.

Another point in favor of using observational learning as a minimal working example is that this kind of learning is very important early in human development: infants cannot act independently, so learning through watching and imitating adults is very important to the acquisition of early motor skills, together with the ability to discern who and what to imitate (Gergely, Bekkering, and Király 2002; Gergely and Csibra 2006; Meltzoff 1988; Wood, Kendal, and Flynn 2013). Developmental research increasingly recognizes that interactivity plays a crucial role in early observational learning, as it relies on the receptivity of the learner to communicative signals from the model and on the model's anticipation and recognition of the learner's needs (Csibra and Gergely 2009; Gergely and Csibra 2005, 2006; Király, Csibra, and Gergely 2013; Gergely and Király, this volume). However, such interactivity is frequently missing from laboratory studies of intergenerational transmission chains with adult participants, where learners' inputs are usually observations of single instances of the to-be-learned behavior with no chance for repeated practices (Caldwell, Renner, and Atkinson 2018; Caldwell et al. 2020; Caldwell and Millen 2008b; Mesoudi and Whiten 2008; Miton and Charbonneau 2018).[1]

To our minds, this treatment is problematic because it carries with it an assumption of unidirectionality: that is, there is no opportunity for mutual exchange of information between the model and the learner. As a result, learners in an observational scenario offer no input to shape the learning outcome. In such accounts, the only control a learner can exert is over where and how they allocate their attention (Heyes 2012). A learner may preselect models to learn from—for example, on the basis of social reputation or prestige (Henrich and Gil-White 2001; Jiménez and Mesoudi 2019). Learners may also discard or down-weight inputs from models who appear incompetent or unreliable (Dautriche et al. 2021; Koenig and Woodward 2010; Poulin-Dubois, Brooker, and Polonia 2011), but they do not have any direct impact on the model's behavior itself. Laboratory studies in turn construct learning scenarios where there can be no interactivity, either because the model demonstrates the action and then leaves before the learner first practices the behavior (Caldwell and Millen 2009) or the learner watches prerecorded videos rather than a live model (Strachan et al. 2021). When laboratory studies of technical transmission do involve interactivity, the dynamics of these interactions—particularly how feedback channels from the learner to the model are exploited during teaching—tend to be overlooked in favor of comparing the production outputs of learners who acquired the skill through teaching against those of learners who learned by imitation or emulation (Morgan et al. 2015). On the other hand, laboratory studies of joint actions (Sebanz, Bekkering, and Knoblich 2006; Sebanz and Knoblich 2009, 2021) indicate that during interactions, dynamic coupling between interaction partners is a complex phenomenon involving sophisticated prediction, planning, and signaling to ensure successful coordination (we discuss this in further detail below in the section on coordination, flexibility, and rigidity).

In cognitive science, this problem of unidirectionality is also known as the "fourth wall" problem, and it has been well documented with regards to eye gaze processing (Gobel, Kim, and Richardson 2015; Risko, Richardson, and Kingstone 2016). The crux of the problem is that within laboratory experiments where participants look at scenes or faces on computer screens, their patterns of gaze fixations are very different from live recordings of participants physically interacting with those scenes or people in real life. When observing another person in real life, the observed individual can detect that they are being watched, which can result in audience effects (Bateson, Nettle, and Roberts 2006; Cañigueral, Ward, and Hamilton 2021; Cañigueral and Hamilton 2019; Hamilton 2016) and even interfere with the performance of skilled behavior (Belletier et al. 2015; DeCaro et al. 2011). Given that audience effects can elicit changes to behaviors that may be context-specific and tailored to other people within the interaction (Krishnan-Barman and Hamilton 2019), their effects on the interaction and learning should be considered in all but truly unidirectional observations (cases where the informational channels are such that a learner can watch a model without any chance of affecting their behavior). Such cases are rare outside of laboratory settings where unidirectionality can be guaranteed through the use of apparatus such as videos or one-way mirrors. In the real world, nearly all social learning interactions are to some degree bidirectional in nature, as a learner *can* affect the behavior of the model and may do so in ways that can be unobservable to those outside the interaction (Stout 2002). An explanation of the mechanisms of social learning should therefore aim at identifying how and when a learner *does* affect the behavior of the model.

We propose to embrace the bidirectionality of social learning interactions by examining the distinction between *know-how* and *information*. In the next section, we clarify this distinction and discuss the implications for broadening the scope of research on social learning.

Know-How versus Information

A social learning episode involves at least two people. One, the model, possesses some technical knowledge (henceforth know-how) that the other, the learner, will acquire as a result of the episode. In this sense, it is a necessary precondition that a social learning interaction has an asymmetry in know-how and a consequent unidirectional flow from the model to the learner. However, know-how is not the only thing communicated in a learning interaction; there is other information that need not flow unidirectionally from the model to the learner. Students can ask questions of their teachers, or learners can make statements about what they are learning or try out ways of using the technique that the model can then react and adapt to. Considering this, it is important to distinguish know-how flow from information flow as dimensions along which interaction contexts can vary. In brief, we use know-how to refer to the context-independent representation of an action or sequence of actions that are necessary to produce a certain outcome, while information is the context-specific feedback from the social interaction partner that allows an actor to make online adjustments that can satisfy either instrumental or communicative goals.

Imagine a situation in which two people—a model (Maria) and a learner (Luisa)—are involved in preparing a meal according to a family recipe of Maria's. Maria knows the recipe but Luisa does not. This is an example of a know-how asymmetry and is a necessary precondition for social learning—the know-how (the family recipe) is held by the model (Maria) and not by the learner (Luisa) who will acquire this know-how over the course of the learning interaction. This know-how (the family recipe) is context-independent because it is not restricted to the immediate context of the interaction in which Maria and Luisa find themselves but instead extends beyond this particular interaction—once learned, Luisa could go on to cook this meal without Maria, and she could do so in a different kitchen with different tools and ingredients of different sizes or quality.

However, while a social learning interaction must have a unidirectional know-how flow through the interaction because of some know-how asymmetry, it can have many different types of *information* flow. This information flow can be unidirectional—Luisa could be learning to cook the recipe by reading a cookbook of family recipes that Maria has published— but it need not be. If Luisa and Maria are in the kitchen together, then Maria can see what Luisa is doing as well and adjust her actions accordingly, perhaps by changing her position at the stove so that Luisa can look over her shoulder. The information flow is also affected by the dynamics of the interaction itself: Luisa could sit in the kitchen and watch Maria prepare the meal, but she could also help Maria in preparing the meal. Any of these situations is an example of social learning, and all can result in the successful transfer of know-how from Maria to Luisa. But, for the two members of the interaction, they are very different scenarios that require very different cognitive mechanisms to act.

These different interaction structures, characterized by different patterns of information flow, highlight a key aspect of social learning interactions, which is the many varied opportunities for coordination between models and learners that are overlooked in many treatments of social learning. Drawing on research in cognitive science in joint action and coordination, we propose that research on social learning can lean into this complexity by considering cognitive mechanisms in social learning episodes that may not be specialized for learning per se but rather for action coordination. This approach centers the social learner not as a spectator or information scrounger but as an actively engaged participant in the behavior.

Learning by Active Engagement

Considering the social learner as an active participant in an interaction that is defined by particular situational constraints opens up opportunities for nonvertical patterns of transmission that researchers describe as collaborative learning (Tomasello, Kruger, and Ratner 1993)—children can play with their peers at grown-up jobs before they are allowed to participate with adults. For example, in Papua New Guinea, Asabano male children who are prohibited from joining adult hunters on the search for dangerous game like feral pigs or cassowaries will often band together and structure their playtime around hunting smaller targets like lizards (Little and Lancy 2016). This gives young children important basic experience running around and throwing sticks at lizards, and it gives older children the valuable opportunity to plan and manage a team of individuals of various ages and skill levels, which can give them much better functional understanding of the mechanics at play in the behavior that will serve them well when they eventually join the adult hunting parties.

Learning through interaction with peers can also be seen in cases where the canonical transmission model of learning through observation of an expert appears otherwise unidirectional. In cases where interaction with elders is discouraged as inappropriate and children must instead congregate to watch experts perform a skill without disturbing them (as in the case of the Akha; Ongaro, this volume), the vertical transmission of skills from adults to children appears unidirectional. However, even in this scenario, transmission is not wholly unidirectional, as the model is undoubtedly aware of the audience and tolerates the observation (see above). Nonetheless, even here there are opportunities for interaction, as learners can take social cues from their peers who are also watching. Collaborative learning can lead to better retention and transfer of information (Craig, Chi, and VanLehn 2009), and this finding echoes laboratory research showing that collaborating partners can efficiently distribute their attention during visual search tasks (Brennan et al. 2008; Wahn et al. 2020). In a case where a pooled group of observational learners consists of different age groups, younger learners may gain a particular benefit by dynamically following the gaze of their more experienced older peers to relevant features of the actions being performed (Richardson and Dale 2005; Williams et al. 2002).

Even in cases where coordination is not strictly necessary to achieve an outcome—as in the case of stone knapping to make tools, for example—established experts will intentionally perform individual actions in social configurations, discussing their tasks with each other in a line or a circle, where each can have perceptual access to what others are doing and where novices can watch multiple experts at work in parallel (Stout 2002, 2005). Beyond ethnographic evidence, there is evidence from the archaeological record of southern African stone tools throughout the Pleistocene that technological innovations can and do occur through bottom-up or learner-driven contributions to the behavior, as opposed to through purely top-down mechanisms of transmission (Wilkins 2020).

We argue that it is this kind of learning through coordination that allows for the flexible transmission of technical traditions. That is, techniques can be learned in a wide variety of ways, and although each may have some idiosyncratic properties favoring certain types of learning interaction over others, an episode of social learning is subject to the specific coordination demands that are dictated by the task constraints, individual motivations, and interaction structure. In turn, learners—and models—recruit specialized coordination mechanisms to address these specific coordination demands, and these mechanisms play a role in determining both what is learned and how it is learned. In the following section, we explore how different coordination mechanisms can lead to systematic behavioral modulations in response to different contextual demands that allow learners and models to adapt to various social learning scenarios. We also offer some speculation as to how rigidity in the way that cultural traditions are practiced (rather than in the way they are transmitted) can be explained as a potential coordination strategy. In doing this, we show how centering interactivity in technical transmission can help to address flexibility and rigidity at the level of cultural variants within a population as well as their transmission.

Coordination, Flexibility, and Rigidity

We now give a brief overview of some literature in joint action and action coordination that describes some of the cognitive mechanisms supporting coordination and demonstrates how they can inform social learning and research into technical evolution and transmission.

Specifically, we describe how a drive to make oneself predictable during coordination can result in rigid and stable patterns of behavior, while having to represent another person's task during a joint action when that actor has different constraints on their actions can lead to dynamic and flexible action modulations to compensate. Finally, we describe the few existing studies that examine social learning and teaching from a joint action perspective.

Coordinating actions between two or more individuals in order to realize some shared or joint outcome is a demanding task subserved by a host of mechanisms (Vesper et al. 2010). A key driving principle behind how these mechanisms are expressed is whether and how they help to facilitate a partner's anticipation of the relevant features of one's actions. Several studies have shown that, when coordinating, actors become less variable from trial to trial than when they act alone as a way of making themselves predictable (Sacheli et al. 2013; Vesper et al. 2011, 2016). While this kind of variability modulation is a very basic way of facilitating coordination, which has even been observed in macaque monkeys (Visco-Comandini et al. 2015), humans are also able to adapt their behavior in more systematic ways.

For example, a study of joint improvisation (Hart et al. 2014) found that expert improvisers who were instructed to synchronize their actions modulated the velocity profiles of their actions such that both partners deviated significantly from how either of them would move individually. Furthermore, rather than simply converging on a pattern of behavior somewhere between the two actors' styles of movements (i.e., averaging out each other's idiosyncrasies), the resulting coordinated actions reflected some universal characteristics across dyads. This suggests that both agents adjusted their behavior in systematic, general ways that they considered easy to predict for any potential interaction partner.

The role of predictability in coordinating joint action is a fundamental one, and one that interacting partners are very sensitive to. This is particularly evident in cases of competition: while coordinating participants modulate their behavior to be predictable, competing participants do the opposite as they try to mislead their opponents' predictions (Glover and Dixon 2017). However, anticipating another's actions is difficult even when both interactants are cooperating; researchers have found that even when participants choose to coordinate with somebody in a cooperative context, this coordination results in systematic performance costs relative to when acting alone (Curioni et al. 2022).

Considering social learning episodes as instances of coordination allows for a great degree of flexibility in the trial-to-trial expression of behaviors and techniques. The mechanisms supporting coordination in joint action are versatile and sensitive to the situational context and local interaction demands. Actors are sensitive to their partners' ecological constraints and embody these in their own movements (Schmitz et al. 2017), and individuals with an easy task will make more adjustments to their behavior to coordinate with a partner who has a more difficult task (Vesper et al. 2013). In cases of unidirectional information flow, coordinating partners will use leader–follower role assignments to dictate the distribution of online adjustments (Curioni et al. 2019). Even specific coordination mechanisms such as adjusting the speed of one's movement are dependent on the local task demands. In cases where partners are trying to synchronize the timing of oscillatory movements such as tapping or swaying, people speed up in an attempt to minimize variability (Vesper et al. 2011; Wolf et al. 2019), while in cases where partners are trying to synchronize the spatial endpoints of actions, they tend to slow down and adjust the ascent-to-descent velocity ratio of their movements to highlight the upcoming target (McEllin, Knoblich, and Sebanz 2018). Actors will even tailor their actions on the basis of their partner's ability in order to ensure that they do not communicate redundantly (Candidi et al. 2015).

Until now, there has been comparatively little work examining how coordination mechanisms are expressed and exploited in instances of social learning and what role these mechanisms might play downstream in the transmission, propagation, and stabilization of cultural traditions. However, early work does show that pedagogical demonstrations share some kinematic characteristics with coordinated joint actions. In one study (McEllin, Knoblich, and Sebanz 2018), where participants learned to play a series of notes on a modified xylophone, the kinematics of the participants who knew the sequence were examined under three conditions: a turn-taking demonstration, where the participant played the piece through for someone who had to learn to play it themselves; an unequal coordination condition, where the knowledgeable participant and a naive participant had to play together at the same time; and an equal coordination condition, where both participants knew the piece and had to play it together at the same time. In all three conditions, participants (demonstrators, leaders, and co-actors) exaggerated the peak height of their movements relative to an individual baseline. When participants had to play together at the same time (unequal and equal coordination conditions), they slowed down the descent phase of their movements in order to facilitate temporal coordination. However, when participants were coordinating with a naive participant, they also slowed down the ascent phase of their movements in order to facilitate their partner's spatial predictions about the upcoming end of the movement. These kinematic signatures, which map clearly to different coordination demands and constraints, are clear evidence of individuals flexibly adapting their behavior to facilitate coordination and communicate information.

In another study (Okazaki, Muraoka, and Osu 2019) the kinematics of participants were monitored during a turn-taking imitation learning task where a teacher demonstrated to a student how to complete the Tower of Hanoi task in order to quantify the interactional dynamics of teaching, which they describe as a type of reciprocal interpersonal coordination. Importantly, they found not just feed-forward information flow from the teacher to the learner but also feedback information flow from the learner to the teacher. Furthermore, this changed over time as teachers interactively scaffolded the learners' behavior by providing more pedagogical information when learners struggled; as learners improved, the transfer of know-how from teacher to learner became a bidirectional exchange of common ground information.

In a later study (Strachan and colleagues 2021), it was examined how learners interpret these pedagogical cues in observed demonstrations (playing a piece of music, this time on a modified set of drums) and specifically whether they incorporate these action modifications into their own reproductions. If participants were observational learning through copying the model's demonstration, then they would reproduce this behavior because these exaggerations are embedded features of the actions. However, as we have shown, under a coordinative framework, it is not always necessary for both actors to adjust their behavior if this is not relevant to the end-goal, so there would be no need to copy the pedagogical modifications that learners observe. This study did indeed find evidence in support for a pragmatic reconstructive learning process whereby learners did not incorporate the modifications they observed into their own productions.

Rigidity as a Coordination Strategy

We have so far discussed the role that coordination mechanisms play in supporting flexibility at the level of social learning interactions. That is, through the use of various cognitive mechanisms specialized for interpersonal action coordination, learners can adapt to a range of situational constraints that allow them to learn a wide array of technical skills. As such, technical skill learning can be made flexible through the use of coordination mechanisms. Coordination mechanisms are also highly important for flexibility at the level of the tradition itself, when those traditions involve joint actions with multiple individuals performing skills together in highly dynamic environments such as team sports; improvised dance, music, or theater; and surgery. However, when dealing with the flexibility and rigidity of technical traditions themselves (rather than the flexibility of the learning interactions), the inverse can also be true: that is, the rigidity of technical traditions as they are practiced can be used as a focal point for successful coordinated action.

Given the cognitive demands of coordination and its objective and measurable costs to performance, participants may use any tools at their disposal to offset these costs and maximize the predictability of their own and their partners' actions. One such tool may be through rigid cultural traditions that transmit a prescribed pattern of behavior from the constituent individuals in an interaction. Strict adherence to an inherited tradition serves a pragmatic coordinative function in that it facilitates the prediction of others' behavior and establishes what people are expected to do. Such mutually manifest information allows any group member with the background knowledge of that tradition to successfully coordinate without having to engage in costly prediction and planning.

If rigid traditions can serve to stabilize coordination, then this may encourage the high-fidelity transmission of cultural practices across generations that involve a component of coordination. Take driving a car as an extreme example. In the United Kingdom, the cultural practice is to drive on the left side of the road. There may have been potential pragmatic reasons for this practice when it first originated, but whatever these were at the time, they are now clearly obsolete as most of the rest of the world drives comfortably on the right side. This obsolete cultural practice is suboptimal: British drivers who drive overseas experience real costs of having to adapt to the other side of the road and having to adapt to different blind spots in their car, as do right-side drivers who visit the United Kingdom. And yet the practice resists innovation or change. Learner drivers in the UK are still taught to drive on the left and adhere to this rigid practice, despite its suboptimality in relation to the rest of the world, because they must still share the road with the previous generation from whom they learned.

When acquiring a technique that involves coordination or joint action, functionally opaque or even transparently obsolete features of the tradition may be preserved purely because they serve to facilitate prediction among coordination partners, particularly if after learning the technique the learner must coordinate with members of the previous generation. The common ground afforded by a rigid tradition allows individuals to partially offset the cognitive costs of coordination, and so any innovation or change that may optimize a technique must first overcome the coordination costs involved with deviating from the shared action plan. Given that breakdowns of coordination are invariably costly in terms of time, materials, or even risk of harm, the usefulness of rigid traditions for coordination may result in strong inertia or resilience to change.

Conclusion

In this chapter, we have outlined an approach for investigating the cognitive mechanisms involved in social learning responsible for the flexibility of technical transmission that emphasizes the interactivity of such interactions. We have described the problem of unidirectionality, which we argue does not place enough emphasis on the interactive context in which the learner is embedded. Specifically, the bidirectional flow of information along mutual channels between models and learners allows for the strategic use of coordination mechanisms to make one's actions more predictable and in turn more learnable.

Acknowledgments

Arianna Curioni was supported by the European Research Council under the European Union's Seventh Framework Program (FP7/2007–2013)/ERC grant agreement no. 609819, SOMICS.

Note

1. A notable exception is the use of closed groups or replacement paradigms (Mesoudi and Whiten 2008). In closed-group paradigms, small groups of participants learn a task with the option to learn (copy) from peer members of their own group. Such paradigms are typically used to explore selection strategies in social learning—*what* and *who* learners copy—or the prevalence of social learning versus individual learning within a population. Replacement group paradigms are similar in that they consist of small groups learning the task, but they also involve an iterated replacement of group members to approximate generational turnover. Both types of paradigms involve multiple opportunities for sustained interaction between participants, setting them apart from transmission chain studies by permitting bidirectional information flow. However, both paradigms (especially replacement groups) are more costly and logistically challenging to run than transmission chains, and they are typically run using decision-making games or mental problem-solving rather than action-based technical skills (e.g., Gürerk, Irlenbusch, and Rockenbach 2006; Insko et al. 1983; Toyokawa, Whalen, and Laland 2019). In the rare cases of exceptions (Caldwell and Millen 2008a, 2009), these paradigms are subject to some of the same criticisms as one-off transmission studies in that they measure transmission and cultural change on the basis of input-output relationships and do not study the dynamic information transaction during learning.

References

Bateson, M., D. Nettle, and G. Roberts. 2006. "Cues of Being Watched Enhance Cooperation in a Real-World Setting." *Biology Letters* 2 (3): 412–414.

Belletier, C., K. Davranche, I. S. Tellier, F. Dumas, F. Vidal, T. Hasbroucq, and P. Huguet. 2015. "Choking under Monitoring Pressure: Being Watched by the Experimenter Reduces Executive Attention." *Psychonomic Bulletin & Review* 22 (5): 1410–1416.

Brennan, S. E., X. Chen, C. A. Dickinson, M. B. Neider, and G. J. Zelinsky. 2008. "Coordinating Cognition: The Costs and Benefits of Shared Gaze during Collaborative Search." *Cognition* 106 (3): 1465–1477.

Caldwell, C. A., M. Atkinson, K. H. Blakey, J. Dunstone, D. Kean, G. Mackintosh, E. Renner, and C. E. H. Wilks. 2020. "Experimental Assessment of Capacities for Cumulative Culture: Review and Evaluation of Methods." *Wiley Interdisciplinary Reviews: Cognitive Science* 11 (1): e1516, 1-15.

Caldwell, C. A., and A. E. Millen. 2008a. "Experimental Models for Testing Hypotheses about Cumulative Cultural Evolution." *Evolution and Human Behavior* 29 (3): 165–171.

Caldwell, C. A., and A. E Millen. 2008b. "Studying Cumulative Cultural Evolution in the Laboratory." *Philosophical Transactions of the Royal Society B: Biological Sciences* 363 (1509): 3529–3539.

Caldwell, C. A., and A. E. Millen. 2009. "Social Learning Mechanisms and Cumulative Cultural Evolution: Is Imitation Necessary?" *Psychological Science* 20 (12): 1478–1483.

Caldwell, C. A., E. Renner, and M. Atkinson. 2018. "Human Teaching and Cumulative Cultural Evolution." *Review of Philosophy and Psychology* 9 (4): 751–770.

Candidi, M., A. Curioni, F. Donnarumma, L. M. Sacheli, and G. Pezzulo. 2015. "Interactional Leader–Follower Sensorimotor Communication Strategies during Repetitive Joint Actions." *Journal of the Royal Society Interface* 12 (110): 20150644.

Cañigueral, R., and A. F. C. Hamilton. 2019. "Effects of Being Watched on Self-Referential Processing, Self-Awareness and Prosocial Behaviour." *Consciousness and Cognition* 76 (November): 102830.

Cañigueral, R., J. A. Ward, and A. F. C. Hamilton. 2021. "Effects of Being Watched on Eye Gaze and Facial Displays of Typical and Autistic Individuals during Conversation." *Autism* 25 (1): 210–226.

Charbonneau, M., and P. Bourrat. 2021. "Fidelity and the Grain Problem in Cultural Evolution." *Synthese* 199 (January): 5815–5836.

Craig, S. D., M. T. H. Chi, and K. VanLehn. 2009. "Improving Classroom Learning by Collaboratively Observing Human Tutoring Videos While Problem Solving." *Journal of Educational Psychology* 101 (4): 779–789.

Csibra, G., and G. Gergely. 2009. "Natural Pedagogy." *Trends in Cognitive Sciences* 13 (4): 148–153.

Curioni, A., C. Vesper, G. Knoblich, and N. Sebanz. 2019. "Reciprocal Information Flow and Role Distribution Support Joint Action Coordination." *Cognition* 187 (June): 21–31.

Curioni, A., P. Voinov, M. Allritz, T. Wolf, J. Call, and G. Knoblich. 2022. "Human Adults Prefer to Cooperate Even When It Is Costly." *Proceedings of the Royal Society B: Biological Sciences* 289 (1973): 20220128.

Dautriche, I., L. Goupil, K. Smith, and H. Rabagliati. 2021. "Knowing How You Know: Toddlers Reevaluate Words Learned from an Unreliable Speaker." *Open Mind* 5 (February): 1–19.

DeCaro, M. S., R. D. Thomas, N. B. Albert, and S. L. Beilock. 2011. "Choking under Pressure: Multiple Routes to Skill Failure." *Journal of Experimental Psychology: General* 140 (3): 390–406.

Gergely, G., H. Bekkering, and I. Király. 2002. "Rational Imitation in Preverbal Infants." *Nature* 415 (6873): 755–755.

Gergely, G., and G. Csibra. 2005. "The Social Construction of the Cultural Mind: Imitative Learning as a Mechanism of Human Pedagogy." *Interaction Studies* 6 (3): 463–481.

Gergely, G., and G. Csibra. 2006. "Sylvia's Recipe: The Role of Imitation and Pedagogy in the Transmission of Cultural Knowledge." In *Roots of Human Sociality: Culture, Cognition, and Human Interaction*, edited by S. C. Levinson and N. J. Enfield, 229–255. New York: Routledge.

Glover, S., and P. Dixon. 2017. "The Role of Predictability in Cooperative and Competitive Joint Action." *Journal of Experimental Psychology: Human Perception and Performance* 43 (4): 644–650.

Gobel, M. S., H. S. Kim, and D. C. Richardson. 2015. "The Dual Function of Social Gaze." *Cognition* 136 (March): 359–364.

Gürerk, Ö., B. Irlenbusch, and B. Rockenbach. 2006. "The Competitive Advantage of Sanctioning Institutions." *Science* 312 (5770): 108–111.

Hamilton, A. F. C. 2016. "Gazing at Me: The Importance of Social Meaning in Understanding Direct-Gaze Cues." *Philosophical Transactions of the Royal Society B: Biological Sciences* 371 (1686): 20150080ne.

Hart, Y., L. Noy, R. Feniger-Schaal, A. E. Mayo, and U. Alon. 2014. "Individuality and Togetherness in Joint Improvised Motion." *PLOS One* 9 (2): e87213.

Henrich, J. 2016. *The Secret of Our Success: How Culture Is Driving Human Evolution, Domesticating Our Species, and Making Us Smarter*. Princeton, NJ: Princeton University Press.

Henrich, J., and F. J. Gil-White. 2001. "The Evolution of Prestige: Freely Conferred Deference as a Mechanism for Enhancing the Benefits of Cultural Transmission." *Evolution and Human Behavior* 22 (3): 165–196.

Henrich, J., and R. McElreath. 2003. "The Evolution of Cultural Evolution." *Evolutionary Anthropology: Issues, News, and Reviews* 12 (3): 123–135.

Heyes, C. M. 1994. "Social Learning in Animals: Categories and Mechanisms." *Biological Reviews* 69 (2): 207–231.

Heyes, C. M. 2001. "Causes and Consequences of Imitation." *Trends in Cognitive Sciences* 5 (6): 253–261.

Heyes, C. M. 2012. "What's Social about Social Learning?" *Journal of Comparative Psychology* 126 (2): 193–202.

Heyes, C. M. 2016. "Blackboxing: Social Learning Strategies and Cultural Evolution." *Philosophical Transactions of the Royal Society B: Biological Sciences* 371 (1693): 20150369.

Hoppitt, W., and K. N. Laland. 2013. *Social Learning: An Introduction to Mechanisms, Methods, and Models.* Princeton, NJ: Princeton University Press.

Insko, C. A., S. Drenan, M. R. Solomon, R. Smith, and T. J. Wade. 1983. "Conformity as a Function of the Consistency of Positive Self-Evaluation with Being Liked and Being Right." *Journal of Experimental Social Psychology* 19 (4): 341–358.

Jiménez, Á. V., and A. Mesoudi. 2019. "Prestige-Biased Social Learning: Current Evidence and Outstanding Questions." *Palgrave Communications* 5 (1): 1–12.

Király, I., G. Csibra, and G. Gergely. 2013. "Beyond Rational Imitation: Learning Arbitrary Means Actions from Communicative Demonstrations." *Journal of Experimental Child Psychology* 116 (2): 471–486.

Kline, M. A. 2015. "How to Learn about Teaching: An Evolutionary Framework for the Study of Teaching Behavior in Humans and Other Animals." *Behavioral and Brain Sciences* 38:e31.

Koenig, M. A., and A. L. Woodward. 2010. "Sensitivity of 24-Month-Olds to the Prior Inaccuracy of the Source: Possible Mechanisms." *Developmental Psychology* 46 (4): 815–826.

Krishnan-Barman, S., and A. F. C. Hamilton. 2019. "Adults Imitate to Send a Social Signal." *Cognition* 187 (June): 150–155.

Little, C. A. J. L., and D. F. Lancy. 2016. "How Do Children Become Workers? Making Sense of Conflicting Accounts of Cultural Transmission in Anthropology and Psychology." *Ethos* 44 (3): 269–288.

McEllin, L., G. Knoblich, and N. Sebanz. 2018. "Distinct Kinematic Markers of Demonstration and Joint Action Coordination? Evidence from Virtual Xylophone Playing." *Journal of Experimental Psychology: Human Perception and Performance* 44 (6): 885–897.

Meltzoff, A. N. 1988. "Infant Imitation after a One-Week Delay: Long-Term Memory for Novel Acts and Multiple Stimuli." *Developmental Psychology* 24 (4): 470–476.

Mesoudi, A. 2011. *Cultural Evolution: How Darwinian Theory Can Explain Human Culture and Synthesize the Social Sciences.* Chicago: University of Chicago Press.

Mesoudi, A., and A. Whiten. 2008. "The Multiple Roles of Cultural Transmission Experiments in Understanding Human Cultural Evolution." *Philosophical Transactions of the Royal Society B: Biological Sciences* 363 (1509): 3489–3501.

Miton, H., and M. Charbonneau. 2018. "Cumulative Culture in the Laboratory: Methodological and Theoretical Challenges." *Proceedings of the Royal Society B: Biological Sciences* 285 (1879): 20180677.

Morgan, T. J. H., N. T. Uomini, L. E. Rendell, L. Chouinard-Thuly, S. E. Street, H. M. Lewis, C. P. Cross, et al. 2015. "Experimental Evidence for the Co-Evolution of Hominin Tool-Making Teaching and Language." *Nature Communications* 6 (1): 6029.

Okazaki, S., Y. Muraoka, and R. Osu. 2019. "Teacher-Learner Interaction Quantifies Scaffolding Behaviour in Imitation Learning." *Scientific Reports* 9 (1): 7543.

Poulin-Dubois, D., I. Brooker., and A. Polonia. 2011. "Infants Prefer to Imitate a Reliable Person." *Infant Behavior and Development* 34 (2): 303–309.

Richardson, D. C., and R. Dale. 2005. "Looking to Understand: The Coupling between Speakers' and Listeners' Eye Movements and Its Relationship to Discourse Comprehension." *Cognitive Science* 29: 1045–1060.

Risko, E. F., D. C. Richardson, and A. Kingstone. 2016. "Breaking the Fourth Wall of Cognitive Science: Real-World Social Attention and the Dual Function of Gaze." *Current Directions in Psychological Science* 25 (1): 70–74.

Sacheli, L. M., E. Tidoni, E. F. Pavone, S. M. Aglioti, and M. Candidi. 2013. "Kinematics Fingerprints of Leader and Follower Role-Taking during Cooperative Joint Actions." *Experimental Brain Research* 226 (4): 473–486.

Schmitz, L., C. Vesper, N. Sebanz, and G. Knoblich. 2017. "Co-Representation of Others' Task Constraints in Joint Action." *Journal of Experimental Psychology: Human Perception and Performance* 43 (8): 1480–1493.

Sebanz, N., H. Bekkering, and G. Knoblich. 2006. "Joint Action: Bodies and Minds Moving Together." *Trends in Cognitive Sciences* 10 (2): 70–76.

Sebanz, N., and G. Knoblich. 2009. "Prediction in Joint Action: What, When, and Where." *Topics in Cognitive Science* 1 (2): 353–367.

Sebanz, N., and G. Knoblich. 2021. "Progress in Joint-Action Research." *Current Directions in Psychological Science* 30 (2): 138–143.

Stout, D. 2002. "Skill and Cognition in Stone Tool Production: An Ethnographic Case Study from Irian Jaya." *Current Anthropology* 43 (5): 693–722.

Stout, D. 2005. "The Social and Cultural Context of Stone-Knapping Skill Acquisition." In *Stone Knapping: The Necessary Conditions for a Uniquely Hominin Behaviour,* edited by V. Roux and B. Bril, 10. McDonald Institute Monographs. Cambridge: McDonald Institute for Archaeological Research.

Strachan, J. W. A., A. Curioni, M. D. Constable, G. Knoblich, and M. Charbonneau. 2021. "Evaluating the Relative Contributions of Copying and Reconstruction Processes in Cultural Transmission Episodes." *PLOS One* 16 (9): e0256901.

Tomasello, M., A. C. Kruger, and H. H. Ratner. 1993. "Cultural Learning." *Behavioral and Brain Sciences* 16 (3): 495–511.

Toyokawa, W., A. Whalen, and K. N. Laland. 2019. "Social Learning Strategies Regulate the Wisdom and Madness of Interactive Crowds." *Nature Human Behaviour* 3 (2): 183–193.

Vesper, C., S. Butterfill, G. Knoblich, and N. Sebanz. 2010. "A Minimal Architecture for Joint Action." *Neural Networks* 23 (8–9): 998–1003.

Vesper, C., L. Schmitz, L. Safra, N. Sebanz, and G. Knoblich. 2016. "The Role of Shared Visual Information for Joint Action Coordination." *Cognition* 153 (August): 118–123.

Vesper, C., R. P. R. D. van der Wel, G. Knoblich, and N. Sebanz. 2011. "Making Oneself Predictable: Reduced Temporal Variability Facilitates Joint Action Coordination." *Experimental Brain Research* 211 (3–4): 517–530.

Vesper, C., R. P. R. D. van der Wel, G. Knoblich, and N. Sebanz. 2013. "Are You Ready to Jump? Predictive Mechanisms in Interpersonal Coordination." *Journal of Experimental Psychology: Human Perception and Performance* 39 (1): 48–61.

Visco-Comandini, F., S. Ferrari-Toniolo, E. Satta, O. Papazachariadis, R. Gupta, L. E. Nalbant, and A. Battaglia-Mayer. 2015. "Do Non-Human Primates Cooperate? Evidences of Motor Coordination during a Joint Action Task in Macaque Monkeys." *Cortex* 70 (September): 115–127.

Wahn, B., A. Czeszumski, M. Labusch, A. Kingstone, and P. König. 2020. "Dyadic and Triadic Search: Benefits, Costs, and Predictors of Group Performance." *Attention, Perception & Psychophysics* 82 (5): 2415–2433.

Whiten, A., N. McGuigan, S. Marshall-Pescini, and L. M. Hopper. 2009. "Emulation, Imitation, Over-Imitation and the Scope of Culture for Child and Chimpanzee." *Philosophical Transactions of the Royal Society B: Biological Sciences* 364 (1528): 2417–2428.

Wilkins, J. 2020. "Learner-Driven Innovation in the Stone Tool Technology of Early Homo Sapiens." *Evolutionary Human Sciences* 2(e40): 1–15.

Williams, A. M., P. Ward, J. M. Knowles, and N. J. Smeeton. 2002. "Anticipation Skill in a Real-World Task: Measurement, Training, and Transfer in Tennis." *Journal of Experimental Psychology: Applied* 8 (4): 259–270.

Wolf, T., C. Vesper, N. Sebanz, P. E. Keller, and G. Knoblich. 2019. "Combining Phase Advancement and Period Correction Explains Rushing during Joint Rhythmic Activities." *Scientific Reports* 9 (1): 9350.

Wood, L. A., R. L. Kendal, and E. G. Flynn. 2013. "Whom Do Children Copy? Model-Based Biases in Social Learning." *Developmental Review* 33 (4): 341–356.

7 Playing with Knives: Children's Learning, Cultural Niche Construction, and the Evolution of Technical Flexibility

Adam Howell Boyette

Introduction

For this project, the other authors and I have been asked to think about flexibility and rigidity in the production and reproduction of technical skills. In this volume, the subject is addressed from a diversity of perspectives and from the scale of individual cognition (Pope-Caldwell, this volume) to the social group or species (Astuti, this volume; Stout, this volume). My contribution lies in a middle ground and focuses on the interplay between the individual and the environment in the context of children's learning of technical skills across childhood (Gauvain 2005b; Greenfield et al. 2003). This approach draws from cultural niche construction theory. Niche construction is a coevolutionary theoretical perspective that links behavior and the environment as interacting forces of evolutionary change (Laland, Odling-Smee, and Feldman 2000; Odling-Smee, Laland, and Feldman 2003). Specifically, organisms are observed to change their environments through their actions in ways that can then change selective pressures acting on them. *Cultural* niche construction refers to the powerful ways in which humans construct spaces, ideas, and artifacts within lifetimes and across generations that shape how we develop and behave, including how we further change the environment (Laland and O'Brien 2011; Sterelny 2004; Stotz 2010). In this chapter, my central claim is that through cultural niche construction, specific regular features of children's developmental contexts can facilitate learning of complex technical skills with great flexibility in the learning mechanisms.

I situate this claim within debates on the evolution of cultural learning and on cultural variation in children's learning contexts. I then draw on data from my fieldwork among BaYaka people in the Congo Basin and describe how stable features of the BaYaka physical and social environment allow children to learn to expertly use blade tools like knives and machetes to perform a huge range of daily tasks with very little direct input from their elders.

First, my understanding of "flexibility" and "rigidity" in regard to technical skill learning is informed by interdisciplinary discussions of the evolution of the human capacity for cumulative culture (Boyd and Richerson 2005; Dean et al. 2014; Laland 2017; Tomasello 1999). In this literature, humans are seen as having constructed a technically complex niche in which adaptation to the environment relies on "high fidelity" social learning mechanisms to reproduce technical skills, some of which have been modified over generations (hence the "cumulative" nature of such culture). Here, "high fidelity" means that a learner can reproduce a behavior with relatively few errors. As such, imitation (Gergely and Csibra 2006; Hoehl et al. 2019; Nielsen and Tomaselli 2010) and teaching (Boyd, Richerson, and Henrich 2011; Csibra and Gergely 2011; Morgan et al. 2015; Strauss and Ziv 2012; Tomasello et al. 2012) are typically regarded as critical "high fidelity" social learning mechanisms for the reproduction of the sometimes-opaque technical skills that have been central to human evolution. While undoubtedly critical to our success as a species, these mechanisms are theoretically and empirically associated with costs in terms of individual- and population-level flexibility. For example, there is the theoretical possibility of imitating behaviors that are maladaptive in current or future environments (Boyd and Richerson 1995; Enquist, Eriksson, and Ghirlanda 2007; Rogers 1988). Additionally, teaching entails opportunity costs for teachers, who might be better off applying their skills instead of sharing them (Fogarty, Strimling, and Laland 2011; Thornton and Raihani 2008). Moreover, while teaching is adaptive because it saves the learner time and energy by directing their attention to specific information to be learned, it has also been shown to constrain learners' exploration and innovation (Bonawitz et al. 2009, 2011).

Thus, in order to build and maintain complex technical traditions, humans must balance the benefits of high-fidelity social learning against the costs of teaching and of filtering out techniques that are poor fits to the current environment. We must also remain flexible in the face of environmental change—being ready to apply technical traditions in new ways or create new ones. In this chapter, I explore how our very capacity for creating material change in the environment through complex technical innovations, along with our unique sociality and life history, creates feedback that can support high-fidelity (i.e., rigid) reproduction of technical skills with great flexibility in how children learn them. I also show how variation in our culturally shaped environments can enhance or constrain this flexibility, with implications for how technical skills are distributed in communities of learners.

Culturally Constructed Contexts of Learning

Cultural niche construction was likely a dominant force in human evolution, perhaps at least since tools became a regular and increasingly diverse feature of hominid economic and social life (Jeffares 2012; Sterelny 2004; Stotz 2010). Theorists tie the use of and reliance on tools (broadly defined) with the evolution of developmental systems of social cognition (Dean et al. 2014; Tomasello 1999), which likely also coevolved with brain size as well as changes in ancestral human social systems that began to favor central-place foraging (Kuhn and Stiner 2019; Stiner 2021), cooperative child-rearing, altricial birth, and an extended childhood period (Burkart and van Schaik 2016; Hill, Barton, and Hurtado 2009; Hrdy 2009; Kramer 2010; Sterelny 2012). As a result of this coevolutionary history, typically developing human children are equipped with the fundamentals that facilitate acquisition of technical skills: the ability to infer the goals and intentions of their tool-using elders and an environment populated by the material basis of learning and a variety of models to learn from.

These fundamentals aside, culturally constructed developmental contexts (Flynn et al. 2013; Kendal 2011; Super and Harkness 1986) vary in the accessibility of materials for technical learning (Lancy 2016b), in how experienced, capable, and willing children's models are (Morelli, Rogoff, and Angelillo 2003), and in the dominant behavioral processes through which children "learn to learn" (Glowacki and Molleman 2017; Mesoudi et al.

2016). Variation in these dimensions influence, at the individual level, children's approach to learning and their familiarity with their society's cultural repertoire, and at the community level, a group's flexibility in the face of environmental change.

For instance, several researchers (Lancy 2012; Lave and Wenger 1991; Mead 1970; Ochs and Izquierdo 2009; Scribner and Cole 1973) compare what Barbara Rogoff and colleagues (Rogoff et al. 2003; Paradise and Rogoff 2009) call learning through "intent participation" with the tradition of formal school-based learning. The latter tradition is dominant in industrialized and postindustrial contexts where households are typically isolated and children's daily experiences are spatially separate from their parent's economic and social activities. In these contexts, learning is conceived of as a formalized, objectified, and decontextualized process; language is the nearly exclusive means of information exchange; learners are age-graded and have little interaction with children of other ages outside the home, and they are situated in hierarchical relationships with teachers, parents, and other adults who are the sole sources of cultural knowledge (Lancy 2016a; Rogoff et al. 2003; Scribner and Cole 1973). The material basis for learning everyday domestic tasks (e.g., using pots, knives, cleaning tools) as well as work in the public sphere (e.g., tools, computers) are often off-limits to children, who are given their own material culture (Lancy 2008; Morelli, Rogoff, and Angelillo 2003). While children certainly learn from observation in such contexts, there may be relatively little regularity in approaches to teaching or in specific cultural knowledge or behavior among models observed across the complex and diverse contexts they experience daily (Ochs and Izquierdo 2009), and there may be limited opportunities to apply what they observe (Morelli, Rogoff, and Angelillo 2003).

More research is needed to understand the specific mechanisms, but children who grow up in these contexts emphasizing formal school-based learning tend to be more extrinsically motivated (Coppens and Alcalá 2015; Ochs and Izquierdo 2009) and attend only to teaching directed at them, following this closely rather than exploring alternative possibilities (López et al. 2010; Shneidman et al. 2016; Silva, Correa-Chávez, and Rogoff 2010). Such a learning style may be effective at preparing children for an environment in which survival is based on specialization within a highly competitive, meritocratic labor market (Bowles and Gintis 1977, 2002). However, there are reasonable concerns that this relatively rigid and costly educational tradition amplifies social inequalities (Croizet et al. 2019; Jackson and Holzman 2020), is inflexible with regard to learning differences (Cainelli and Bisiacchi 2019; Shifrer 2013), and leads to decreases in mental health and children's well-being (Boyce et al. 2012; Gray 2013; Narvaez et al. 2012).

In contrast, the "intent participation" tradition is typical of small-scale subsistence societies where children are embedded in the daily economic and social activities of their communities. As such, children can learn the norms, practices, skills, and knowledge essential to life in their community through observation and legitimate participation, with teaching used sparingly for abstract or specialized tasks or symbolic knowledge (Gaskins 2000; Lancy 2010; Maynard 2002; Rogoff et al. 2003). Through intent participation, children's learning is embodied and affectively linked to its performance alongside close family, friends, and neighbors, and it is often scaffolded within collaborative learning within mixed-age child groups (Whiting and Edwards 1988). These social learning processes seem to facilitate the development of both an internal motivation to learn through helping others in their work and an open attentional stance in which children learn readily from observation, collaboration with others, and listening to others being taught (Correa-Chávez and Rogoff 2009; Gaskins and Paradise 2010; López et al. 2010; Mejía-Arauz et al. 2007; Silva, Correa-Chávez, and Rogoff 2010).

Barry Hewlett and I (Boyette and Hewlett 2017) note that there is also variation across intent participation communities in the organization of learning. In particular, among many small-scale farming and pastoralists societies, children are typically expected to be obedient to their parents and other elders (Erchak 1980; LeVine et al. 1994), such that their learning is within the context of task assignments by elders oriented toward supporting the family economy. As obedience and correct behavior is normatively valued in such contexts, parents use more direct instruction and rigid behavioral control in their teaching. For example, Boyette and Hewlett (2017) measured a small but statistically significant difference in the amount of direct instruction received by farmer children compared to Aka forager children. Indeed, forager parents highly endorse their children's autonomous learning in most domains (Boyette and Lew-Levy 2021; Briggs 1991; Naveh 2016), and though they often assign tasks as a means of teaching, compliance is not expected (Boyette and Lew-Levy 2021; Johnson 2003).[1] My claim here is that such variations in the learning context (e.g., cultural models of learning) are part of the culturally constructed niche in which technical skills are socialized (table 7.1).

Of course, there is enormous variation in cultural models of learning within school-based learning societies as well (Kusserow 2004), and some alternatives to the stereotypical pattern I described above parallel aspects of the intent participation tradition (Gray and Feldman 2004; Lillard and Taggart 2019). As suggested in table 7.1, a relative reliance on rigid versus flexible social learning mechanisms may scale up to influence how flexible communities are able to be when confronted with environmental change. A full discussion of this variation is beyond the scope of this chapter; however, I would suggest that an investigation of how such variation is related to flexibility in response to environmental change is an area in need of investigation. For now, the case study that follows focuses on the ways in which cultural niche construction by BaYaka foragers flexibly facilitates learning of complex technical skills.

Learning Blade Tool Use

Unfortunately, there are few empirical studies of cultural niche construction as such. In the following case study, I focus on children's acquisition of nonhunting blade tool techniques to illustrate how the processes of cultural niche construction can facilitate high-fidelity learning. I focus on nonhunting blades for several reasons. Hunting tools and the techniques involved in their use are, arguably, not as flexible as the use of other types of blade tools, as they have been crafted for more specific purposes (e.g., the pursuit of specific types of game in specific ecologies). Additionally, hunting techniques are also typically learned later in life and not necessarily mastered until the third decade of life (Koster et al. 2020). Furthermore, hunting techniques are typically learned by males, with young male foragers already showing a greater interest in hunting than females from early childhood. Adults endorse this gendered division of activity interest (which supports the gendered division of labor) by giving males toy hunting tools and girls toy foraging tools (Lew-Levy et al. 2018).[2] Finally, nonhunting blade tools are perhaps more evolutionarily relevant, since it is likely that early hominid stone tools were general purpose and not used for the types of pursuit hunting common today (Plummer 2004). Thus, a focus on the acquisition of nonhunting

Table 7.1
Cultural learning contexts and their hypothetical implications for learning technical flexibility

Culturally constructed learning context	Accessibility of materials	Accessibility of models	Dominant social learning processes	Hypothetical implications for cultural learning of technical flexibility
Formal schooling (industrial model, Western, decontextualized)	Few materials freely accessible; toys and learning tools (often standardized) used to enhance cognitive flexibility, not practical, embodied know-how	Limited; parents in domestic spaces; teachers in school; variable regularity across contexts	Teaching, rigid adult-led scaffolding, competitive motivation, play	Limited opportunities for embodied learning; high specialization; *individuals learn narrow technical repertoire; may limit community flexibility with adaptation requiring coordination across multiple, specialized institutions*
Intent participation—hierarchical (e.g., small-scale farmers, pastoralists)	Highly accessible; child-sized tools as early scaffolds	Highly accessible; extended family most available; little separation from adult and child spaces; high regularity across contexts	Observation, play, adult-mediated participation, collaboration, teaching of specialized skills	Many opportunities for embodied learning of critical skills; open learning stance; *individually flexible; community may be less flexible as result of subsistence specialization*
Intent participation—egalitarian (e.g., some contemporary mobile foragers)	Highly accessible; child-sized tools also used	Most accessible; residential community available; little separation from adult and child spaces; high regularity in social and cultural contexts; physical contexts flexible with regular features	Observation, play, autonomous participation, collaboration, teaching of specialized skills	Many opportunities for embodied learning of all skills; open learning stance; *most flexible—individually autonomous learning of broad repertoire; flexible subsistence strategy; with skills transferable to new tasks and domains*

Source: All sources used to compile this table are cited in the text.

techniques by children is a better model of how such cultural learning may have been involved in the evolution of human tool use. However, I will come back to such cases of more specialized techniques later.

Importantly, I am focusing my analysis simply on the availability of nonhunting blade tools (hereafter "blade tools") and the opportunities to observe, be taught, and practice their use. In other words, I do not examine blade tool construction. Undoubtedly, the construction and use of blade tools for a variety of subsistence tasks was a fundamental cultural adaptation for our species (McBrearty and Brooks 2000; Morgan et al. 2015; Stout 2011) and remains so today. It also may have involved more concentrated learning and teaching than, I will argue, use of nonhunting blade tools does (Morgan et al. 2015; Stout 2011). However, in the region where I work, as in much of the world today, blades are rarely locally made but imported. Among the BaYaka subjects of this case study, there are those with knowledge of iron forging who can craft axe heads and knife blades, but these tools typically come from nonlocal sources and are acquired through trade. Thus, the vast majority of other blades are made of steel, and their construction is not part of the daily learning experiences of interest here. It should be noted, though, that novice toolmakers' presence near and participation in the construction of blade tools is inferred from sites across the world and deep into human evolutionary history (e.g., Grimm 2000; Takakura 2013). While we cannot be sure of the ages of the novice toolmakers, we can infer that with the emergence of the "domestic space"—a central place for food sharing and cooperative child-rearing—juveniles were at least present to observe various uses of stone tools beginning as early as 450 thousand years ago (Kuhn and Stiner 2019). Today, while the use of blade tools by children tends to be discouraged in families from modern, large-scale industrial and postindustrial societies, this is not the case across contemporary small-scale subsistence societies, where children begin to manipulate knives, machetes, and other tools long before they can productively use them (Lancy 2016b). This is the context to which I now turn.

Case Study: BaYaka Children's Experience with Blade Tools

The BaYaka are a group of several populations of tropical forest foragers living across the Congo Basin (Lewis 2002). "BaYaka" is a general term for several groups, including those referred to in publications as the Aka (Bahuchet 1985; Boyette 2016; Boyette and Lew-Levy 2019; B. L. Hewlett 2005; B. S. Hewlett 1991), the Mbendjele (Lewis 2002), the BaYaka (Boyette et al. 2020; Lew-Levy et al. 2020) or Mbendjele-BaYaka (Jang et al. 2019; Sonoda, Bombjaková, and Gallois 2018) who live in the northern Republic of the Congo and south-western Central African Republic. Here, I'll be drawing on qualitative and quantitative data from two BaYaka populations I have worked with: the Aka of the Lobaye Province in the Central African Republic and the BaYaka of the Motaba River region in the Likouala Province of the Republic of the Congo. In my analysis, I'll use the term "BaYaka" to refer to these two populations because the data I will draw from can be reasonably generalized to both, who intermarry and share the same language, values, expressive culture, and sharing norms, with some regional variation.

BaYaka Cultural Niche Construction

As noted above, the processes of cultural niche construction can support cultural learning through lowering costs to learning, such as making it easier to find models to observe, teachers to teach, or opportunities to practice. In this way, BaYaka settlements are highly supportive of blade tool use learning. Barry S. Hewlett and colleagues (2019) describe BaYaka spaces and spatial use as intimate, even in comparison to a general trend toward close living among foragers as compared to food-producing peoples. For instance, Aka typically live in a one-room dwelling, 4.8-square-meters in size, accommodating an average of 3.1 people or about one-square-meter of space per person, on average. Importantly, these houses are, on average, placed within 4.3 meters of the nearest neighbors, and very little is done inside the dwellings other than sleeping, with much of life being conducted outdoors in sight of all other members of the community. BaYaka settlements are generally of two types: smaller forest settlements with traditional *mongoulou* domed houses and village settlements with larger, mud-brick houses. However, the public nature of most activities is consistent across both types of settlements, and children in particular are never barred from entering any dwelling, no matter who is currently sleeping therein. Hewlett and colleagues (2019) find consistencies between Aka spatial use patterns and cultural models of social relationships—especially between parents and children and between marital partners—as valuing physical closeness among social partners. Anecdotally, when joining a group of people sitting together, BaYaka people will often sit so as to be in physical contact with one another, no matter how much space is available.

These spatial aspects of cultural niche construction are consistent with a larger emphasis on sharing that is common to mobile foraging peoples. Construed widely, this sharing includes not only food but also other material resources, time, and knowledge (Lavi and Friesem 2019). Here, I will emphasize in particular that these spaces facilitate both sharing of material bases for learning techniques, such as blade tool use, as well as the knowledge and skills embodied in others. In terms of the former, as noted, most work is done outside the house, and most household objects are also stored outside. Blades are prominent among these objects. For example, a brief survey of items used in food preparation of a random sample of 14 Aka households at a village settlement in Central African Republic (CAR) in 2012 shows that most households had at least two knives or machetes, making these blade tools the second most numerous of 12 items named, after plates and tied with the number of cooking pots per household (table 7.2). Note that the traditional BaYaka axe, the *djumbi*, was not counted among these items, nor were, for example, razor blades, or *gileti*, which are commonly in circulation though disposed of after one or two uses. Thus, table 7.2 reflects a bare minimum of children's opportunities to observe and interact with blades throughout daily life, and these blades are used to chop wood, clear the ground around camp, build a house, make baskets and mats, crack nuts, peel manioc, skin game, divide meat, cut edible leaves, dig for wild yams, style hair, and cut fingernails.

Blade Tools and Early Childhood

In terms of the specific ways that BaYaka cultural niche construction enhances social and individual learning opportunities, I will now discuss how such knowledge and skill sharing occurs across childhood to gradually (re)produce technically competent members of the community. In general, respect for autonomy is a core cultural model—or foundational cultural schema (Boyette 2019; B. S. Hewlett et al. 2011)—among the BaYaka, and this is true for all individuals regardless of age (Boyette and Lew-Levy 2021). As such, BaYaka parenting is considered indulgent. For instance, infants nurse on demand and are responded

Table 7.2
Aka kitchen item inventory (n = 14 households)

	Object	Sum	Mean
1	Plate	32	3.56
2	Cooking pot	30	2.31
3	**Knife/machete**	**30**	**2.31**
4	Spoon	21	1.62
5	Jerry can	14	1.56
6	Sieve	9	1.12
7	Basin	8	1.60
8	Bucket	4	1.33
9	Mortar + pestle	3	1.00
10	Cup	1	1.00
11	Cutting board	1	1.00
12	Stirring stick	1	1.00

to immediately when distressed by any nearby caregiver, typically but not always a parent, who nurses or sooths them (B. S. Hewlett et al. 1998). One aspect of the autonomy granted to young children is that they are permitted to engage with any objects they find around camp, including blade tools. Parents and other caretakers are not being negligent when allowing children access to these tools. Indeed, children being cut or cutting others with blades is a common concern among BaYaka parents (Boyette, unpublished data). However, they indulge children's autonomous interest and even encourage interaction with these tools. For example, when the hafts of knives used in food preparation break, these now "useless" blades are given to young children to play with. In the parlance of social learning, this is a classic example of opportunity scaffolding. In a sample of 10 Aka infants aged 12 to 14 months, Barry S. Hewlett and Casey Roulette (2016) observed opportunity scaffolding nearly once every hour, on average, in their sample of 10 hours of infant-focused video. While they did not code the objects used, knives were prominent across the range of teaching behaviors they observed, including instances of natural pedagogy, moving a child's body, and demonstration. These observations are not surprising, given the spatial prominence of these blades and their regular use by people across a wide range of everyday tasks.

Blade Tools from Middle-Childhood to Adolescence

Until around age 7, BaYaka children typically stay nearby the settlement, typically under the casual supervision of at least one elder who remains in camp if others are away. As BaYaka children transition from toddlerhood and weaning into middle childhood, across the ages of four to six years old, children start spending their time with a multiage, all-child play group, and as a group, they may venture into the forest to play or autonomously forage (Bombjaková 2018; Boyette 2016; B. S. Hewlett et al. 2011; Lew-Levy, Kissler, et al. 2020). This context offers significant opportunities for social learning and individual practice with blade tools through scaffolded interactions with other, slightly older children (Gauvain 2005a; Lew-Levy, Kissler, et al. 2020). As a quantitative demonstration of such opportunities, I present data collected during my study of Aka children's time allocation and social learning in CAR in 2010.

More details of the methods used to collect these data can be found elsewhere (Boyette 2016; Boyette and Hewlett 2017; Lew-Levy and Boyette 2018). Briefly, I used systematic behavior coding during focal follows of 50 individual Aka children aged 4 through 17, with a mean of 238.62 observations per child.[3] These children were from eight different forest settlements. While I did not specifically set out to record blade use, early during my observations, I started noting such use in the margins of my data sheets, as I was struck by how capable children were with machetes, axes, and knives. These notes are not as systematic as the behavioral coding that was the focus of my study, but they do reflect a bare minimum estimate of the presence of blade tools in children's everyday contexts across my observations.

To quantify blade tool use in these data, I performed a search of the notes I kept intermittently in a column of my minute-by-minute data coding sheets. Specifically, I searched for the following text strings: "chop," "cut," "dig," "kni," "machete," "yebe," "ax," "jumbi," and "blade." Note that *yebe* is the word for "knife," *djumbi* is the traditional axe, and machetes (and knives during children's play) are often used for digging. All notes that were found through the text search were then reviewed so that irrelevant observations could be excluded (e.g., if "cut" was in the word "cute" or the note did not reference the ongoing activity). Overall, one or more of these text strings was noted in 74 observations spread across 26 of the 50 children (16 female, 10 male), with an average of 2.8 observations per child. These included 22 mentions of "chop," 15 of "dig," 17 of "cut," and 25 mentions of a blade (five observations had mentions of a blade and an action, and these were counted as a single observation). It is also important to note that I included in these data mentions of blade use by the child *or* by someone with the child. For the sake of my argument, this distinction is not critical, as again, these data were not systematically collected and thus provide only a reasonable, minimal account of the opportunities to learn blade tool techniques in these children's lives.

First, while not noted that frequently overall, blades were noted persistently throughout my observations. For example, blade use was noted in focal follows of children from seven of the eight Aka forest communities, or "camps," in which I lived. Additionally, there was no obvious concentration of observations in any one camp, with 10.6 observations noted per camp. Furthermore, as plotted in figure 7.1a, these observations were not concentrated at any particular time during the study period, which extended from April until September (the gap in time in figure 7.1a was the time that I worked with Ngandu farmer children, who also used blade tools but are excluded from the current analysis). Nor was blade use noted among children of any one age (figure 7.1b), and observations were consistently distributed across boys and girls of different ages and throughout the study period, although use of the tools by girls was noted twice as frequently: 50 times for girls versus 24 times for boys.

Second, blade use was observed across a broad range of children's activities. In fact, as can be seen in table 7.3, it was *not* observed in only the three least frequently observed activities: childcare, music, and hygiene. Notably, blades were overrepresented in three activities relative to how frequent those activities were observed in the full dataset. In particular, while work-themed pretense play was observed in only 4.7 percent of all observations, 29.7 percent of the observations with blades were during this activity. This reflects 3.9 percent of all observations of work-themed pretense play, which was coded when a child reenacted traditional subsistence activities, such as digging for wild yams, making a playhouse, or chopping trees (Lew-Levy and Boyette 2018). While less frequently noted across children's activities, blades were also noted during child-initiated work and work at a higher frequency than was observed in the full

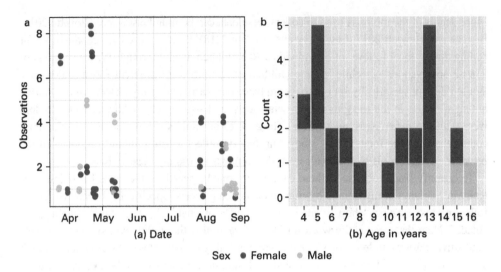

Sex ● Female ● Male

Figure 7.1
Counts of observations of blade tool use noted in field notes during focal follows of Aka children by (a) date of observation and (b) age of children in years. Each plot also shows the distribution of observations of females versus males across the study period and among the ages sampled. *Source*: Data collected by author. See references in text.

Table 7.3
Relative representation of blade tool use across all activities coded compared to the relative representation of each activity in the full dataset

Activity	Percent of full dataset	Percent of blade-use dataset	Percent of observations of activity in which blade tool use was observed[1]
Work-themed pretense play	4.70	29.70	3.90
Child-initiated work	10.30	17.60	1.06
Work	7.50	10.80	0.90
Other	4.80	4.10	0.52
Rest	27.70	20.30	0.45
Play	20.90	12.20	0.36
Travel	5.50	2.70	0.31
Eats	5.10	1.40	0.17
Visit	9.60	1.40	0.09
Childcare	1.30	NA	NA
Music	1.40	NA	NA
Hygiene	1.10	NA	NA

1. Percentages are calculated as count of all observations of activity when blade use was noted divided by the count of all observations of activity times 100.

dataset. The former was coded when the focal child was involved in subsistence work without involvement from adults, while the latter was coded when the child worked with an adult or the subsistence activity was initiated by an adult (whether or not the adult and child worked together at the task).

Third, I also examined the presence of adults and the settings in which blade use was observed. During my observations, I coded whether mother, father, both parents, parents and other adults, or no adults were within visual range of the focal child. When blade tool use was observed, 66 percent of the time the focal child was in sight of parents and others; 9 percent of the time either the children's mother or both parents were present; and 24 percent of the time no adults were in sight of the child. In terms of settings, the observations of blade tool use were fairly similar in their distribution to where children generally spent their time. In particular, 68 percent of the observations were in the domestic space of the forest camp, whereas 20 percent were in the forest. The other 12 percent were either in or near gardens or at the forest camp of an Ngandu farmer family with whom they spent time.

General Discussion

The analysis presented here is a relatively coarse-grained account of BaYaka children's experience with blade tools during their daily lives. A finer-grained analysis could only further support my conclusions. In brief, every BaYaka household owns multiple blade tools, which young children observe in use every day and seek to interact with. While parents worry about accidental cuts, they still begin to scaffold children's use of blades early, through opportunity scaffolding and occasional direct instruction. This in and of itself demonstrates that parents see the benefits of early acquisition of blade techniques as outweighing any costs. As children grow, blades remain present when children are in the domestic space and travel with them as they go into the forest to forage or play, whether or not adults are present. These tools are used across most contexts of daily life, especially during work-themed pretense play as well as legitimate productive subsistence work. Previous work has shown that BaYaka children work more and play less from early childhood through adolescence (Boyette 2016), including play that is an imitation of work (Lew-Levy and Boyette 2018). We have interpreted this pattern to indicate that play serves as a context in which children learn so that they increasingly become legitimate participants in daily economic life, consistent with the intent participation tradition (Lave and Wenger 1991; Lew-Levy and Boyette 2018; Paradise and Rogoff 2009; Rogoff et al. 2003). The complex techniques associated with blade tool use are critical to this learning. Through developmental feedback from the social and physical environment to their bodies and minds, BaYaka children will inevitably learn how to use knives, axes, machetes, and razor blades to perform the wide variety of technical tasks that their elders do with these tools.

Other ecocultural theories of child development emphasize how settings and the company children keep shape children's cultural learning opportunities (Markus and Kitayama 2010; Weisner 1984; Whiting 1963; Whiting and Edwards 1988). Consistent with such approaches, the cultural niche construction perspective emphasizes that regularities of children's developmental context have supported adaptation to particular social and ecological environments and can continue to serve such a role. For the BaYaka, the distributed nature of expertise in blade tool use likely contributes to the overall high level of coordination of subsistence activities and cooperation in general that characterizes their society. Barry S. Hewlett and colleagues emphasize that for BaYaka children, their settings are open to free exploration and that the entire community is available to learn from, such that few social roles or specific skills are hidden from young cultural learners (B. S. Hewlett 2014; B. S. Hewlett, Berl, and Roulette 2016). From infancy, responsive caregiving by many others and respect for autonomy to explore spaces and artifacts leads to security and trust and highly self-motivated learners, who see learning technical skills as part of belonging to the group (Over 2016). Such features of the BaYaka cultural niche enhance the cultural learning of technical knowledge and skills through multiple learning processes and pathways through-out children's development.

Of these processes, top-down instruction in blade tool techniques does not appear to be as important as others. Multiple studies have demonstrated that teaching is an important part of BaYaka children's daily lives (Boyette and Hewlett 2017; B. S. Hewlett and Roulette 2016; Lew-Levy, Kissler, et al. 2020). However, teaching is far less frequent than observational learning (Boyette 2016). The importance of observational learning is clear in the data on middle-childhood that I present here. For instance, it is noteworthy that 20 percent of the observations were when children's activities were coded as "rest" (table 7.3), when children were simply watching others use blades or lying around fiddling with blades themselves. Such observations are then honed through individual learning by playing and working, as shown here by the disproportionate observations of blade use during these types of activities. Through this process, children become able to flexibly apply their skills across the wide range of subsistence tasks for which they are needed. Moreover, the openness of the BaYaka learning environment and diversity of others available to be observed leads to "concerted" or "many-to-one" cultural transmission (B. S. Hewlett and Cavalli-Sforza 1986). This mode of cultural transmission, through which cultural learners learn from many other, more experienced individuals, theoretically leads to low variation between individuals in the population and high conservation of skills across generations. At the same time, such conservation of skills is counterbalanced by a high degree of autonomy and mobility, the basis of which is, in part, to seek new information (MacDonald and Hewlett 1999), such that innovations are also learned and spread (Lew-Levy, Milks, et al. 2020).

Importantly, there are examples among the BaYaka of specialized tools or techniques that require some degree of teaching and demonstration. For instance, hunting nets and snares and other tools that require specific types of knots (e.g., a basket to fetch honey from high trees) seem to require some degree of careful demonstration. For instance, I once watched a BaYaka friend put his two children in front of him on the ground as he laid his unfinished porcupine hunting nets on the mat and said, "Watch me. You're going to learn how to hunt." This was all the teaching that occurred, but he clearly marked the pedagogical nature of the context (Csibra and Gergely 2011) and placed the children in a position where they might best observe (Strachan, Curioni, and McEllin, this volume). The children were then given the opportunity to watch him finish tying. While I have not seen it, I predict that as a child begins to try to tie the cords together to weave the net, the method of teaching may change to some degree to include correction. Such teaching has been documented in the technique of weaving mats as described by Bonnie Hewlett (2013). Specifically, she describes being a novice learning to weave a basket and how her Aka teacher (another adult woman) would tell her "No, not like that" and correct her actions. As another example, during our observations of Aka nut-cracking (Boesch et al. 2017), we watched as BaYaka women and older girls would place a machete or an axe blade-side-up on the soft forest floor, stabilize it with their feet, then hold a nut carefully on the edge of the blade with its seam oriented vertically, so that they could hit the nut onto the blade with a wooden hammer to crack it open. Obviously, a miscalculation could result in severe injury if she gets a finger in between the hammer and the blade. However, even in this situation (where we did see adolescents cut themselves multiple times), the extent of teaching that we observed was a woman pointing to the nut's seam where it should be placed on the blade's edge. The teaching moment lasted only a few seconds, and the learner, a preadolescent girl, stopped trying after a few attempts.

Even in the context of learning these specialized techniques—in which learners must already be of a certain developmental readiness and which require careful instruction from older individuals—the relatively minimal role of teaching lies in stark contrast with the cultural model of teaching represented by the formal school-based education tradition described in the introduction (see table 7.1). For BaYaka learners, learning is by their own autonomous design; the techniques of production, their material basis, and their purpose are accessible; the social environment is dense with familiar, trusted people from whom to learn. Techniques may be partially learned, played with, practiced, and honed, cognitively internalized and embodied over time, with minimal constraint. Such an educational context is highly flexible with respect to the needs of learners and likely robust to changes in the state of the environment, as new technical knowledge is easily integrated and shared (Astuti, this volume; Pope-Caldwell, this volume). Indeed, this may represent the learning context of many human foraging groups throughout history that spread across every environment on Earth.

To conclude, I do not suggest the BaYaka cultural niche or its effects on children's learning are representative of other foraging groups or of past human societies per se. However, I do think there are aspects of their social structure and ecological adaptations that can permit some inferences about how similar contexts of cultural learning would support the gradual and inevitable acquisition of blade tool techniques during individual lifetimes and the reproduction of such expertise across generations. In particular, with the coevolution of the human family, cooperative childcare, and the domestic space of the hearth, which becomes the center of social activity and sharing well before the evolution of *Homo sapiens* (Kuhn and Stiner 2019), the cultural niche construction perspective focuses the analytical lens on the opportunities afforded for the (re)production of at least some common techniques that have been critical to the evolution of human cumulative culture and cognition, such as the use of nonhunting blade tools.

Acknowledgments

I thank Senay Cebioğlu, Daša Bombjaková and Mathieu Charbonneau for their insightful comments on earlier versions of this chapter. Additionally, I thank my fellow contributors for the stimulating discussion throughout the development of the volume. Finally, I must also thank the many children from Central African Republic and Congo who let me learn in context alongside them.

Notes

1. I thank Daša Bomjaková for pointing out that the domain of learning is important here. For instance, in ritual or religious education, compliance is expected and disobedience could lead to a refusal to teach (personal communication, July 2021).

2. Of course, this may also be an example of the coevolution of biology (e.g., sexual dimorphism in physiology, human mating, and family systems) and culture (e.g., gender socialization) that is supported by cultural niche construction. However, such an analysis is outside the scope of this chapter.

3. The "focal follow" is a method for systematically measuring time allocation—how much time is spent doing specific behaviors or activities. In this case, individual children were observed continuously during three two-hour

blocks of time, and their behavior was coded every minute according to a predetermined list of behaviors of interest (see Boyette and Hewlett 2017 for more details).

References

Bahuchet, S. 1985. *Les Pygmées Aka et La Forêt Centrafricaine*. Paris: SELAF.

Boesch, C., D. Bombjaková, A. H. Boyette, and A. Meier. 2017. "Technical Intelligence and Culture: Nut Cracking in Humans and Chimpanzees." *American Journal of Physical Anthropology* 163 (2): 339–355.

Bombjaková, D. 2018. "The Role of Public Speaking, Ridicule, and Play in Cultural Transmission among Mbendjele BaYaka Forest Hunter-Gatherers." PhD diss., University College London.

Bonawitz, E., P. Shafto, H. Gweon, N. D. Goodman, E. S. Spelke, and L. Schulz. 2009. "The Double-Edged Sword of Pedagogy: Modeling the Effect of Pedagogical Contexts on Preschoolers' Exploratory Play." In *Proceedings of the 31st Annual Meeting of the Cognitive Science Society*, XX. Netherlands: VU University Amsterdam.

Bonawitz, E., P. Shafto, H. Gweon, N. D. Goodman, E. Spelke, and L. Schulz. 2011. "The Double-Edged Sword of Pedagogy: Instruction Limits Spontaneous Exploration and Discovery." *Cognition* 120 (3): 322–330.

Bowles, S., and H. Gintis. 1977. *Schooling in Capitalist America: Educational Reform and the Contradictions of Economic Life*. New York: Basic Books.

Bowles, S., and H. Gintis. 2002. "Schooling in Capitalist America Revisited." *Sociology of Education* 75 (1): 1–18.

Boyce, W. T., J. Obradovi, N. R. Bush, J. Stamperdahl, Y. S. Kim, and N. Adler. 2012. "Social Stratification, Classroom Climate, and the Behavioral Adaptation of Kindergarten Children." *Proceedings of the National Academy of Sciences* 109 (S2): 17168–17173.

Boyd, R., and P. J. Richerson. 1995. "Why Does Culture Increase Human Adaptability." *Ethology and Sociobiology* 16:125–143.

Boyd, R., and P. J. Richerson. 2005. *The Origin and Evolution of Cultures*. Oxford: Oxford University Press.

Boyd, R., P. J. Richerson, and J. Henrich. 2011. "Colloquium Paper: The Cultural Niche: Why Social Learning Is Essential for Human Adaptation." *Proceedings of the National Academy of Sciences* 108 (S2): 10918–10925.

Boyette, A. H. 2016. "Children's Play and Culture Learning in an Egalitarian Foraging Society." *Child Development* 87 (3): 759–769.

Boyette, A. H. 2019. "Autonomy, Cognitive Development, and the Socialisation of Cooperation in Foragers: Aka Children's Views of Sharing and Caring." *Hunter Gatherer Research* 3 (3): 475–500.

Boyette, A. H., and B. S. Hewlett. 2017. "Autonomy, Equality, and Teaching among Aka Foragers and Ngandu Farmers of the Congo Basin." *Human Nature* 28 (3): 289–322.

Boyette, A. H., and S. Lew-Levy. 2019. "Variation in Cultural Models of Resource Sharing between Congo Basin Foragers and Farmers: Implications for Learning to Share." In *Inter-Disciplinary Perspectives on Sharing among Hunter-Gatherers in the Past and Present*, edited by D. Friesem and N. Lavi, 171–184. Cambridge: McDonald Institute for Archaeological Research.

Boyette, A. H., and S. Lew-Levy. 2021. "Socialization, Autonomy, and Cooperation: Insights from Task Assignment among the Egalitarian BaYaka." *Ethos* 48 (3): 400–418.

Boyette, A. H., S. Lew-Levy, M. S. Sarma, M. Valchy, and L. T. Gettler. 2020. "Fatherhood, Egalitarianism, and Child Health in Two Small-Scale Societies in the Republic of the Congo." *American Journal of Human Biology* 32 (4): e23342. https://doi-org.offsitelib.eva.mpg.de/10.1002/ajhb.23342.

Briggs, J. L. 1991. "Expecting the Unexpected: Canadian Inuit Training for an Experimental Lifestyle." *Ethos* 19 (3): 259–287.

Burkart, J. M., and C. P. van Schaik. 2016. "The Cooperative Breeding Perspective Helps in Pinning Down When Uniquely Human Evolutionary Processes Are Necessary." *Behavioral and Brain Sciences* 39: e34. doi:10.1017/S0140525X15000072.

Cainelli, E., and P. S. Bisiacchi. 2019. "Diagnosis and Treatment of Developmental Dyslexia and Specific Learning Disabilities: Primum Non Nocere." *Journal of Developmental & Behavioral Pediatrics* 40 (7): 558–562.

Coppens, A. D., and L. Alcalá. 2015. "Supporting Children's Initiative." *Advances in Child Development and Behavior*, 49, 91–112.

Correa-Chávez, M., and B. Rogoff. 2009. "Children's Attention to Interactions Directed to Others: Guatemalan Mayan and European American Patterns." *Developmental Psychology* 45 (3): 630–641.

Croizet, J.-C., F. Autin, S. Goudeau, M. Marot, and M. Millet. 2019. "Education and Social Class: Highlighting How the Educational System Perpetuates Social Inequality." In *The Social Psychology of Inequality*, edited by J. Jetten and K. Peters, 139–152. Cham: Springer International Publishing.

Csibra, G., and G. Gergely. 2011. "Natural Pedagogy as Evolutionary Adaptation." *Philosophical Transactions of the Royal Society B: Biological Sciences* 366 (1567): 1149–1157.

Dean, L. G., G. L. Vale, K. N. Laland, E. Flynn, and R. L. Kendal. 2014. "Human Cumulative Culture: A Comparative Perspective: Human Cumulative Culture." *Biological Reviews* 89 (2): 284–301.

Enquist, M., K. Eriksson, and S. Ghirlanda. 2007. "Critical Social Learning: A Solution to Rogers's Paradox of Nonadaptive Culture." *American Anthropologist* 109 (4): 727–734.

Erchak, G. M. 1980. "The Acquisition of Cultural Rules by Kpelle Children." *Ethos* 8 (1): 40–48.

Flynn, E. G., K. N. Laland, R. L. Kendal, and J. R. Kendal. 2013. "Target Article with Commentaries: Developmental Niche Construction." *Developmental Science* 16 (2): 296–313.

Fogarty, L., P. Strimling, and K. N. Laland. 2011. "The Evolution of Teaching." *Evolution* 65 (10): 2760–2770.

Gaskins, S. 2000. "Children's Daily Activities in a Mayan Village: A Culturally Grounded Description." *Cross-Cultural Research* 34 (4): 375–389.

Gaskins, S., and R. Paradise. 2010. "Learning through Observation in Daily Life." In *The Anthropology of Learning in Childhood*, edited by D. F. Lancy, J. Bock, and S. Gaskins, 85–118. Walnut Creek, CA: AltaMira Press.

Gauvain, M. 2005a. "Scaffolding in Socialization." *New Ideas in Psychology* 23:129–139.

Gauvain, M. 2005b. "Sociocultural Contexts of Learning." In *Learning in Cultural Context: Family, Peers, and School*, edited by A. E. Maynard and M. I. Martini, 11–40. New York: Kluwer Academic/Plenum Publishers.

Gergely, G., and G. Csibra. 2006. "Sylvia's Recipe: The Role of Imitation and Pedagogy in the Transmission of Human Culture." In *Roots of Human Sociality: Culture, Cognition, and Human Interaction*, edited by S. C. Levinson and N. J. Enfield, 229–255. Oxford: Berg Publishers.

Glowacki, L., and L. Molleman. 2017. "Subsistence Styles Shape Human Social Learning Strategies." *Nature Human Behaviour* 1 (5): 0098.

Gray, P. 2013. *Free to Learn: Why Unleashing the Instinct to Play Will Make Our Children Happier, More Self-Reliant, and Better Students for Life*. New York: Basic Books.

Gray, P., and J. Feldman. 2004. "Playing in the Zone of Proximal Development: Qualities of Self-Directed Age Mixing between Adolescents and Young Children at a Democratic School." *American Journal of Education* 110 (2): 108–146.

Greenfield, P. M., H. Keller, A. Fuligni, and A. Maynard. 2003. "Cultural Pathways through Universal Development." *Annual Review of Psychology* 54 (1): 461–490.

Grimm, L. 2000. "Apprentice Flintknapping: Relating Material Culture and Social Practice in the Upper Palaeolithic." In *Children and Material Culture*, edited by J. S. Derevenski, 53–71. London: Routledge.

Hewlett, B. L. 2005. "Vulnerable Lives: The Experience of Death and Loss among the Aka and Ngandu Adolescents of the Central African Republic." In *Hunter-Gatherer Childhoods: Evolutionary, Developmental, and Cultural Perspectives*, edited by B. S. Hewlett and M. E. Lamb, 322–342. New Brunswick, NJ: Aldine Transaction.

Hewlett, B. L. 2013. *Listen, Here Is a Story: Ethnographic Life Narratives from Aka and Ngandu Women of the Congo Basin*. Oxford: Oxford University Press.

Hewlett, B. S. 1991. *Intimate Fathers: The Nature and Context of Aka Pygmy Paternal Infant Care*. Ann Arbor: University of Michigan Press.

Hewlett, B. S. 2014. "Hunter-Gatherer Childhoods in the Congo Basin." In *Hunter-Gatherers of the Congo Basin*, edited by B. S. Hewlett, 245–275. New Brunswick, NJ: Transaction Publishers.

Hewlett, B. S., R. E. W. Berl, and C. J. Roulette. 2016. "Teaching and Overimitation among Aka Hunter-Gatherers." In *Social Learning and Innovation in Contemporary Hunter-Gatherers*, edited by H. Terashima and B. S. Hewlett, 35–45. Tokyo: Springer Japan.

Hewlett, B. S., and L. L. Cavalli-Sforza. 1986. "Cultural Transmission among Aka Pygmies." *American Anthropologist* 88 (4): 922–934.

Hewlett, B. S., H. N. Fouts, A. H. Boyette, and B. L. Hewlett. 2011. "Social Learning among Congo Basin Hunter-Gatherers." *Philosophical Transactions of the Royal Society B* 366:1168–1178.

Hewlett, B. S., J. Hudson, A. H. Boyette, and H. N. Fouts. 2019. "Intimate Living: Sharing Space among Aka and Other Hunter-Gatherers." In *Towards a Broader View of Hunter-Gatherer Sharing*, edited by D. Friesem and N. Lavi, 39–56. Cambridge: McDonald Institute for Archaeological Research.

Hewlett, B. S., M. E. Lamb, D. Shannon, B. Leyendecker, and A. Schölmerich. 1998. "Culture and Early Infancy among Central African Foragers and Farmers." *Developmental Psychology* 34 (4): 653–661.

Hewlett, B. S., and C. J. Roulette. 2016. "Teaching in Hunter-Gatherer Infancy." *Royal Society Open Science* 3 (1): 150403.

Hill, K., M. Barton, and A. Magdalena Hurtado. 2009. "The Emergence of Human Uniqueness: Characters Underlying Behavioral Modernity." *Evolutionary Anthropology: Issues, News, and Reviews* 18 (5): 187–200.

Hoehl, S., S. Keupp, H. Schleihauf, N. McGuigan, D. Buttelmann, and A. Whiten. 2019. "'Over-Imitation': A Review and Appraisal of a Decade of Research." *Developmental Review* 51:90–108.

Hrdy, S. B. 2009. *Mothers and Others: The Evolutionary Origins of Mutual Understanding.* Cambridge, MA: Harvard University Press.

Jackson, M., and B. Holzman. 2020. "A Century of Educational Inequality in the United States." *Proceedings of the National Academy of Sciences* 117 (32): 19108–19115.

Jang, H., C. Boesch, R. Mundry, S. D. Ban, and K. R. L. Janmaat. 2019. "Travel Linearity and Speed of Human Foragers and Chimpanzees during Their Daily Search for Food in Tropical Rainforests." *Scientific Reports* 9 (1): 11066.

Jeffares, B. 2012. "Thinking Tools: Acquired Skills, Cultural Niche Construction, and Thinking with Things." *Behavioral and Brain Sciences* 35 (4): 228–229.

Johnson, A. 2003. *Families of the Forest: The Matsigenka Indians of the Peruvian Amazon.* Berkeley: University of California Press.

Kendal, J. R. 2011. "Cultural Niche Construction and Human Learning Environments: Investigating Sociocultural Perspectives." *Biological Theory* 6 (3): 241–250.

Koster, J., R. McElreath, K. Hill, D. Yu, G. Shepard, N. van Vliet, M. Gurven, et al. 2020. "The Life History of Human Foraging: Cross-Cultural and Individual Variation." *Science Advances* 6 (26): eaax9070.

Kramer, K. L. 2010. "Cooperative Breeding and Its Significance to the Demographic Success of Humans." *Annual Review of Anthropology* 39:417–436.

Kuhn, S. L., and M. C. Stiner. 2019. "Hearth and Home in the Middle Pleistocene." *Journal of Anthropological Research* 75 (3): 305–327.

Kusserow, A. 2004. *American Individualisms: Child Rearing and Social Class in Three Neighborhoods.* New York: Palgrave Macmillan.

Laland, K. N. 2017. *Darwin's Unfinished Symphony: How Culture Explains the Evolution of the Human Mind.* Princeton, NJ: Princeton University Press.

Laland, K. N., and M. J. O'Brien. 2011. "Cultural Niche Construction: An Introduction." *Biological Theory* 6 (3): 191–202.

Laland, K. N., F. J. Odling-Smee, and M. W. Feldman. 2000. "Niche Construction, Biological Evolution, and Cultural Change." *Behavioral and Brain Sciences* 23:131–175.

Lancy, D. F. 2008. *The Anthropology of Childhood: Cherubs, Chattel, Changelings.* Cambridge: Cambridge University Press.

Lancy, D. F. 2010. "Learning 'From Nobody': The Limited Role of Teaching in Folk Models of Children's Development." *Childhood in the Past* 3:79–106.

Lancy, D. F. 2012. "The Chore Curriculum." In *African Children at Work: Working and Learning in Growing Up for Life,* edited by G. Spittler and M. Bourdillon, 23–56. Berlin: LIT Verlag.

Lancy, D. F. 2016a. "Teaching: Natural or Cultural?" In *Evolutionary Perspectives on Education and Child Development,* edited by D. Berch and D. Geary, 33–65. Heidelberg: Springer.

Lancy, D. F. 2016b. "Playing with Knives: The Socialization of Self-Initiated Learners." *Child Development* 87 (3): 654–665.

Lave, J., and E. Wenger. 1991. *Situated Learning: Legitimate Peripheral Participation.* Cambridge: Cambridge University Press.

Lavi, N., and D. Friesem, eds. 2019. *Towards a Broader View of Hunter-Gatherer Sharing.* Cambridge: McDonald Institute for Archaeological Research.

LeVine, R. A., S. Dixon, S. LeVine, A. Richman, P. H. Leiderman, C. Keefer, and S. Harkness. 1994. *Child Care and Culture: Lessons from Africa.* Cambridge: Cambridge University Press.

Lewis, J. 2002. "Forest Hunter-Gatherers and Their World: A Study of the Mbendjele Yaka Pygmies of Congo-Brazzaville and Their Secular and Religious Activities and Representations." PhD diss., London School of Economics and Political Science.

Lew-Levy, S., and A. H. Boyette. 2018. "Evidence for the Adaptive Learning Function of Work and Work-Themed Play among Aka Forager and Ngandu Farmer Children from the Congo Basin." *Human Nature* 29 (2): 157–185.

Lew-Levy, S., S. M. Kissler, A. H. Boyette, A. N. Crittenden, I. A. Mabulla, and B. S. Hewlett. 2020. "Who Teaches Children to Forage? Exploring the Primacy of Child-to-Child Teaching among Hadza and BaYaka Hunter-Gatherers of Tanzania and Congo." *Evolution and Human Behavior* 41 (1): 12–22.

Lew-Levy, S., N. Lavi, R. Reckin, J. Cristóbal-Azkarate, and K. Ellis-Davies. 2018. "How Do Hunter-Gatherer Children Learn Social and Gender Norms? A Meta-Ethnographic Review." *Cross-Cultural Research* 52 (2): 213–255.

Lew-Levy, S., A. Milks, N. Lavi, S. M. Pope, and D. E. Friesem. 2020. "Where Innovations Flourish: An Ethnographic and Archaeological Overview of Hunter-Gatherer Learning Contexts." *Evolutionary Human Sciences* 2:e31.

Lillard, A. S., and J. Taggart. 2019. "Pretend Play and Fantasy: What If Montessori Was Right?" *Child Development Perspectives* 13 (2): 85–90.

López, A., M. Correa-Chávez, B. Rogoff, and K. Gutiérrez. 2010. "Attention to Instruction Directed to Another by U.S. Mexican-Heritage Children of Varying Cultural Backgrounds." *Developmental Psychology* 46 (3): 593–601.

MacDonald, D. H., and B. S. Hewlett. 1999. "Reproductive Interests and Forager Mobility." *Current Anthropology* 40 (4): 501–523.

Markus, H. Rose, and S. Kitayama. 2010. "Cultures and Selves: A Cycle of Mutual Constitution." *Perspectives on Psychological Science* 5 (4): 420–430.

Maynard, A. E. 2002. "Cultural Teaching: The Development of Teaching Skills in Maya Sibling Interactions." *Child Development* 73 (3): 969–982.

McBrearty, S., and A. S. Brooks. 2000. "The Revolution That Wasn't: A New Interpretation of the Origin of Modern Human Behavior." *Journal of Human Evolution* 39 (5): 453–563.

Mead, M. 1970. "Our Educational Emphases in Primitive Perspective." In *From Child to Adult: Studies in the Anthropology of Education*, edited by J. Middleton, 1–13. American Museum Sourcebooks in Anthropology. Garden City, NY: Natural History Press.

Mejía-Arauz, R., B. Rogoff, A. Dexter, and B. Najafi. 2007. "Cultural Variation in Children's Social Organization." *Child Development* 78 (3): 1001–1014.

Mesoudi, A., L. Chang, S. R. X. Dall, and A. Thornton. 2016. "The Evolution of Individual and Cultural Variation in Social Learning." *Trends in Ecology & Evolution* 31 (3): 215–225.

Morelli, G., B. Rogoff, and C. Angelillo. 2003. "Cultural Variation in Young Children's Access to Work or Involvement in Specialised Child-Focused Activities." *International Journal of Behavioral Development* 27: 264–274.

Morgan, T. J. H., N. T. Uomini, L. E. Rendell, L. Chouinard-Thuly, S. E. Street, H. M. Lewis, C. P. Cross, et al. 2015. "Experimental Evidence for the Co-Evolution of Hominin Tool-Making Teaching and Language." *Nature Communications* 6 (January): 6029.

Narvaez, D., J. Panksepp, A. N. Schore, and T. R. Gleason. 2012. "The Value of Using an Evolutionary Framework for Gauging Children's Well-Being." In *Evolution, Early Experience and Human Development*, edited by D. Narvaez, J. Panksepp, A. N. Schore, and T. R. Gleason, 3–30. Oxford: Oxford University Press.

Naveh, D. 2016. "Social and Epistemological Dimensions of Learning among Nayaka Hunter-Gatherers." In *Social Learning and Innovation in Contemporary Hunter-Gatherers: Evolutionary and Ethnographic Perspectives*, edited by B. S. Hewlett and H. Terashima, 125–134. Tokyo: Springer Japan.

Nielsen, M., and K. Tomaselli. 2010. "Overimitation in Kalahari Bushman Children and the Origins of Human Cultural Cognition." *Psychological Science* 21 (5): 729–736.

Ochs, E., and C. Izquierdo. 2009. "Responsibility in Childhood: Three Developmental Trajectories." *Ethos* 37 (4): 391–413.

Odling-Smee, F. J., K. N. Laland, and M. W. Feldman. 2003. *Niche Construction: The Neglected Process in Evolution*. Princeton, NJ: Princeton University Press.

Over, H. 2016. "The Origins of Belonging: Social Motivation in Infants and Young Children." *Philosophical Transactions of the Royal Society B: Biological Sciences* 371 (1686): 20150072.

Paradise, R., and B. Rogoff. 2009. "Side by Side: Learning by Observing and Pitching in." *Ethos* 37 (1): 102–138.

Plummer, T. 2004. "Flaked Stones and Old Bones: Biological and Cultural Evolution at the Dawn of Technology." *American Journal of Physical Anthropology* 125 (S39): 118–164.

Rogers, A. R. 1988. "Does Biology Constrain Culture?" *American Anthropologist* 90 (4): 819–831.

Rogoff, B., R. Paradise, R. Mejía Arauz, M. Correa-Chávez, and C. Angelillo. 2003. "Firsthand Learning through Intent Participation." *Annual Review of Psychology* 54 (1): 175–203.

Scribner, S., and M. Cole. 1973. "Cognitive Consequences of Formal and Informal Education: New Accommodations Are Needed between School-Based Learning and Learning Experiences of Everyday Life." *Science* 182 (4112): 553–559.

Shifrer, D. 2013. "Stigma of a Label: Educational Expectations for High School Students Labeled with Learning Disabilities." *Journal of Health and Social Behavior* 54 (4): 462–480.

Shneidman, L., H. Gweon, L. E. Schulz, and A. L. Woodward. 2016. "Learning from Others and Spontaneous Exploration: A Cross-Cultural Investigation." *Child Development* 87 (3): 723–735.

Silva, K. G., M. Correa-Chávez, and B. Rogoff. 2010. "Mexican-Heritage Children's Attention and Learning from Interactions Directed to Others: Mexican-Heritage Children's Attention." *Child Development* 81 (3): 898–912.

Sonoda, K., D. Bombjaková, and S. Gallois. 2018. "Cultural Transmission of Foundational Schemas among Congo Basin Hunter-Gatherers." *African Study Monographs, Supplement* 54:155–169.

Sterelny, K. 2004. "Externalism, Epistemic Artefacts and the Extended Mind." In *The Externalist Challenge: New Studies on Cognition and Intentionality*, edited by R. Schantz, 239–254. Berlin: de Gruyter.

Sterelny, K. 2012. *The Evolved Apprentice: How Evolution Made Humans Unique*. Jean Nicod Lectures. Cambridge, MA: MIT Press.

Stiner, M. C. 2021. "The Challenges of Documenting Coevolution and Niche Construction: The Example of Domestic Spaces." *Evolutionary Anthropology: Issues, News, and Reviews* 30, no. 1 (2021): 63–70.

Stotz, K. 2010. "Human Nature and Cognitive–Developmental Niche Construction." *Phenomenology and the Cognitive Sciences* 9 (4): 483–501.

Stout, D. 2011. "Stone Toolmaking and the Evolution of Human Culture and Cognition." *Philosophical Transactions of the Royal Society B: Biological Sciences* 366 (1567): 1050–1059.

Strauss, S., and M. Ziv. 2012. "Teaching Is a Natural Cognitive Ability for Humans." *Mind, Brain, and Education* 6 (4): 186–196.

Super, C. M., and S. Harkness. 1986. "The Developmental Niche: A Conceptualization at the Interface of Child and Culture." *International Journal of Behavioral Development* 9:545–569.

Takakura, J. 2013. "Using Lithic Refitting to Investigate the Skill Learning Process: Lessons from Upper Paleolithic Assemblages at the Shirataki Sites in Hokkaido, Northern Japan." In *Dynamics of Learning in Neanderthals and Modern Humans Volume 1*, edited by T. Akazawa, Y. Nishiaki, and K. Aoki, 151–171. Tokyo: Springer Japan.

Thornton, A., and N. J. Raihani. 2008. "The Evolution of Teaching." *Animal Behaviour* 75 (6): 1823–1836.

Tomasello, M. 1999. *The Cultural Origins of Human Cognition*. Cambridge, MA: Harvard University Press.

Tomasello, M., A. P. Melis, C. Tennie, E. Wyman, and E. Herrmann. 2012. "Two Key Steps in the Evolution of Human Cooperation." *Current Anthropology* 53 (6): 673–692.

Weisner, T. S. 1984. "Ecocultural Niches of Middle Childhood: A Cross-Cultural Perspective." In *Development During Middle Childhood: The Years From Six to Twelve*, 335–369. Washington, DC: National Academy Press.

Whiting, B. B. 1963. *Six Cultures: Studies of Child Rearing*. New York: Wiley.

Whiting, B. B., and Carolyn P. Edwards. 1988. *Children of Different Worlds: The Formation of Social Behavior*. Cambridge, MA: Harvard University Press.

8 Stability and Change in Paleolithic Toolmaking

Dietrich Stout

Introduction

Culture is remarkable in its capacity to sustain both rapid change and enduring traditions. The rise of social media has occasioned a tectonic shift in cultural norms, language, economics, and politics in less than 20 years, and yet schoolchildren still sing nursery rhymes that are hundreds of years old and learn about farm animals few of them will ever encounter. Different cultural evolutionary research traditions have tended to focus on explaining either the adaptive flexibility or the stable rigidity of cultural traits (Sterelny 2017) and thus emphasized either processes of information transmission and incremental modification (Richerson and Boyd 2005; Henrich 2016; Laland 2018) or stabilizing factors of convergent reconstruction (Sperber 1996; Scott-Phillips, Blancke, and Heintz 2018; Strachan et al. 2021). Each of these is clearly relevant to understanding human culture, and many of the disagreements between the two traditions may be more apparent than real (Sterelny 2017). Nevertheless, theoretical emphases do play an important role in generating research questions and framing expectations. This chapter considers how the expectation that humanlike culture is characterized by change and diversification rather than stability and convergence has influenced interpretations of the Paleolithic archaeological record and shaped big-picture hypotheses about human evolution.

Stability and Change in the Paleolithic

The early archaeological record is widely thought to present a dilemma for human evolutionary studies. On the one hand, even the simplest known Paleolithic tools appear to be outside the behavioral range of modern nonhuman primates (Harmand et al. 2015, 142; Toth, Schick, and Semaw 2006; Braun et al. 2019; Stout et al. 2019) and are thus evocative of humanlike cultural capacities (Holloway 1969; Gärdenfors and Högberg 2017). This includes the reliable reproduction of particular tool forms across vast spans of time and space (Mithen 1999). On the other hand, many scholars consider rates of change in Paleolithic tools to be inhumanly slow and lacking the patterns of diversification and incremental improvement expected from full-fledged cultural evolution (Tennie et al. 2017; Richerson and Boyd 2005; Corbey et al. 2016; Foley 1987). This conundrum has led to various special explanations for the "monotonous" (Isaac 1976) and "bewildering" (Richerson and Boyd 2005) lack of Lower Paleolithic spatiotemporal variation, including proposals that stone toolmaking behaviors were genetically encoded rather than learned (Richerson and Boyd 2005; Corbey et al. 2016) or that Lower Paleolithic hominids lacked the cognitive capacities for innovation (Mithen 1999), imitation (Tennie et al. 2017), or effective teaching (Morgan et al. 2015) necessary for cumulative culture evolution.

These conjectures offer solutions to the problem of Paleolithic invariance, but is this really a problem we need to solve? How strong are expectations defining "humanlike" rates of culture change, and where do they come from? This is an important question insofar as it concerns underlying assumptions that guide hypothesis generation and evaluation. For example, Claudio Tennie and colleagues (2016, 2017) offer slow rates of change as one of several reasons to assume (until proven otherwise) that high-fidelity social transmission was absent in the Paleolithic. Accepting such a "null hypothesis" would make the presence of cultural learning in the Paleolithic an exceptional claim requiring exceptional evidence, and it would have important implications for our ability to recognize and study the borderline or intermediate cases critical to gradualist evolutionary accounts (Stout 2017; Stout et al. 2019).

Similarly, the suggestion that the detailed form of Acheulean bifaces was genetically constrained (Richerson and Boyd 2005), for example, by a "predisposition toward the basic behavioral routines involved, such as invasive bifacial reduction while realizing cutting edges in the secant plane, working from the tip down, and keeping symmetry" (Corbey et al. 2016, 14), potentially offers a cure that is worse than the disease insofar as it posits a level of detail in the genetic instruction of behavior that is theoretically questionable (Laland et al. 2015) and unlike anything known in modern primate tool behavior (Wynn and Gowlett 2018). These concerns echo earlier debates about the plausibility of a genetically specified "universal grammar" underlying human language (e.g., Deacon 1997), an idea that has arguably acted as a major impediment to progress in the study of language evolution (Christiansen and Chater 2017). Considering the theoretical and disciplinary stakes of attempts to resolve the dilemma of Paleolithic stability, it seems wise to make sure that there really is a problem in the first place.

In this chapter, I will argue that the dilemma of Paleolithic stability is more apparent than real and that this misapprehension highlights problems with a number of underlying assumptions about the nature of human culture and technology. These include the prevailing "information transmission" paradigm in cultural evolutionary studies, the dichotomization of individual versus social learning, a taxonomic approach that ranks social learning mechanisms in terms of intrinsic fidelity, and the view that human cultural evolution is broadly directional. Implicitly or explicitly, these assumptions have been built into the highly successful "cumulative cultural evolution" concept that has framed the debate around the origin of human technology in terms of the emergence of particular social learning mechanisms such as imitation or teaching. As an alternative, I advocate a complex, contingent, and variable vision of Paleolithic technological change and stability as resulting from the evolutionary and developmental interaction of a wide range of ecological, cognitive, material, and social factors (Stout 2021a). These factors include, but are not limited to, the rigidity or flexibility of individual cognitive processes supporting technological innovation and social reproduction (Gergely and Király, this volume; Strachan, Curioni, and McEllin, this volume; Roux et al., this volume).

What Is Technology?

Before proceeding, it is necessary to specify what we mean by "technology." Perhaps surprisingly, this is not a simple question. According to Leo Marx (1997), the term *technology* has its origins in the seventeenth century but did not come into widespread popular usage until the 1930s. In this process, its original meaning as a field of study focusing on the "useful" or "mechanic" arts shifted to a far more sweeping label for the new configurations of materials, institutions, and socioeconomic relations of the industrial revolution that were so radically transforming society. As such, technology emerged as a concept intimately entwined with notions of power, progress, and value and was used to both promote and critique an emerging capitalist system (Marx 1997). Arguably, this contributed to an even further broadening of meaning in which technology can refer to anything from a collection of physical hardware (tools and machines) to an abstract system of rules (Dusek 2006) and is commonly used as a marker of prestige or importance rather than a theoretically meaningful label.

For example, even the simplest Paleolithic stone tools are conventionally referred to as products of a "lithic technology" for little apparent reason other than the fact that they were tools. This terminology implicitly asserts the evolutionary importance, cultural nature, and human uniqueness of the earliest stone artifacts by linking them to a broader narrative of technological progress over human evolution (cf. Kuhn 2021). It is not usually seen to require explicit justification despite ongoing debates about precisely these points (e.g., Tennie et al. 2017; Wynn et al. 2011). Conversely, authors wishing to emphasize continuity with nonhuman primates have applied the label "technology" to ape and monkey tool use (Wynn et al. 2011), again without explicit justification. These usages may or may not be appropriate—the key point is that absence of any justification or discussion essentially reduces technology to a prestigious synonym for "tool" (see also Osiurak and Heinke 2018).

At the other end of the spectrum, the meaning of technology has been expanded to include everything from music (Patel 2019) to behavior modification (Skinner 2002). For example, French philosopher Michel Foucault wrote extensively about modern "technologies of power." While these could implicate material elements (e.g., bodily techniques, spatial arrangements), they are technological primarily in the sense of being a system of social rules or practices (Behrent 2013). This allowed Foucault (1988) to enumerate technologies of production, sign systems, social control, and the self without worrying about whether any actual tools, machines, or equipment were involved. Foucault's approach to technology as a social phenomenon seems to have arisen out of contemporary interest in industrial organization and reflects a deep ambivalence about the progressive versus dehumanizing effects of emerging technological systems (Behrent 2013). However, it also risks robbing the concept of technology of any specific meaning by expanding it to encompass almost any aspect of human culture and cognition. Are kinship systems a form of social technology? Do traditional meditative practices constitute a technology of the self? Are languages and number systems cognitive technologies? These questions are evocative but present a technology concept that is arguably too broad and amorphous for systematic theoretical analysis.

A reasonable middle path between over-reduction and over-expansion is supplied by the *technological systems* approach, which defines a technology as an integrated system of hardware, people, skills, knowledge, social relations, and institutions applied to practical tasks (Dusek 2006; Hughes 1987). This conceptualization is specific in a way that respects the history (Marx 1997) and colloquial usage of the term, thus avoiding confusion while retaining key features that made technology seem like an interesting topic in the first place. However, it leaves the critical question of what exactly constitutes a "practical task" unaddressed—essentially asking readers to make a subjective and value-laden judgment about what qualifies. To address this, I (Stout 2021a) proposed an evolutionary grounding of the concept in which technological tasks are practical in the specific sense that they involve the kind of material production necessary to support distinctive human life history and reproductive strategies.

In this "technological niche" (Stout and Khreisheh 2015; Stout and Hecht 2017), surplus production funds a biocultural reproductive strategy (Bogin, Bragg, and Kuzawa 2014) in which community members who may not be close biological kin donate resources (e.g., time, effort, food) to support offspring. This strategy allows costly investments in extended development and reduced offspring mortality without reducing fertility rates, thus supporting the protracted learning periods, extended life span, and large brain size that enable further surplus production (Kaplan et al. 2000) through the innovation and intergenerational reproduction of technological skills, knowledge, and equipment (Henrich 2016; Richerson and Boyd 2005; Laland 2018). Critically, this framework emphasizes that technological skills and knowledge are not simply "transmitted" or copied across social networks but must be actively reproduced (cf. "reconstructed" in Strachan et al. 2021) by individuals through extended learning processes (Lew-Levy et al. 2017) involving a complex mix of social interaction and individual trial-and-error practice (Sterelny, this volume; Ericsson, Krampe, and Tesch-Romer 1993; Lave and Wenger 1991). Building on the technological systems approach, this evolutionary framing supports a theoretical definition of technology as a behavioral domain of socially reproduced, collaborative activities involving the manipulation and modification of objects to enact changes in the physical environment.

According to this definition, technologies extend beyond simple tool use to encompass longer causal chains (e.g., use of a tool to construct a mechanism to harvest a resource to make a product) involving (1) the coordinated activity of many individuals, (2) the use of objects and materials in a wide range of roles other than as handheld instruments, and (3) the processes of social reproduction that sustain and elaborate technological systems. At the same time, this definition constrains technology to *materially instrumental* activities primarily intended to produce physical effects and excludes *communicative* activities primarily intended to affect the thoughts, behaviors, or experiences of the self or others. Whereas culture is a broader term that clearly encompasses both kinds of activity, technology is here defined by its focus on the former.

This definition recognizes that communication is an inherent part of the collaboration and social reproduction characteristic of technological systems but also specifies that it is not their ultimate goal or product. Thus, there could be technologies for the production of physical communications tools such as books, musical instruments, or computer networks, but there would be no "technologies" for training people in these skills, nor for effective storytelling, musical composition, rhetoric, or social messaging. Systematic approaches in these communicative domains might then be termed "arts" or "sciences" rather than technologies. This semantic distinction is important because materially instrumental versus communicative goals tend to present different functional demands, design constraints, and cultural evolutionary dynamics. This reflects the fact that the former are shaped by physical situations and materials, which may be relatively invariant across time and space, whereas the latter address human psychology in the context of specific cultural systems of meaning. Of course, some instrumental goals (e.g., production of particular artifact shapes, textures, or colors) are themselves occasioned by superordinate communicative goals. The stance advanced here is that it is nevertheless both possible and useful to distinguish the material constraints relevant to the former from the culturally situated psychological constraints on the latter. More to the point, this definition is not intended to lead to hair-splitting arguments about whether liminal cases are or are not proper "technologies" but rather to a focus on what is theoretically interesting about technologies (Stout 2013) and a more careful analysis of interwoven technological and nontechnological aspects of cultural phenomena (Stout 2021a).

Paleolithic Technologies?

By this definition, it is not immediately clear that various Paleolithic stone toolmaking behaviors actually qualify as technologies. Rather than arguing over the label, however, this should serve to focus our attention on assessing the expression of the three core features of technology: production, collaboration, and reproduction. Critically, these are not traits that can simply be ticked off as present or absent; rather, they are dimensions of variation. Focusing on these aspects encourages us to take a closer look at the archaeological evidence from multiple perspectives. Although quite a bit of recent attention has focused on assessing the nature of social reproduction mechanisms indicated by early stone tools and the implications for rates of change (Tennie et al. 2017; Morgan et al. 2015; Gärdenfors and Högberg 2017; Stout et al. 2019), this has been largely independent of attention to the complexity of production systems or demands for collaborative action. And yet these dimensions are clearly related to one another. The theoretical definition of technology adopted here directs our attention to this broader set of interacting factors when attempting to understand stability and change in the Paleolithic.

For example, what indications are there for longer technological production chains extending beyond immediate handheld tool use? In the earliest record, we might consider resource procurement, transport, and management (Reeves et al. 2021; Linares-Matás and Clark 2021) as well as the possibility that lithic artifacts were themselves part of a larger system for the production and use of wooden hunting and digging tools (Hayden 2015). More direct evidence for expanding technological systems comes later—for example, in the production of shaped antler and bone percussors that were themselves used as equipment for stone toolmaking (Stout et al. 2014; Moigne et al. 2016) and with the subsequent appearance of hafted, multicomponent tools (Barham 2013). In addition to placing increased demands on mechanisms of social reproduction (Pargeter, Khreisheh, and Stout 2019), such increases in complexity may themselves be inherently destabilizing and tend to favor increased rates of change. This is because (1) highly connected, interdependent systems are more susceptible to rapid transitions, including collapse (Scheffer et al. 2012), and (2) increasing technological complexity may have autocatalytic effects on innovation and diversification as more elements become available for recombination (Stout 2011; Lewis and Laland 2012). These potential effects of size and complexity on rates of change and stability further direct our attention to the collaboration needed to support elaborated technological systems.

Technology is perhaps most clearly distinguished from simple tool use by its mobilization of collaborative action by many individuals over extended periods of time. This is allowed in part by the concrete materiality of technology, which provides a durable medium for interaction across time and space, but it also requires underlying cognitive and social mechanisms for interactive coordination (Stout 2021a). Despite the fact that this capacity for collaboration is regarded as a distinctive and critically important human social cognitive trait (Tomasello and Gonzalez-Cabrera 2017), it has received relatively little attention in the study of Paleolithic stone tools. Evolutionary accounts have instead focused on zooarchaeological evidence of cooperative hunting (e.g., Stiner, Barkai, and Gopher 2009) or an inferred context of cooperative breeding (Hrdy 2009; Hawkes 2014). This perhaps reflects an implicit perception of stone toolmaking as a solitary activity or the pragmatic difficulty of identifying signatures of collaboration that would be visible in the lithic record. However, modern human lithic technologies are often highly collaborative (Reynolds 1993; Stout 2002; Apel 2001), and the possibility of Paleolithic collaboration warrants further attention. Theory (Stout 2021a) and ethnographic evidence suggest that collaboration is increasingly likely as lithic technological systems become more complex—for example, in the production of hafted tools (Stout 2002; Reynolds 1993). Activities such as the quarrying of large flake blanks (Shipton 2013) for Acheulean handaxe making, the production of complex adhesives for hafting (Wadley, Hodgskiss, and Grant 2009), and the heat treatment of lithic raw materials (Brown et al. 2009) seem to be promising subjects of study in this regard.

In sum, social reproduction is clearly a core feature of technology, but it is unlikely to tell the whole story of why and how fast Paleolithic technologies emerge, change, and disappear. Critically, increasing technological complexity and the more extensive collaboration required to support it come with costs as well as benefits. Longer production chains obviously require more immediate effort but also increased investments in the skill acquisition (Pargeter, Khreisheh, and Stout 2019) and social coordination (Currie et al. 2021) required to maintain them. For technological innovation to occur, benefits must outweigh these costs in the currently prevailing socioecological setting. This creates the possibility for fitness valleys separating highly stable local optima from theoretically possible increases. A classic example is the spread of agriculture, which was not a simple process of invention and adoption but rather involved complex coevolutionary interactions of changing crops, environment, and social institutions (Bowles and Choi 2013). Conditions of Paleolithic technological change are less well known but are likely to have involved multiple factors in addition to hominin social learning capacities (Tennie et al. 2017) and the population dynamics of information transmission (Powell, Shennan, and Thomas 2009; Derex and Mesoudi 2020). In order to create theoretical space for these more diverse causal pathways, it may be helpful to reconsider the core "culture as information" paradigm that dominates cultural evolutionary theory (Stout 2020).

Culture as Information

By the mid-twentieth century, the Modern Synthesis (MS) of evolutionary theory had firmly established evolutionary biology as a mathematical and statistical enterprise dealing in abstract genetic units (Pigliucci 2009). The central "genes as information" paradigm embodied by the MS was succinctly expressed by Eörs Szathmary and John Maynard Smith (1995, 231), who wrote: "Developmental biology can be seen as the study of how information in the genome is translated into adult structure, and evolutionary biology of how the information came to be there in the first place." This informational metaphor was rapidly appropriated to describe human culture as "an information-holding system with functions similar to that of cellular DNA" such that "the instructions needed for coping with the environment and performing specialized roles are provided by learned information, which is symbolically encoded and culturally transmitted" (d'Andrade 1984, 198). In anthropology, such symbolic informational approaches soon fell out of favor and were replaced by more enactive (Geertz 1973), dialectical (Giddens 1976), and embodied (Bourdieu 1977) conceptions of culture as something people *do* (e.g., practice theory). However, the informational conception has remained dominant in the study of cultural evolution. As Peter Richerson and Robert Boyd (2005, 5) specify: "Culture is information capable of affecting individuals' behaviors that they acquire from other members of their species through teaching, imitation, and other forms of social transmission."

This is somewhat ironic, as the population-based gene-culture coevolutionary thinking pioneered by scholars like Richerson and Boyd has now become an important part of an "Extended Evolutionary Synthesis" (EES) specifically questioning the MS conception of biological evolution as the transmission and expression of genetic information (Laland et al. 2015). The EES proposes a more inclusive and materially grounded conception of evolution in terms of dynamic, multidirectional processes such as reciprocal organism-environment causation, constructive development of adaptive phenotypes, and inclusive inheritance through nongenetic social, cultural, ecological, physiological, and epigenetic mechanisms. As a result, a wider range of evolutionary causes beyond mutation, selection, drift, and gene flow are emphasized. There is some controversy as to whether the processes highlighted by the EES are actually novel or important for evolutionary theory (e.g., Futuyma 2017), but there does seem to be broad agreement that they are worthy of further investigation. The suggestion here is that the cultural evolutionary component of inclusive inheritance could use a similar rethink. As with the EES more broadly, this should not be construed as a repudiation of past work or even as presenting previously unrecognized mechanisms and empirical findings. In fact, one of the reasons the EES has been so controversial is that its primary contribution is theoretical or even philosophical rather than empirical (Pigliucci and Finkelman 2014). The EES takes a stance on the nature and goals of evolutionary explanation whose relevance and appeal will depend on the questions and objectives of different research programs (Welch 2017). This is equally true with respect to the study of cultural evolution.

Richerson and Boyd (2005, 259) explicitly state that their definition of culture as information is a pragmatic one, intended to promote productive research, rather than the only possible one. In this respect, it has clearly been successful and has generated an ever-growing body of literature elucidating everything from the influence of population size and structure on cultural evolution (Henrich 2004; Derex and Mesoudi 2020) to the relevance of variation in learning strategies (Kendal et al. 2018; Miu et al. 2020). As with gene-centered approaches to biological evolution (cf. Welch 2017), the power of this informational approach stems from its relative simplicity, broad generalizability, and amenability to formal modeling. However, these broad strengths may be less well suited to explaining the precise causal-historical details of particular cases (Stout 2018), especially when variables employed in formal models are difficult to relate to empirical measures of real-world data. This parallels the case with the EES, which may be most relevant and helpful to researchers interested in explaining particular evolutionary histories (cf., Welch 2017).

Transmission or Causation?

According to the EES, "phenotypes are not inherited, they are reconstructed in development" (Laland et al. 2015, 1–14). The same might easily be said of cultural concepts and practices. In fact, the need for such individual "reconstruction" is a core premise of Cultural Attraction Theory (CAT) (see Sperber 1996; Sterelny 2017; Strachan et al. 2021; Scott-Phillips, Blancke, and Heintz 2018; Claidière, Scott-Phillips, and Sperber 2014). Illustrative examples of CAT tend to focus on communicative (e.g., songs, jokes, stories) rather than technological culture. They thus emphasize psychological rather than ecological and material explanations (Scott-Phillips, Blancke, and Heintz 2018) and often continue to describe culture as transmitted information (e.g., Strachan et al. 2021). Critically, however, CAT explicitly extends to practical skills and material mechanisms of attraction and more precisely theorizes cultural reconstruction and attraction as products of complex causal chains rather than information transmission. This conceptual shift from information transmission to causal relations is subtle but important and mirrors the EES move to a more inclusive vision of inheritance involving more than just genes. While a case can be made that DNA codes information (Maynard Smith 2000), and intentional communication certainly does, material conditions influencing evolution (e.g., constructed niches) or skill learning (e.g., equipment) can be considered "information" in only a loose metaphorical sense. Causal relations provide a more robust and inclusive framework for thinking about the diverse mechanisms potentially involved in reproducing and altering the patterns of behavior that constitute culture (cf. Roepstorff, Niewöhner, and Beck 2010).

This is especially true of technology. Causal mechanisms potentially contributing to technological stability and change extend beyond learning processes per se to include relative costs and benefits in particular behavioral systems and ecologies (Režek et al. 2018; Pargeter, Khreisheh, and Stout 2019), social structure (Derex and Mesoudi 2020) and institutions (Bowles and Choi 2013; Brahm and Poblete 2021; Roux 2010), intrinsic features of (Stout 2021a) and interactions between (Kolodny, Creanza, and Feldman 2015) technologies, and potential coevolutionary relationships between these diverse factors (Strassberg and Creanza 2021; Kolodny, Creanza, and Feldman 2016). Indeed, the cultural evolution literature is already replete with examples of material and other causes of technological stability and change, including functional design demands, inflexible production processes, technological entrenchment, innovation cascades, market integration, and environmental change (reviewed by Mesoudi et al. 2013, table 11.2). In the information transmission paradigm, however, such particular features are viewed as proximate mechanisms inflecting local rates and patterns of change rather than ultimate explanations for the origin of cultural diversity and adaptation. As in gene-centered approaches to biological evolution, the latter are expected to be expressed in purely terms of the population dynamics of information variation, transmission, and selection.

An alternative approach would be to emphasize the causal power of such "proximate" mechanisms to actually drive evolutionary change (cf. Laland et al. 2015). For example, there is some debate in the cultural evolution literature over whether technological innovation is usually blind and random (i.e., like genetic mutation), with optimization purely due to selective retention, or whether individual learning commonly acts to "guide" variation toward desired outcomes and allow for optimization even in the absence of selection (Mesoudi 2021). A largely neglected third possibility in this debate is that material and social conditions can also guide variation and affect retention. A simple nonhuman example is the way in which the durability of artifacts associated with primate tool use can facilitate the reproduction of tool behavior (Fragaszy et al. 2013). In humans, ecology, ideology, and economics can affect the nature, frequency, and retention of innovations (Lew-Levy et al. 2020; Greenfield 2003; Macfarlane and Harrison 2000), and particular technologies may be more or less evolvable due to the modularity versus interdependence of component parts or procedures (Martin 2000; Mesoudi and O'Brien 2008; Charbonneau 2016). The famously accidental discovery of penicillin by Alexander Fleming was made a lot more likely by the physical and institutional infrastructure of his bacteriology lab at St. Mary's Hospital, and even then, purification, clinical application, and mass production of penicillin took more than a decade of effort from a large community of people receiving government support in a wartime setting (Gaynes 2017). The further one zooms in on particular histories of technological change and stability, the more helpful it becomes to consider diverse causes beyond information variation, transmission, and selection. As argued in the previous section, the utility of such particularism may often depend on the nature and scale of the questions being asked. Growing appreciation of the complex, multilineal, intermittent, asynchronous course of human evolution (Antón and Kuzawa 2017; Falk et al. 2005; Holloway et al. 2018; d'Errico and Stringer 2011; Vaesen and Houkes 2021) suggests that greater resolution may be needed to adequately explain the origins, tempo, and mode of cultural evolution in the Paleolithic.

Cumulative Cultural Evolution?

Another foundational premise of the informational paradigm is that there exists a strict dichotomy between social and individual learning, with the former providing the information transmission mechanisms necessary for adaptive cultural evolution to occur (e.g., Boyd, Richerson, and Henrich 2011). Early on, this was perceived to produce a paradox insofar as a mixed population of individual and social learners has an equilibrium mean fitness no higher than a population of pure individual learners (Rogers 1988). This is because the benefits of social learning are frequency dependent under the assumption that copying is cheap but fails to track environmental change or produce fitness-enhancing innovations. However, Robert Boyd and Peter Richerson (1996) showed that allowing individuals to pursue mixed strategies of individual and social learning can indeed produce incremental improvement in a fitness-enhancing skill and thus increase population mean fitness. Eventually, this can lead to the emergence of skills that would have been beyond the inventive capacity (the "reaction norm") of individuals in the first generation, a process that Boyd and Richerson dubbed "cumulative cultural evolution" (CCE).

This model assumed that social learning requires specialized psychological mechanisms that come at some cost. This stance was inspired by the cultural learning concept (Tomasello, Kruger, and Ratner 1993) that sought to explain human cognitive uniqueness as a product of enhanced social learning mechanisms including imitation, instruction, and collaboration. High-fidelity reproduction by these mechanisms was proposed to support a cultural evolutionary "ratchet effect" that allowed the accumulation of modifications over generations, later described as "improvement" (Tomasello 1999, 5) and attributed to one key biological innovation that put humans on a qualitatively different cultural path of evolution. The assumed costs of high-fidelity social learning were found to create a barrier to the initiation of CCE (Boyd and Richerson 1996), potentially explaining its rarity in nature. According to this argument, accumulation must begin with simple skills that are within the inventive potential of individuals. Insofar as the benefits of learning these skills socially (i.e., decreased cost) remain small relative to the costs of enhanced social learning mechanisms, this creates an "adaptive valley" that must be crossed before CCE can start to produce the body of complex, difficult-to-learn, and useful cultural content that would allow these expensive mechanisms to pay for themselves and initiate sustained biocultural feedback. These ideas lead to a picture of CCE as a unique and characteristic human trait—a unitary capacity that emerged as a key event (Boyd, Richerson, and Henrich 2011; Tomasello 1999) or a crossing of a coevolutionary Rubicon (Henrich 2016) that put humans on a novel evolutionary trajectory.

However, there is substantial evidence that the cognitive processes and neural systems supporting social and individuals learning actually overlap extensively (Heyes 2018; Olsson, Knapska, and Lindström 2020). Considering them as independent traits in evolutionary modeling may thus be misleading. More conceptually, CAT collapses the social and individual dichotomy by focusing on complex processes of reconstruction rather than straightforward transmission or copying of information (Claidière, Scott-Phillips, and Sperber 2014; Sterelny 2017). This is again particularly relevant and convincing in the case of technology. The materiality of technology often demands precise control of physical contingencies in pursuit of complex goals, and this in turn demands a protracted, collaborative learning process, including dedicated practice in supportive material and social contexts (Stout 2021a; Stout and Hecht 2017). Such intertwined social, individual, and contextual mechanisms are clearly problematic for any strict dichotomy between social and individual learning. Because different technologies are also expected to vary substantially in their particular cognitive and motor demands (Stout 2021a), this also problematizes any attempt to characterize particular learning mechanisms as inherently "high" or "low" fidelity.

The informational paradigm posits a diverse taxonomy of social learning mechanisms ranked by transmission capacity and fidelity. From low to high, the three most widely used categories are *stimulus enhancement* (direction of attention), *emulation* (copying action goals or outcomes), and *imitation* (reproducing specific behavioral means). Although there is some confusion in the literature over the precise meaning of "imitation" (Heyes 2021; Stout et al. 2019), it is commonly argued to be critical to the ratchet effect of CCE (e.g., Tennie, Call, and Tomasello 2009). However, real-world learning and behavior occur on multiple levels of organization and across extended periods of time. Distinctions between goals and means (Stout et al. 2019) and measures of reproductive fidelity (Charbonneau and Bourrat 2021) thus depend on the scale of analysis, and the relevant scale depends on details and objectives of the behavior in question (Stout 2021a; Legare and Nielsen 2015). Similarly, it is expected that the utility of learning strategies ranging from independent exploration to end-state emulation or body movement reenactment will depend on the specific skills to be reproduced rather than some inherent information transmission capacity (Heyes 2021; Stout and Hecht 2017). For example, under even moderately variable conditions, increased reliance on individual trial-and-error learning can actually lead to higher reproductive fidelity than precise behavior-copying (Truskanov and Prat 2018). Experimental work with transmission chains has similarly shown that the importance of different learning mechanisms depends on the particular task and context being studied (Caldwell 2020). Even greater diversity and variability can be expected in the real world, including the Paleolithic.

The recognition of a distinct form of "cumulative" culture evolution emerged as a useful marker in a debate over the possibility of fitness-enhancing gene-culture coevolution (Boyd and Richerson 1996) and has led to much work on the importance of reproductive fidelity and innovation (Lewis and Laland 2012), social learning strategies (Kendal et al. 2018), and population size and structure (Henrich 2004; Powell, Shennan, and Thomas 2009; Derex and Mesoudi 2020) on cultural adaptation. In hindsight, however, it is not clear that the concept of cumulative culture captures anything that is not already encompassed by concepts of inheritance, adaptation, and persistent evolutionary trends that have already been extensively theorized in evolutionary biology (Stout 2021b). Insofar as the concept of fitness is undertheorized in cultural evolutionary studies and invariably represented by some kind of proxy variable (Mesoudi and Thornton 2018), the core CCE criterion of "iterative improvement" carries substantial risk of progressivist misinterpretation. It might thus be preferable to revert to the more precise and the less value-laden terminology from evolutionary biology.

Directionality in Human Evolution

Taken together, these critiques of the informational paradigm highlight a causal diversity that decenters social learning mechanisms as the key factor determining the stability and evolvability of technologies. In so doing, they undermine the influential idea that cumulative culture capacity is a unitary trait dependent on one (Tomasello 1999) or several (Boyd, Richerson, and Henrich 2011; Henrich 2016) key psychological adaptations for learning from other people. If cumulative culture is not a species typical characteristic of *Homo sapiens* (Vaesen and Houkes 2021) that emerged at some key point in our past, then the long-term persistence of some Paleolithic technologies appears less in need of special explanation (or at least not more so than any other case of technological stability or change).

Of course, all this theorizing is well and good, but isn't it an *empirical* fact that human evolution has followed a unique and consistent path in need of some unifying explanation? There are many unknowns in human evolution, but one thing that has long seemed clear is that our evolutionary history has been characterized by long-term parallel trends toward increasing brain size (Du et al. 2018) and technological sophistication (Stout 2011). The apparent consistency of these trends over so much time and space, and presumably very different selective contexts, has suggested to many that they reflect some kind of intrinsic biocultural feedback dynamic (Washburn 1960; Isler and van Schaik 2014; Holloway 1981; Miller, Barton, and Nunn 2019). Indeed, such feedback lies at the heart of the technological niche concept used to motivate the theoretical definition of technology adopted here (Stout 2021a). Such logic leads one to expect that once this powerful coevolutionary process gets started by some "initial kick" (Holloway 1981), key cognitive adaptation (Tomasello 1999), or crossed threshold (Henrich 2016), it should become self-sustaining across diverse conditions and thus produce the long-term directional trends thought to characterize human evolution on a macro scale.

But what if human evolution is not actually characterized by such trends? For example, Krist Vaesen and Wibo Houkes (2021) have argued that there is actually very little evidence that human cultural evolution has been predominantly cumulative, showing iterative improvement in performance over time (Mesoudi and Thornton 2018). This is perhaps unsurprising for communicative aspects of human culture (e.g., art, religion, ritual) in which performance is always relative to specific cultural and psychological contexts and long-term "improvement" can be an elusive concept. More surprisingly, it may also be true of technology. Technology is the domain most commonly referenced in studies of cumulative culture because its material goals, means, and payoffs appear more easily comparable across time and contexts (cf. Derex and Mesoudi 2020; Vaesen and Houkes 2021). Even so, it is not clear that technology is characteristically cumulative. For example, I argued (Stout 2011) that Paleolithic technological change was cumulative because maximum expressed complexity increased over time by adding levels of hierarchical structure to previously established technologies. While this establishes that iterative increases *sometimes* occurred, it fails to show that they were a characteristic or predominant pattern in technological evolution because it ignores patterns of stability and change in simpler technologies that also existed throughout the Paleolithic (Vaesen and Houkes 2021; Režek et al. 2018). In fact, even a random walk (increase and decrease equally likely) could produce this pattern if complexity has a lower bound constraining variation in that direction (cf. the "zero force evolutionary law" of McShea and Brandon 2010). The absence of a clear and persistent overall trend toward increasing technological complexity or performance suggests that more particularistic explanations of specific instances of Paleolithic change and stability may be needed (Stout 2018).

Could this also be the case for hominin brain evolution? It is, of course, indisputable that the endocranial volume of *Homo sapiens* is more than three times that of *Australopithecus*. Furthermore, quantitative analyses of fossil hominin endocranial volumes over time generally support a pattern of gradual (Du et al. 2018) and accelerating (Miller, Barton, and Nunn 2019) increase over time. These findings are consistent with a microevolutionary process of long-term directional brain size selection on hominin populations within lineages (anagenesis) driven by accelerating biocultural feedback (Miller, Barton, and Nunn 2019). However, there are problems. Most importantly, these analyses do not include the relatively recent small-brained species *Homo floresiensis* (Falk et al. 2005) and *Homo naledi* (Holloway et al. 2018). These species represent either episodes of secondary brain size reduction or the survival of smaller-brained lineages much later in the fossil record than previously thought. Considering that both of these discoveries have occurred within the past 20 years, we must seriously consider that more such examples may remain to be found. If this turns out to be the case, then patterns of hominin brain size evolution would start to look more like diversification above a minimum "ape grade" level, rather than a gradual and persistent directional trend.

A second problem arising from the incomplete nature of fossil record is that it is difficult to be sure that the apparent trend toward brain size increase actually reflects microevolutionary processes within lineages versus macroevolutionary processes of lineage splitting and extinction. Andrew Du and colleagues (2018) found that microevolutionary anagenesis accounted for 64 to 88 percent of hominin brain size change; however, this result is obviously dependent on the taxonomic classification of fossils as well as known first and last appearance dates of these hypothetical species. Unfortunately, neither the taxonomic diagnosis of hominin species (Athreya and Hopkins 2021) in biologically real terms (i.e., as separately evolving lineages in de Queiroz 2005) nor our knowledge of their actual temporal range appear to be particularly reliable, especially for species represented by a small number of fossil specimens. This uncertainty matters because macroevolutionary processes of lineage formation and extinction may involve very different mechanisms (Gould 2002) from the gradual, microevolutionary selection envisioned by biocultural coevolutionary models (Du et al. 2018). For example, an apparently unitary trend could actually reflect the average of a relatively small number of idiosyncratic lineage-level events or a tendency for brain size (or its body size, life history, or ecological correlates) to be associated with increased speciation or reduced extinction rates rather than individual fitness.

These issues in our understanding of the broad patterning of human brain and technological evolution weaken the empirical motivation for positing a key threshold or capacity that initiated a uniquely human pattern of persistent and cumulative biocultural evolution. This may be a tough pill to swallow, as expectations of a hard animal and human boundary (Cartmill 1990) and human evolutionary progress (Ruse 1996) are deeply ingrained in paleoanthropological thought. Indeed, the implications of such a shift in perspective are many and deep and extend to paleoanthropology's tragic role in establishing and perpetuating racist colonial hierarchies (Athreya and Ackermann 2019). For current purposes, they significantly affect our expectations for patterns of technological stability and change in the Paleolithic. In the absence of a robust overall trend, there would be no reason to view stability as anomalous or to expect a single key factor such as cultural transmission capacities (Tennie et al. 2017; Lewis and Laland 2012) and strategies (Lycett and Gowlett 2008; Kendal et al. 2018) or the population dynamics of transmission (Powell, Shennan, and Thomas 2009; Henrich 2016; Derex and Mesoudi 2020) to fully explain rates of change.

Importantly, this does not invalidate biocultural feedback hypotheses (e.g., Henrich 2016) or their use (Stout 2021a) as theoretical grounding for a definition of technology. It does emphasize that the *potential* feedback effects envisioned would only be triggered in particular circumstances, are unlikely to be indefinitely self-sustaining once initiated, and could produce complex dynamics other than continuously accelerating increase (Stout and Hecht 2017). Indeed, such feedback hypotheses for human evolution are derived from comparative (primate and other) data (e.g., Isler and van Schaik 2014) that identify potential coevolutionary relationships while simultaneously showing that they often do not result in runaway feedback. In other words, biocultural feedback is not a unitary, monocausal explanation for human evolution but rather constitutes a set of general principles or relationships that will behave differently across different contexts. Attempts to understand particular instances of technological stability or change must combine such general comparative principles with particular historical details from disciplines like archaeology and paleontology in order to generate specific explanations (Stout 2018). Sometimes variation in hominin cognition or social learning may indeed be critical; in other cases, alternative considerations such as context-specific costs and benefits (Pargeter, Khreisheh, and Stout 2019; Režek et al. 2018), inheritance of material infrastructure (Pradhan, Tennie, and van Schaik 2012), social arrangements (Powers, van

Schaik, and Lehmann 2016; Currie et al. 2021; Derex and Mesoudi 2020), or interactions between technologies (Kolodny, Creanza, and Feldman 2015) may be key.

Conclusion

Stephen Jay Gould and Niles Eldredge (1977) famously argued that "stasis is data" and needs explanation. This certainly applies to the stability and persistence of Paleolithic toolmaking behaviors such a simple core-and-flake ("Oldowan") or large cutting tool ("Acheulean") production. Yet what is it about this stability that needs to be explained? The so-called California program arising from the work of Robert Boyd and Peter Richerson focuses on uniquely human capacities for cultural accumulation and diversification (Sterelny 2017) and sees stasis as something of an anomaly or even deficit to be explained in terms of cognitive limitations, low population size, or other problems (Morgan 2016; Henrich 2016; Richerson and Boyd 2005). In contrast, the CAT "Paris program" arising from the work of Dan Sperber seeks to explain cultural stability in a variable world (Sterelny 2017) and so focuses on causal mechanisms that create stable attractors (Scott-Phillips, Blancke, and Heintz 2018).

This focus may provide a more felicitous frame for discussing Paleolithic stability, and all of the major proposals can be construed as suggesting different causal mechanisms leading to observed technological attractors. These include psychological factors of attraction such as genetically evolved biases toward certain tool forms or procedures (Richerson and Boyd 2005; Corbey et al. 2016), reliance on particular learning mechanisms (Tennie et al. 2017), or more general perceptual biases (Wynn and Gowlett 2018), as well as ecological factors such as population structure (Powell, Shennan, and Thomas 2009) or characteristic modes of social organization and transmission (Lycett and Gowlett 2008). Archaeologists often emphasize functional forces of attraction, such as design constraints (Wynn and Gowlett 2018) or the role of tools in larger behavioral ecological strategies (Režek et al. 2018). These perspectives generally expect successful strategies to be stable and thus focus more on explaining episodes of change in terms of extrinsic causes of change such as climatically driven habitat shifts (Antón, Potts, and Aiello 2014). Finally, there are evolving organismal factors of attraction such as more general perceptual-motor and cognitive capacities (Stout et al. 2019; Pargeter et al. 2020) or biomechanics and manipulative capacities (Karakostis et al. 2021) that might affect the relative costs and benefits of particular technologies. Most likely, each of these mechanisms and more have been relevant at different times and places in the Paleolithic and would have interacted in complex and historically contingent ways to produce the observed archaeological record (Stout 2018).

This is perhaps unfortunate, as prime mover explanations of human uniqueness as arising from CCE and biocultural feedback dynamics resulting from enhanced social learning capacities are attractive for their parsimony and synthetic power. Humans are exceptional in so many ways at the same time that more piecemeal adaptive accounts do seem to be missing a bigger picture. How likely is it that everything from dexterous manipulation to theory of mind and metacognitive learning strategies would just happen to occur together in one species? By analogy, if one person flipped a coin and got heads 10 times in a row, we would expect some kind of general explanation for this highly unlikely outcome (e.g., biased coin). However, if one person out of 2,000 flipped 10 heads in a row, we would recognize that this is simply an expected result of probability. The appropriate level of explanation for why *this particular* person got that result would then focus on the dynamics of each individual flip. If it is true that human evolution is better characterized by diversification rather than a single directional trend toward increasing brain size and technological complexity, then our explanatory task is more similar to the latter case. We would thus need to focus on particular causes for particular instances of stability and change rather than positing single, overarching explanations. For this perspective, stasis would simply be data rather than an anomaly or exception to a more general evolutionary process.

Importantly, this is not to say that some degree of theoretical synthesis cannot be achieved. As outlined above, comparative and modeling studies of biocultural evolution have identified potential interactions between diverse life history, neural, cognitive, social, and technological variables (Isler and van Schaik 2014; Kolodny, Creanza, and Feldman 2016; Morgan 2016; Kaplan et al. 2000; Muthukrishna et al. 2018) that may allow synthesis of complex evolutionary phenomena in terms of a smaller number of recurring relationships and processes (Stout 2021a). However, the relevant causal relations in any particular case may still vary depending on specific context. Moving away from the information transmission paradigm of cultural evolution and the presumption of an overarching directionality to human biocultural evolution opens theoretical space for considering this wider range of causal relations and explanations for Paleolithic stability and change.

Acknowledgments

I would like to thank Mathieu Charbonneau and Dan Sperber for the opportunity to participate in this project, workshop participants for stimulating contributions and discussion, and Mathieu Charbonneau for helpful comments on a draft of this chapter.

References

Antón, S. C., and C. W. Kuzawa. 2017. "Early Homo, Plasticity and the Extended Evolutionary Synthesis." *Interface Focus* 7 (5): 20170004.

Antón, S. C., R. Potts, and L. C. Aiello. 2014. "Evolution of Early Homo: An Integrated Biological Perspective." *Science* 345 (6192): 1236828.

Apel, J. 2001. *Daggers, Knowledge and Power*. PhD diss. Uppsala: Acta Universitatis Upsaliensis, 2001.

Athreya, S., and R. Rogers Ackermann. 2019. "Colonialism and Narratives of Human Origins in Asia and Africa." In *Interrogating Human Origins*, 72–95. New York: Routledge.

Athreya, S., and A. Hopkins. 2021. "Conceptual Issues in Hominin Taxonomy: Homo Heidelbergensis and an Ethnobiological Reframing of Species." *American Journal of Physical Anthropology* 175 (2021): 4–26.

Barham, L. 2013. *From Hand to Handle: The First Industrial Revolution*. Oxford: Oxford University Press.

Behrent, M. C. 2013. "Foucault and Technology." *History and Technology* 29 (1): 54–104.

Bogin, B., J. Bragg, and C. Kuzawa. 2014. "Humans Are Not Cooperative Breeders but Practice Biocultural Reproduction." *Annals of Human Biology* 41 (4): 368–380.

Bourdieu, P. 1977. *Outline of a Theory of Practice*. Cambridge: Cambridge University Press.

Bowles, S., and J.-K. Choi. 2013. "Coevolution of Farming and Private Property during the Early Holocene." *Proceedings of the National Academy of Sciences* 110 (22): 8830–8835.

Boyd, R., and P. J. Richerson. 1996. "Why Culture Is Common but Cultural Evolution Is Rare." In *Evolution of Social Behaviour Patterns in Primates and Man*, edited by W. G. Runciman, J. Maynard Smith, and R. I. M. Dunbar, 77–93. Oxford: Oxford University Press.

Boyd, R., P. J. Richerson, and J. Henrich. 2011. "The Cultural Niche: Why Social Learning Is Essential for Human Adaptation." *Proceedings of the National Academy of Sciences* 108 (S2): 10918–10925.

Brahm, F., and J. Poblete. 2021. "The Evolution of Productive Organizations." *Nature Human Behaviour* 5 (1): 39–48.

Braun, D. R., V. Aldeias, W. Archer, J. Ramon Arrowsmith, N. Baraki, C. J. Campisano, A. L. Deino, et al. 2019. "Earliest Known Oldowan Artifacts at >2.58 Ma from Ledi-Geraru, Ethiopia, Highlight Early Technological Diversity." *Proceedings of the National Academy of Sciences* 116 (24): 11712.

Brown, K. S., C. W. Marean, A. I. R. Herries, Z. Jacobs, C. Tribolo, D. Braun, D. L. Roberts, et al. 2009. "Fire as an Engineering Tool of Early Modern Humans." *Science* 325 (5942): 859–862.

Caldwell, C. A. 2020. "Using Experimental Research Designs to Explore the Scope of Cumulative Culture in Humans and Other Animals." *Topics in Cognitive Science* 12 (2): 673–689.

Cartmill, M. 1990. "Human Uniqueness and Theoretical Content in Paleoanthropology." *International Journal of Primatology* 11 (3): 173–192.

Charbonneau, M. 2016. "Modularity and Recombination in Technological Evolution." *Philosophy & Technology* 29 (4): 373–392.

Charbonneau, M., and P. Bourrat. 2021. "Fidelity and the Grain Problem in Cultural Evolution." *Synthese* 199 (3–4): 5815–5836.

Christiansen, M. H., and N. Chater. 2017. "Towards an Integrated Science of Language." *Nature Human Behaviour* 1 (8): 1–3.

Claidière, N., T. C. Scott-Phillips, and D. Sperber. 2014. "How Darwinian Is Cultural Evolution?" *Philosophical Transactions of the Royal Society B: Biological Sciences* 369 (1642): 20130368.

Corbey, R., A. Jagich, K. Vaesen, and M. Collard. 2016. "The Acheulean Handaxe: More Like a Bird's Song than a Beatles' Tune?" *Evolutionary Anthropology: Issues, News, and Reviews* 25 (1): 6–19.

Currie, T. E., M. Campenni, A. Flitton, T. Njagi, E. Ontiri, C. Perret, and L. Walker. 2021. "The Cultural Evolution and Ecology of Institutions." *Philosophical Transactions of the Royal Society B* 376 (1828): 20200047.

d'Andrade, R. G. 1984. "Cultural Meaning Systems." In *Behavioral and Social Science Research: A National Resource: Part II*, edited by R. A. Shweder and R. A. Le Vine, 197–236. Washington, DC: The National Academies Press. https://doi.org/10.17226/63.

d'Errico, F., and C. B. Stringer. 2011. "Evolution, Revolution or Saltation Scenario for the Emergence of Modern Cultures?" *Philosophical Transactions of the Royal Society B: Biological Sciences* 366 (1567): 1060.

de Queiroz, K. 2005. "Ernst Mayr and the Modern Concept of Species." *Proceedings of the National Academy of Sciences* 102 (S1): 6600–6607.

Deacon, T. W. 1997. *The Symbolic Species: The Co-evolution of Language and the Brain.* New York: W. W. Norton.

Derex, M., and A. Mesoudi. 2020. "Cumulative Cultural Evolution within Evolving Population Structures." *Trends in Cognitive Sciences* 24 (8): 654-667.

Du, A., A. M. Zipkin, K. G. Hatala, E. Renner, J. L. Baker, S. Bianchi, K. H. Bernal, et al. 2018. "Pattern and Process in Hominin Brain Size Evolution Are Scale-Dependent." *Proceedings of the Royal Society B: Biological Sciences* 285 (1873): 20172738.

Dusek, V. 2006. *Philosophy of Technology: An Introduction.* Malden, MA: Blackwell Publishing.

Ericsson, K. A., R. T. Krampe, and C. Tesch-Romer. 1993. "The Role of Deliberate Practice in the Acquisition of Expert Performance." *Psychological Review* 100 (3): 363–406.

Falk, D., C. Hildebolt, K. Smith, M. J. Morwood, T. Sutikna, P. Brown, Jatmiko, et al. 2005. "The Brain of LB1, Homo Floresiensis." *Science* 308 (5719): 242–245.

Foley, R. 1987. "Hominid Species and Stone-Tool Assemblages: How Are They Related?" *Antiquity* 61 (233): 380–392.

Foucault, M. 1988. *Technologies of the Self: A Seminar with Michel Foucault.* Amherst: University of Massachusetts Press.

Fragaszy, D. M., D. Biro, Y. Eshchar, T. Humle, P. Izar, B. Resende, and E. Visalberghi. 2013. "The Fourth Dimension of Tool Use: Temporally Enduring Artefacts Aid Primates Learning to Use Tools." *Philosophical Transactions of the Royal Society B: Biological Sciences* 368 (1630): 20120410.

Futuyma, D. J. 2017. "Evolutionary Biology Today and the Call for an Extended Synthesis." *Interface Focus* 7 (5): 20160145.

Gärdenfors, P., and A. Högberg. 2017. "The Archaeology of Teaching and the Evolution of *Homo docens*." *Current Anthropology* 58 (2): 188–208.

Gaynes, R. 2017. "The Discovery of Penicillin—New Insights after More than 75 Years of Clinical Use." *Emerging Infectious Diseases* 23 (5): 849–853.

Geertz, C. 1973. *The Interpretation of Cultures*. New York: Basic Books.

Giddens, A. 1976. *New Rules of Sociological Method: A Positive Critique of Interpretative Sociologies*. New York: Basic Books.

Gould, S. J. 2002. *The Structure of Evolutionary Theory*. Cambridge, MA: Harvard University Press.

Gould, S. J., and N. Eldredge. 1977. "Punctuated Equilibria: The Tempo and Mode of Evolution Reconsidered." *Paleobiology* 3 (2): 115–151.

Greenfield, P. M. 2003. "Historical Change, Cultural Learning, and Cognitive Representation in Zinacantec Maya Children." *Cognitive Development* 18 (4): 455.

Harmand, S., J. E. Lewis, C. S. Feibel, C. J. Lepre, S. Prat, A. Lenoble, X. Boës, et al. 2015. "3.3-Million-Year-Old Stone Tools from Lomekwi 3, West Turkana, Kenya." *Nature* 521 (7552): 310–315.

Hawkes, K. 2014. "Primate Sociality to Human Cooperation." *Human Nature* 25 (1): 28–48.

Hayden, B. 2015. "Insights into Early Lithic Technologies from Ethnography." *Philosophical Transactions of the Royal Society of London. Series B: Biological Sciences* 370 (1682): 20140356.

Henrich, J. 2004. "Demography and Cultural Evolution: How Adaptive Cultural Processes Can Produce Maladaptive Losses: The Tasmanian Case." *American Antiquity* 69 (2): 197–214.

Henrich, J. 2016. *The Secret of Our Success: How Culture Is Driving Human Evolution, Domesticating Our Species, and Making Us Smarter*. Princeton, NJ: Princeton University Press.

Heyes, C. M. 2018. *Cognitive Gadgets: The Cultural Evolution of Thinking*. Cambridge, MA: Harvard University Press.

Heyes, C. M. 2021. "Imitation." *Current Biology* 31:R1–R6.

Holloway, R. 1969. "Culture: A Human Domain." *Current Anthropology* 10:395–412.

Holloway, R. 1981. "Culture, Symbols and Human Brain Evolution: A Synthesis." *Dialectical Anthropology* 5:287–303.

Holloway, R. L., S. D. Hurst, H. M. Garvin, P. T. Schoenemann, W. B. Vanti, L. R. Berger, and J. Hawks. 2018. "Endocast Morphology of Homo Naledi from the Dinaledi Chamber, South Africa." *Proceedings of the National Academy of Sciences* 115 (22): 5738–5743.

Hrdy, S. B. 2009. *Mothers and Others: The Evolutionary Origins of Mutual Understanding*. Cambridge, MA: Harvard University Press.

Hughes, T. P. 1987. "The Evolution of Large Technological Systems." *The Social Construction of Technological Systems: New Directions in the Sociology and History of Technology* 82:51–82.

Isaac, G. L. 1976. "Stages of Cultural Elaboration in the Pleistocene: Possible Archaeological Indicators of the Development of Language Capabilities." *Annals of the New York Academy of Sciences* 280 (1): 275–288.

Isler, K., and C. P. van Schaik. 2014. "How Humans Evolved Large Brains: Comparative Evidence." *Evolutionary Anthropology: Issues, News, and Reviews* 23 (2): 65–75.

Kaplan, H., K. Hill, J. Lancaster, and A. M. Hurtado. 2000. "A Theory of Human Life History Evolution: Diet, Intelligence, and Longevity." *Evolutionary Anthropology: Issues, News, and Reviews* 9 (4): 156–185.

Karakostis, F. A., D. Haeufle, I. Anastopoulou, K. Moraitis, G. Hotz, V. Tourloukis, and K. Harvati. 2021. "Biomechanics of the Human Thumb and the Evolution of Dexterity." *Current Biology* 31 (6): 1317–1325.

Kendal, R. L., N. J. Boogert, L. Rendell, K. N. Laland, M. Webster, and P. L. Jones. 2018. "Social Learning Strategies: Bridge-Building between Fields." *Trends in Cognitive Sciences* 22 (7): 651–665.

Kolodny, O., N. Creanza, and M. W. Feldman. 2015. "Evolution in Leaps: The Punctuated Accumulation and Loss of Cultural Innovations." *Proceedings of the National Academy of Sciences* 112 (49): 6762–6769.

Kolodny, O., N. Creanza, and M. W. Feldman. 2016. "Game-Changing Innovations: How Culture Can Change the Parameters of Its Own Evolution and Induce Abrupt Cultural Shifts." *PLOS Computational Biology* 12 (12): e1005302.

Kuhn, S. L. 2021. *The Evolution of Paleolithic Technologies*. London: Routledge.

Laland, K. 2018. *Darwin's Unfinished Symphony: How Culture Made the Human Mind*. Princeton, NJ: Princeton University Press.

Laland, K. N., T. Uller, M. W. Feldman, K. Sterelny, G. B. Müller, A. Moczek, E. Jablonka, et al. 2015. "The Extended Evolutionary Synthesis: Its Structure, Assumptions and Predictions." *Proceedings of the Royal Society of London B: Biological Sciences* 282 (1813): 1–14, 20151019.

Lave, J., and E. Wenger. 1991. *Situated Learning: Legitimate Peripheral Participation*. Cambridge: Cambridge University Press.

Legare, C. H., and M. Nielsen. 2015. "Imitation and Innovation: The Dual Engines of Cultural Learning." *Trends in Cognitive Sciences* 19 (11): 688–699.

Lewis, H. M., and K. N. Laland. 2012. "Transmission Fidelity Is the Key to the Build-Up of Cumulative Culture." *Philosophical Transactions of the Royal Society of London. Series B: Biological Sciences* 367 (1599): 2171–2180.

Lew-Levy, S., A. Milks, N. Lavi, S. M. Pope, and D. E. Friesem. 2020. "Where Innovations Flourish: An Ethnographic and Archaeological Overview of Hunter-Gatherer Learning Contexts." *Evolutionary Human Sciences* 2: e31.

Lew-Levy, S., R. Reckin, N. Lavi, J. Cristóbal-Azkarate, and K. Ellis-Davies. 2017. "How Do Hunter-Gatherer Children Learn Subsistence Skills?" *Human Nature* 28 (4): 367–394.

Linares-Matás, G. J., and J. Clark. 2021. "Seasonality and Oldowan Behavioral Variability in East Africa." *Journal of Human Evolution* 164: 103070. https://doi.org/10.1016/j.jhevol.2021.103070.

Lycett, S., and J. Gowlett. 2008. "On Questions Surrounding the Acheulean 'Tradition.'" *World Archaeology* 40:295–315.

Macfarlane, A., and S. Harrison. 2000. "Technological Evolution and Involution: A Preliminary Comparison of Europe and Japan." In *Technological Innovation as an Evolutionary Process*, edited by J. Ziman, 77–89. Cambridge: Cambridge University Press.

Martin, G. 2000. "Stasis in Complex Artefacts." In *Technological Innovation as an Evolutionary Process*, edited by J. Ziman, 90–100. Cambridge: Cambridge University Press.

Marx, L. 1997. "Technology: The Emergence of a Hazardous Concept." *Social Research* 64 (3): 965–988.

Maynard Smith, J. 2000. "The Concept of Information in Biology." *Philosophy of Science* 67 (2): 177–194.

McShea, D. W., and R. N. Brandon. 2010. *Biology's First Law: The Tendency for Diversity and Complexity to Increase in Evolutionary Systems*. Chicago: University of Chicago Press.

Mesoudi, A. 2021. "Blind and Incremental or Directed and Disruptive? On the Nature of Novel Variation in Human Cultural Evolution." *American Philosophical Quarterly* 58 (1): 7–20.

Mesoudi, A., K. N. Laland, R. Boyd, B. Buchanan, E. Flynn, R. N. McCauley, J. Renn, et al. 2013. "The Cultural Evolution of Technology and Science." In *Cultural Evolution: Society, Technology, Language, and Religion*, edited by P. J. Richerson and M. H. Christiansen, 193–216. Cambridge, MA: MIT Press.

Mesoudi, A., and M. J. O'Brien. 2008. "The Learning and Transmission of Hierarchical Cultural Recipes." *Biological Theory* 3 (1): 63–72.

Mesoudi, A., and A. Thornton. 2018. "What Is Cumulative Cultural Evolution?" *Proceedings of the Royal Society B: Biological Sciences* 285 (1880): 20180712.

Miller, I. F., R. A. Barton, and C. L. Nunn. 2019. "Quantitative Uniqueness of Human Brain Evolution Revealed through Phylogenetic Comparative Analysis." *eLife* 8:e41250.

Mithen, S. 1999. "Imitation and Cultural Change: A View from the Stone Age, with Specific Reference to the Manufacture of Handaxes." In *Mammalian Social Learning: Comparative and Ecological Perspectives*, edited by H. O. Box and K. Rita Gibson, 389–413. Cambridge: Cambridge University Press.

Miu, E., N. Gulley, K. N. Laland, and L. Rendell. 2020. "Flexible Learning, Rather than Inveterate Innovation or Copying, Drives Cumulative Knowledge Gain." *Science Advances* 6 (23): eaaz0286.

Moigne, A.-M., P. Valensi, P. Auguste, J. García-Solano, A. Tuffreau, A. Lamotte, C. Barroso, et al. 2016. "Bone Retouchers from Lower Palaeolithic Sites: Terra Amata, Orgnac 3, Cagny-l'Epinette and Cueva del Angel." *Quaternary International* 409:195–212.

Morgan, T. J. H. 2016. "Testing the Cognitive and Cultural Niche Theories of Human Evolution." *Current Anthropology* 57 (3): 370–377.

Morgan, T. J. H., N. T. Uomini, L. E. Rendell, L. Chouinard-Thuly, S. E. Street, H. M. Lewis, C. P. Cross, et al. 2015. "Experimental Evidence for the Co-evolution of Hominin Tool-Making Teaching and Language." *Nature Communications* 6:6029.

Muthukrishna, M., M. Doebeli, M. Chudek, and J. Henrich. 2018. "The Cultural Brain Hypothesis: How Culture Drives Brain Expansion, Sociality, and Life History." *PLOS Computational Biology* 14 (11): e1006504.

Olsson, A., E. Knapska, and B. Lindström. 2020. "The Neural and Computational Systems of Social Learning." *Nature Reviews Neuroscience* 21 (4): 197–212.

Osiurak, F., and D. Heinke. 2018. "Looking for Intoolligence: A Unified Framework for the Cognitive Study of Human Tool Use and Technology." *American Psychologist* 73 (2): 169.

Pargeter, J., N. Khreisheh, J. J. Shea, and D. Stout. 2020. "Knowledge vs. Know-How? Dissecting the Foundations of Stone Knapping Skill." *Journal of Human Evolution* 145:102807.

Pargeter, J., N. Khreisheh, and D. Stout. 2019. "Understanding Stone Tool-Making Skill Acquisition: Experimental Methods and Evolutionary Implications." *Journal of Human Evolution* 133:146–166.

Patel, A. D. 2019. "Music as a Transformative Technology of the Mind: An Update." In *The Origins of Musicality*, edited by H. Honing, 113–126. Cambridge: MIT Press.

Pigliucci, M. 2009. "An Extended Synthesis for Evolutionary Biology." *Annals of the New York Academy of Sciences* 1168:218–228.

Pigliucci, M., and L. Finkelman. 2014. "The Extended (Evolutionary) Synthesis Debate: Where Science Meets Philosophy." *BioScience* 64 (6): 511–516.

Powell, A., S. Shennan, and M. G. Thomas. 2009. "Late Pleistocene Demography and the Appearance of Modern Human Behavior." *Science* 324 (5932): 1298–1301.

Powers, S. T., C. P. van Schaik, and L. Lehmann. 2016. "How Institutions Shaped the Last Major Evolutionary Transition to Large-Scale Human Societies." *Philosophical Transactions of the Royal Society B: Biological Sciences* 371 (1687): 20150098.

Pradhan, G. R., C. Tennie, and C. P. van Schaik. 2012. "Social Organization and the Evolution of Cumulative Technology in Apes and Hominins." *Journal of Human Evolution* 63 (1): 180–190.

Reeves, J. S., D. R. Braun, E. M. Finestone, and T. W Plummer. 2021. "Ecological Perspectives on Technological Diversity at Kanjera South." *Journal of Human Evolution* 158:103029.

Reynolds, P. C. 1993. "The Complementation Theory of Language and Tool Use." In *Tools, Language and Cognition in Human Evolution*, edited by K. R. Gibson and T. Ingold, 407–428. Cambridge: Cambridge University Press.

Režek, Ž., H. L. Dibble, S. P. McPherron, D. R. Braun, and S. C. Lin. 2018. "Two Million Years of Flaking Stone and the Evolutionary Efficiency of Stone Tool Technology." *Nature Ecology & Evolution* 2 (4): 628–633.

Richerson, P. J., and R. Boyd. 2005. *Not by Genes Alone: How Culture Transformed Human Evolution*. Chicago: Chicago University Press.

Roepstorff, A., J. Niewöhner, and S. Beck. 2010. "Enculturing Brains through Patterned Practices." *Neural Networks* 23 (8): 1051–1059.

Rogers, A. R. 1988. "Does Biology Constrain Culture?" *American Anthropologist* 90 (4): 819–831.

Roux, V. 2010. "Technological Innovations and Developmental Trajectories: Social Factors as Evolutionary Forces." In *Innovation in Cultural Systems: Contributions from Evolutionary Anthropology*, edited by M. J. O'Brien and S. J. Shennan, 217–234. Cambridge, MA: MIT Press.

Ruse, M. 1996. *Monad to Man: The Concept of Progress in Evolutionary Biology*. Cambridge, MA: Harvard University Press.

Scheffer, M., S. R. Carpenter, T. M. Lenton, J. Bascompte, W. Brock, V. Dakos, J. van de Koppel, et al. 2012. "Anticipating Critical Transitions." *Science* 338 (6105): 344–348.

Scott-Phillips, T., S. Blancke, and C. Heintz. 2018. "Four Misunderstandings about Cultural Attraction." *Evolutionary Anthropology: Issues, News, and Reviews* 27 (4): 162–173.

Shipton, C. B. Kersey. 2013. *A Million Years of Hominin Sociality and Cognition: Acheulean Bifaces in the Hunsgi-Baichbal Valley, India*. Oxford: Archaeopress.

Skinner, B. F. 2002. *Beyond Freedom and Dignity*. Cambridge: Hackett Publishing.

Sperber, D. 1996. *Explaining Culture: A Naturalistic Approach*. Oxford: Blackwell.

Sterelny, K. 2017. "Cultural Evolution in California and Paris." *Studies in History and Philosophy of Science Part C: Studies in History and Philosophy of Biological and Biomedical Sciences* 62:42–50.

Stiner, M. C., R. Barkai, and A. Gopher. 2009. "Cooperative Hunting and Meat Sharing 400–200 kya at Qesem Cave, Israel." *Proceedings of the National Academy of Sciences* 106 (32): 13207–13212.

Stout, D. 2002. "Skill and Cognition in Stone Tool Production: An Ethnographic Case Study from Irian Jaya." *Current Anthropology* 45 (3): 693–722.

Stout, D. 2011. "Stone Toolmaking and the Evolution of Human Culture and Cognition." *Philosophical Transactions of the Royal Society B: Biological Sciences* 366 (1567): 1050–1059.

Stout, D. 2013. "Neuroscience of Technology." In *Cultural Evolution: Society, Technology, Language, and Religion*, edited by P. J. Richerson and M. H. Christiansen, 157–173. Cambridge, MA: MIT Press.

Stout, D. 2017. "Comment." *Current Anthropology* 58 (5): 661–662.

Stout, D. 2018. "Human Evolution: History or Science?" In *Rethinking Human Evolution*, edited by J. H. Schwartz, 297–317. Cambridge, MA: MIT Press.

Stout, D. 2020. "Culture, Mind, and Brain in Human Evolution: An Extended Evolutionary Perspective on Paleolithic Toolmaking as Embodied Practice." In *Culture, Mind, and Brain: Emerging Concepts, Models, and Applications*, edited by L. J. Kirmayer, C. M. Worthman, S. Kitayama, R. Lemelson, and C. Cummings. Cambridge: Cambridge University Press.

Stout, D. 2021a. "The Cognitive Science of Technology." *Trends in Cognitive Sciences* 25 (11): 964–977.

Stout, D. 2021b. "What Is 'Cumulative' Evolution? Comment." *Current Anthropology* 62 (2): 230–231.

Stout, D., J. Apel, J. Commander, and M. Roberts. 2014. "Late Acheulean Technology and Cognition at Boxgrove, UK." *Journal of Archaeological Science* 41:576–590.

Stout, D., and E. E. Hecht. 2017. "Evolutionary Neuroscience of Cumulative Culture." *Proceedings of the National Academy of Sciences* 114 (30): 7861–7868.

Stout, D., and N. Khreisheh. 2015. "Skill Learning and Human Brain Evolution: An Experimental Approach." *Cambridge Archaeological Journal* 25 (4): 867–875.

Stout, D., M. J. Rogers, A. V. Jaeggi, and S. Semaw. 2019. "Archaeology and the Origins of Human Cumulative Culture: A Case Study from the Earliest Oldowan at Gona, Ethiopia." *Current Anthropology* 60 (3): 309–340.

Strachan, J. W. A., A. Curioni, M. D. Constable, G. Knoblich, and M. Charbonneau. 2021. "Evaluating the Relative Contributions of Copying and Reconstruction Processes in Cultural Transmission Episodes." *PLOS One* 16 (9): e0256901.

Strassberg, S. Saxton, and N. Creanza. 2021. "Cultural Evolution and Prehistoric Demography." *Philosophical Transactions of the Royal Society B: Biological Sciences* 376 (1816): 20190713.

Szathmary, E., and J. Maynard Smith. 1995. "The Major Evolutionary Transitions." *Nature* 374 (6519): 227–232.

Tennie, C., D. R. Braun, L. S. Premo, and S. P. McPherron. 2016. "The Island Test for Cumulative Culture in the Paleolithic." In *The Nature of Culture: Based on an Interdisciplinary Symposium "The Nature of Culture,"* *Tübingen, Germany*, edited by M. Haidle, N. Conard, and M. Bolus, 121–133. Dordrecht: Springer Netherlands.

Tennie, C., J. Call, and M. Tomasello. 2009. "Ratcheting Up the Ratchet: On the Evolution of Cumulative Culture." *Philosophical Transactions of the Royal Society B: Biological Sciences* 364 (1528): 2405–2415.

Tennie, C., L. S. Premo, David R. Braun, and S. P. McPherron. 2017. "Early Stone Tools and Cultural Transmission: Resetting the Null Hypothesis." *Current Anthropology* 58 (5): 652–654.

Tomasello, M. 1999. *The Cultural Origins of Human Cognition*. Cambridge, MA: Harvard University Press.

Tomasello, M., and I. Gonzalez-Cabrera. 2017. "The Role of Ontogeny in the Evolution of Human Cooperation." *Human Nature* 28 (3): 274–288.

Tomasello, M., A. C. Kruger, and H. H. Ratner. 1993. "Cultural Learning." *Behavioral and Brain Sciences* 16:495–552.

Toth, N., K. D. Schick, and S. Semaw. 2006. "A Comparative Study of the Stone Tool-Making Skills of *Pan*, *Australopithecus*, and *Homo sapiens*." In *The Oldowan: Case Studies into the Earliest Stone Age*, edited by N. Toth and K. D. Schick, 155–222. Gosport: Stone Age Institute Press.

Truskanov, N., and Y. Prat. 2018. "Cultural Transmission in an Ever-changing World: Trial-and-Error Copying May Be More Robust than Precise Imitation." *Philosophical Transactions of the Royal Society B: Biological Sciences* 373 (1743): 20170050.

Vaesen, K., and W. Houkes. 2021. "Is Human Culture Cumulative?" *Current Anthropology* 62 (2): 218–238.

Wadley, L., T. Hodgskiss, and M. Grant. 2009. "Implications for Complex Cognition from the Hafting of Tools with Compound Adhesives in the Middle Stone Age, South Africa." *Proceedings of the National Academy of Sciences* 106 (24): 9590–9594.

Washburn, S. L. 1960. "Tools and Human Evolution." *Scientific American* 203 (3): 3–15.

Welch, J. J. 2017. "What's Wrong with Evolutionary Biology?" *Biology & Philosophy* 32 (2): 263–279.

Wynn, T., and J. Gowlett. 2018. "The Handaxe Reconsidered." *Evolutionary Anthropology: Issues, News, and Reviews* 27 (1): 21–29.

Wynn, T., R. A. Hernandez-Aguilar, L. F. Marchant, and W. C. Mcgrew. 2011. "'An Ape's View of the Oldowan' Revisited." *Evolutionary Anthropology* 20 (5): 181–197.

III EXOGENOUS FACTORS OF TECHNICAL RIGIDITY AND FLEXIBILITY

9 How Variability, Predictability, and Harshness Shape Cognitive Flexibility

Sarah Pope-Caldwell

Introduction

Flexible problem-solving has long been revered as a hallmark of human ingenuity. Our global expansion and the sheer diversity of contemporaneous human cultures is a testament to our ability to flexibly adapt existing techniques and seek alternative strategies. Yet, humans' predilection for "sticking to what we know" is evident in a wide range of behaviors and biases. The question is not are humans flexible or rigid problem-solvers but rather when do the benefits of flexibly changing tack exceed the costs? The *constrained flexibility framework* (CFF) seeks to understand how variability, predictability, and harshness work in tandem to shape flexible versus inflexible strategy use. The CFF distinguishes between strategy switching that occurs in response to failure and strategy switching that involves searching for better alternatives, here referred to as responsive and elective flexibility, respectively. Efforts to understand human cognitive flexibility have almost exclusively focused on how well individuals switch when they have to, rather than when, and under what contexts, they choose to. Understanding the conditions in which changing strategies is most beneficial may help explain how our species strikes the balance between maintaining working strategies and finding or creating better ones.

Adaptive Flexibility

Francy Ntamboudila (2020) tells a Congolese fable of how the antelope, a weak but wily animal, became the king of the jungle. It had been decreed that whichever animal was strong enough to shoot an arrow through the trunk of the largest tree would win the throne. The elephant and the leopard, the strongest of the animals, practiced and practiced. But on the day of the competition, their arrows barely pierced its bark. They laughed when the antelope stepped up to take his turn. However, knowing that he would never be strong enough, the antelope had come up with a different strategy. Before the event, the antelope convinced his friend, the bee, to drill a hole through the tree in a place that only he would know. On the day of the competition, the antelope took aim and shot the arrow through the hole—thereby winning the throne. The leopard and the elephant failed because they relied solely on their might, whereas the antelope's flexible thinking was his true strength.

Flexible behavior is an adaptive response to dynamic environments. As challenges arise or opportunities shift, strategies must be updated in order to meet these changing demands. This chapter will focus on the cognitive mechanisms underlying flexible behavior, also known as cognitive flexibility. *Cognitive flexibility* is an individual's ability to adaptively select between known solutions and innovated or acquired novel solutions in response to relevant environmental changes (Laureiro-Martínez and Brusoni 2018; Pope 2018; Ueltzhöffer, Armbruster-Genç, and Fiebach 2015). In other words, cognitive flexibility is characterized by the contextually mediated optimal balance between repeating or returning to a familiar strategy and searching for or switching to an alternative, with suboptimal strategy-selection occurring as a result of either *inflexibility* (an inability to switch from one strategy to another) or *overflexibility* (an inability to maintain a working strategy).

Environments fluctuate predictably, with daily or seasonal cycles, as well as unpredictably, in response to stochastic events like earthquakes or cyclones. Moreover, an animal's lived environment is also a product of its local habitat, which may span multiple or even micro climates, and its ranging patterns, which might regularly extend into novel territory. Flexible behavior is especially important for long-lived, mobile species, like humans, who must cope with changing environments, both climatic and social, throughout their lifetimes (Dingemanse and Wolf 2010; Sol et al. 2016; van Schaik 2013; Vicente and Wang 1998). Indeed, humans exhibit an unprecedented degree of behavioral diversity (Fogarty, Creanza, and Feldman 2015; Hill, Barton, and Magdalena Hurtado 2009) that allows us, for better or for worse, to inhabit and modify over half of the Earth's landmass (Henrich and McElreath 2003; Vitousek et al. 1997), suggesting that, at least as a species, humans are quite capable of updating existing strategies to meet exogenous demands. However, on an individual level, humans' flexibility is not always evident.

Like the elephant and the leopard, humans are often blinded by familiar strategies or known solutions. In a classic example, children and adults who learned to solve a set of arithmetic "water jar" problems using a somewhat tedious four-step solution were later unable to replace this familiar strategy with a more efficient one-step alternative (Luchins 1942). Abraham Luchins tested thousands of Americans under various conditions and consistently found that people ignored the better alternative in favor of their learned strategy (Luchins 1942; Luchins and Luchins 1950). Termed *cognitive set*,[1] the propensity for familiar strategies to occlude other (even more efficient) alternatives has been demonstrated across a wide variety of contexts, including strategic reasoning (Bilalić, McLeod, and Gobet 2008), design and engineering (Chrysikou and Weisberg 2005; Jansson and Smith 1991), mathematics (Crooks and McNeil 2009; Lemaire and Leclère 2014; Wertheimer 1945), sequential problem-solving (Jacobson and Hopper 2019; Pope et al. 2015, 2019, 2020; Watzek, Pope, and Brosnan 2019), insight problem-solving (Hanus et al. 2011; Öllinger, Jones, and Knoblich 2008), and functional fixedness paradigms (Adamson 1952; Duncker and Lees 1945; German and Barrett 2005). This gives rise to a contradiction. On the one hand, humans are profoundly adaptive, the inventors and modifiers behind a technological revolution. On the other hand, we can be dismally conservative, either unwilling or unable to move beyond the tried and true. How is it that humans can be simultaneously flexible and inflexible? The answer lies in a different question altogether: *When do the benefits of being flexible exceed the costs?*

In this chapter, I begin by discussing the relative risks and benefits of repeating a current strategy versus switching to an alternative, while highlighting that the optimal balance between the two is dependent on exogenous demands. Next, building on existing theories of human cognitive evolution, I propose the CFF for how variability, predictability, and harshness might work in concert with one another to shape cognitive flexibility.

Costs and Benefits of Staying, Switching, and Searching

Why Change Strategies?

Changing strategies is beneficial when an alternative strategy is better than the current strategy. Under *forced-switch conditions*, when a current strategy no longer works, a shift in goal or approach is clearly adaptive because the value of the current strategy has become zero. Switching that occurs in response to strategy failure, or even predicted failure, will be henceforth referred to as *responsive flexibility* (see also Tenpas, Schweinfurth, and Call, this volume, who describe individuals' *innovative* behavior and groups' *advancement*). Strategies can become temporarily or even permanently ineffective for a number of reasons. Capabilities, goals, and opportunities shift throughout development and in response to exogenous changes in the environment, like predation or other mortality risks, seasonal or spatial fluctuations in weather and resource availability, as well as downstream effects of climatic variability or anthropogenic disturbance. When a current strategy stops working, adaptive behavior might entail the discovery or innovation of a new technique or falling back on another known strategy. The majority of human cognitive flexibility research makes use of forced-switch metrics, wherein responsive flexibility, or rather inflexibility, is measured by the extent to which individuals perseverate with an acquired strategy after it stops working (Doebel and Zelazo 2015; Meiran 2010; Monsell 2003; Sakai 2008; Zelazo 2004). Changing strategies under forced-switch conditions is considered an integral part of executive functioning, the suite of cognitive skills that regulate our thoughts and actions (Doebel and Zelazo 2015; Friedman et al. 2008; Miyake et al. 2000).

The other context in which flexible behavior can be beneficial is when a current strategy is or becomes less efficient than an alternative. Under *optional-switch conditions*, when a current strategy continues to work but alternatives are available, rather than avoiding failure, changing strategies can simply reduce inefficiency—but not always. Switching between known solutions or searching for new ones that occurs under optional-switch conditions will be henceforth referred to as *elective flexibility* (see also Tenpas, Schweinfurth, and Call, this volume, and their description of individuals' *creative* behavior and groups' *shifting*). Elective flexibility offers an important mechanism for understanding human ingenuity and cumulative culture, which necessarily builds on existing techniques. However, within psychology, flexibility under optional-switch conditions is rarely measured. Existing metrics include the water jar (described above) and other learned-sequence tasks (Luchins 1942; Luchins and Luchins 1950; Pope et al. 2020; Watzek, Pope, and Brosnan 2019) along with some task shifting (Ardiale and Lemaire 2012; Arrington and Logan 2004; Lemaire and Leclère 2014), token exchange (Hopper et al. 2011; van Leeuwen et al. 2013; van Leeuwen and Call 2017), and extractive foraging tasks (Davis et al. 2019; Jacobson and Hopper 2019; Price et al. 2009). However, the balance between exploiting useful strategies and exploring alternatives has been extensively studied within foraging (Charnov 1976; Cohen, McClure, and Yu 2007; Stephens and Krebs 1986) and decision-making research (Acuna and Schrater 2007;

Fischhoff and Broomell 2020; Gittins and Jones 1979; Payne, Bettman, and Johnson 1994; Peterson and Verstynen 2019), wherein it is measured using a range of reinforcement learning and sequential decision-making paradigms like patch-leaving and multiarmed bandit tasks (see Averbeck 2015 for a review). In optional-switch conditions, the benefit of selecting a better strategy is pitted against the computational and search efforts required to compare alternatives combined with the opportunity costs incurred by learning or switching delays.

Why Repeat a Strategy?

Failure to switch strategies occurs in many contexts and is not always maladaptive. In forced-switch conditions, *perseveration* is inflexibility arising from the continued use of a previously successful strategy, despite evidence that it no longer works (Floresco 2011; Schillemans 2011). Perseveration can be adaptive if the failure is transient or the causal mechanisms are unclear (i.e., the failure might have been a fluke). In optional-switch conditions, *conservatism* is the disinclination to explore or adopt novel strategies when a productive one is already known (Brosnan and Hopper 2014; Hrubesch, Preuschoft, and van Schaik 2009). When a familiar strategy still works, changing to another is only valuable if it leads to better outcomes. In other words, it would be maladaptive to switch if the chosen alternative was not more efficient than the current strategy—or if the efficiency benefit of the chosen alternative failed to exceed the time invested in finding, selecting, and honing it. Another benefit of inflexibility is that it can help maintain or refine useful strategies, reallocating the effort that might be spent searching toward skill practice and eventually specialization (Dingemanse and Wolf 2013). Additionally, learning and maintaining a new strategy is neurologically expensive (Laughlin and Sejnowski 2003). It seems likely that repeating a strategy is an adaptive default approach, but one that should be deviated from when the benefits of finding an alternative exceed the costs (Duckworth 2010).

Elective Flexibility under Ambiguity

When the outcomes of all possible strategies are known, optimal behavior is simply a matter of switching to or maintaining the best one. However, we do not occupy nor have we evolved within unambiguous environments. Under conditions of uncertainty, the relative risks and rewards of potential strategies are often unclear. One could search for all possible strategies and test their payouts, but the time and effort spent gathering these data may quickly outweigh the benefit of using the better alternative, at least in the short-term. Heuristics, such as "always choose the second-cheapest wine" (McFadden, Machina, and Baron 1999), are guiding principles that support people's decision-making under uncertainty, often by systematically ignoring some of the available information (Gigerenzer, Hertwig, and Pachur 2011). The computational requirements of perfect decision-making are immense, and heuristics can be useful cognitive shortcuts, but they lead to predictable biases and errors (Kahneman et al. 1982; Tversky and Kahneman 1974), especially when cognitive capacity is limited (Cash-Padgett and Hayden 2019). Behavior guided by heuristics is not often optimal, but on average, it should reach an acceptable level of efficiency, at least within the environment in which it was formed (Fawcett, Hamblin, and Giraldeau 2013). Under conditions of ambiguity, elective flexibility (i.e., the decision to either repeat a strategy or find or switch to another) is guided by heuristics that are, ideally, tuned to optimize decision-making in that environment

(Todd and Gigerenzer 2007; Vicente and Wang 1998). Thus, to understand how humans balance rigidity versus flexibility, we must first consider how exogenous pressures, like variability, predictability, and harshness, shape both responsive and elective flexibility.

The Constrained Flexibility Framework

When considering how exogenous forces shape cognitive flexibility, we can make use of existing hypotheses aimed at understanding the evolutionary origins of adaptive cognition (see box 9.1). The specific catalysts that gave rise to human cognition is a topic of considerable debate, but the most prominent hypotheses posit that our perhaps uniquely adaptive cognition was heavily influenced by environmental *variability*, the extent to which the environment changes over time or space; *predictability*, the temporal regularity of changes or the degree to which they are correlated; and *harshness*, which refers to exposure to factors that increase morbidity and mortality or the level of consequence elicited by strategy failure (Riotte-Lambert and Matthiopoulos 2020; Young, Frankenhuis, and Ellis 2020). The hypotheses described in box 9.1 offer important perspectives; however, they either conflate variability, predictability and harshness or consider their influence in isolation. Here, I propose that recognizing how variability, predictability, and harshness work in concert with one another to regulate responsive and elective flexibility is integral to predicting when strategies should be rigidly maintained or when they should be flexibly adjusted.

The CFF unifies existing hypotheses regarding the impacts of exogenous variability, predictability, and harshness on adaptive behavior, taking into account their unique and combined effects in order to predict the circumstances under which elective and responsive flexibility are most beneficial. Recall that responsive flexibility consists of switching as a result of strategy failure, and elective flexibility is switching that occurs despite already having a working strategy. Elective flexibility is risky, but here I argue that under certain circumstances, the adaptive benefit of potentially finding or innovating a better solution is enough to offset the costs. The CFF posits that (1) responsive flexibility is adaptive whenever failure is reliably detected or predicted, regardless of environmental variability or harshness, and (2) elective flexibility is suppressed in harsh, stable, or predictably variable environments but may be a valuable tool in unpredictable environments, so long as harshness is low.

Specifically, in stable environments, the usefulness of a strategy does not change over time, so responsive flexibility is not needed and elective flexibility is not pragmatic. Once an optimal strategy is found, there is little benefit derived from maintaining, using, or seeking alternatives. Thus, under stable conditions, both responsive and elective flexibility are maladaptive—and should be suppressed—so long as a good strategy can be maintained.

In variable environments, the usefulness of a strategy can either change regularly, as in predictably variable environments, or irregularly, as in unpredictably variable environments. In variable environments, regardless of harshness, responsive flexibility is critical for handling situations when a previously effective strategy stops (or will soon stop) working. However, the conditions that predict elective flexibility are more complex. In predictably variable environments, such as those governed by reliable seasonal changes, elective flexibility is not useful because even though optimal strategies rotate over time, the set of useful strategies is unchanged. Thus, cognitive resources should be devoted to detecting or predicting when changes occur and to maintaining and honing one's repertoire of working strategies, rather than searching for alternatives. For example, to compensate for predictable traffic jams during rush hour, one might switch from a normal route to an alternate. This is an example of responsive flexibility between two known strategies.

In unpredictably variable environments, previous strategies may or may not work well or even at all. Here, increased elective flexibility might serve to minimize delays in updating to new optimal strategies or, in some cases, result in the discovery or innovation of a novel, more efficient technique. To elaborate on our driving example, if traffic patterns are unpredictable, perhaps because of frequent accidents, one may need to reroute at any

Box 9.1.
Hypotheses regarding Environmental Impacts on Adaptive Cognition

The idea that environmental challenges, such as variability, predictability, and harshness, act as major selection forces in shaping cognition is certainly not new. Many researchers have proposed a role of exogenous variation, or novelty, in cognitive evolution. For example, according to the *behavioral drive* hypothesis (Wyles, Kunkel, and Wilson 1983), expansion into novel habitats requires larger brains that are capable of handling the accompanying challenges, such as locating, identifying, and procuring unknown resources or avoiding risks that stem from novel climates, landscapes, and predators. And the *adaptive flexibility* hypothesis proposed that an initial period of high behavioral diversity is essential to adjusting to a novel habitat, with a subsequent decline as successful behavioral variants are honed (Wright et al. 2010).

Other hypotheses have focused on the role of encephalization in adaptive behavior. For instance, the *cognitive buffer* hypothesis (Allman, McLaughlin, and Hakeem 1993; Deaner, Barton, and van Schaik 2003) and the *brain size–environmental* hypothesis (Sol et al. 2005) suggest that increased cognitive capacity, supported by larger brains, acts as a buffer or as a means of handling changing environments. To support these claims, there is evidence that in primates, carnivorous mammals, and birds, neocortex size correlates with rates of innovative problem-solving (Benson-Amram et al. 2016; Lefebvre, Reader, and Sol 2004; Reader and Laland 2002; Reader and MacDonald 2003).

The role of harsh environments in promoting adaptive change (reviewed by van de Pol et al. 2017) is prominent in many well-known hypotheses aimed at understanding animal behavior and speciation, including *habitat theory* (Vrba 1992, 1995), the *savanna* hypothesis (Dart and Salmons 1925; deMenocal 1995), and *niche construction* (Odling-Smee 1988; also see Laland, Matthews, and Feldman 2016). For example, in the *necessity is the mother of invention* hypothesis (Laland and Reader 1999; Reader and Laland 2001), innovative adaptations are thought to only arise when needed. However, this stands in opposition to the *spare time* hypothesis, which suggests that innovation occurs in times of low stress because of the lower consequences of failure (Kummer and Goodall 1985).

Taking into account the role of environmental variation rather than just extremes, the *variability selection* hypothesis (Potts 1996) suggests that in response to fluctuations in temperature, aridity, and water abundance, early hominins developed a suite of cognitive traits that allowed them to not only move to more suitable environments but also develop flexible approaches to buffer local environmental instability (Grove 2011; see also Potts 2012; Boyd and Richerson 2005, 66–82; Richerson and Boyd 2013). And in support of this, through a combined analysis of paleoenvironmental and archaeological records of the last 5 million years, Richard Potts and J. Tyler Faith (2015) demonstrate considerable overlap between periods of high environmental variability and key behavioral adaptations and speciation events in the evolution of Hominins (see also Snell-Rood 2013).

point. Here, it is beneficial to have many alternative routes to select from. Elective flexibility, or trying alternatives before they are needed, serves to increase familiarity with available strategies. Thus, in unpredictably variable environments, elective flexibility minimizes search costs, allowing responsive flexibility to be more efficient.

Importantly, both predictably and unpredictably variable environments can exist partly or wholly in states of harshness, which places a penalty on elective flexibility commensurate with the severity of the consequences of failure. If you had an important meeting to get to, you should postpone trying an unknown route, which could result in significant delay, until conditions are more forgiving. By contrast, responsive flexibility is indifferent to harshness, as it only occurs when a previous strategy stops (or will soon stop) working, and thus no amount of harshness will outweigh the benefit of switching to or searching for an alternative. For example, if your normal route is suddenly impeded, it is better to use an alternative and risk delay than to not arrive at all.

The following sections will describe existing evidence supporting the CFF. See figure 9.1 for a graphic representation of these dynamics.

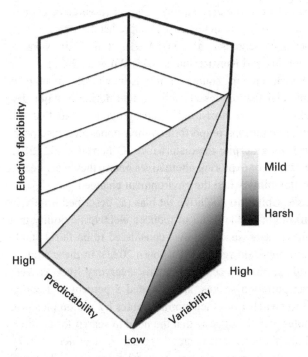

Figure 9.1
Putative impacts of variability, predictability, and harshness on elective flexibility as described by the CFF. The height of the object (z-axis) corresponds to the predicted likelihood of elective flexibility occurring. As variability increases (x-axis) from low to high or from stable to variable, elective flexibility also increases because the relative value of a current strategy is more likely to change. However, this effect is mediated by predictability (y-axis), such that unpredictable environments require more elective flexibility because the best strategy at any given point is unclear. In high-predictability environments, resources are better spent honing a repertoire of best strategies and recognizing when to switch between them. Harshness, which is indicated by gradient intensity, suppresses elective flexibility to limit the risk (and therefore the consequences) of failure at all levels of variability and predictability. *Source*: Provided by author.

The CFF and Cognitive Flexibility

In an uncertain and changing environment, where values of all potential options are unknown or the values of these options change over time, one must adapt by flexibly alternating between exploration and exploitation in order to maintain efficient performance over time and to keep track of the state of the environment (Addicott et al. 2017, 1932).

Predictable versus Unpredictable Environments

Both predictable and unpredictable variation are common in the natural world. Predictable variation can be seen in the many cycles that govern our climate on daily, monthly, and annual bases. Seasonal fluctuations, especially, might be drastic and require a range of behaviors throughout the year. However, seasonal transitions occur smoothly and predictably. Louise Riotte-Lambert and Jason Matthiopoulos (2020) suggest that when environmental predictability is high, learning and memory are advantageous because they allow existing strategies to be recalled and activated when needed, reducing the need for strategy search and its accompanying costs (also see Colwell 1974; Milton 1981). Indeed, several decision-making studies, which required participants to either exploit a current resource or explore alternatives, found increased exploratory behavior in response to unpredictable or stochastic, compared with stable, reward structures (Behrens et al. 2007; Meder et al. 2020; Navarro, Newell, and Schulze 2016; Speekenbrink and Konstantinidis 2015; Wu et al. 2021).

Even within mostly predictable environments, some volatility is inevitable. An analysis of the positional records of commercial fishing vessels showed that during a major disturbance, which resulted in the closure of the most-used fishing grounds, boats that had explored more waters during normal conditions outperformed their nonexploratory peers and were more likely to continue fishing despite the disturbance (O'Farrell et al. 2019). In other words, higher elective flexibility, or exploring alternatives before they were necessary, resulted in better responsive flexibility when the environment changed.

It seems likely that humans' susceptibility to cognitive set bias (as described briefly in this chapter's introduction section) might be the result of reduced elective flexibility that occurs when the context is—in these cases, inaccurately—considered to be fairly stable (Bilalić et al. 2008; Luchins 1942; Pope-Caldwell and Washburn 2022). In these studies, participants underexplored, and this seems to occur in humans' everyday lives as well. Following two days of public transportation strikes, an estimated 5 percent of London commuters, who had been forced to switch from their normal routes by station closures, did not return to their previous routes, which suggests that the need to search for an alternative strategy resulted in their finding a better one (Larcom, Rauch, and Willems 2017).

Inaccurate stability judgments may also help explain cross-cultural differences in humans' susceptibility to cognitive set. We (Pope et al. 2019) found that compared with their American study subjects, Namibian Himba participants were approximately four times more likely to use a more efficient "shortcut" strategy than Americans on a touch-screen task. It is possible that unfamiliarity with computer games led Himba participants to consider the game less predictable than their American counterparts. However, another intriguing possibility is that if decision-making heuristics are somewhat domain-general, unpredictability in other areas of the Himba participants' lives might have resulted in a lowered threshold for elective flexibility. This possibility aligns with other research reporting that younger children exhibited increased exploration and elective flexibility compared with adolescents and adults (German and Defeyter 2000; Gopnik, Griffiths, and Lucas 2015; Pope et al. 2015). Indeed, young children may be more prone to trying alternatives as they are in a constant state of strategy acquisition, wherein predictive decision-making, based on reasoning and executive functioning, may still be out of reach (see Gopnik 2020; Ionescu 2017).

Harsh Environments

Elective flexibility is a valuable tool for handling environmental changes, especially those that are unpredictable, but it comes with a price. Searching for another strategy can take time and may result in failure. Even if a viable alternative is found, there are likely to be learning delays before it can be enacted, or else the attempts to learn it could fail. Finally, the net benefit of using the alternative strategy might be lower than the previous strategy, based on the time invested in finding, learning, and using it.

In harsh environments, the consequences of failure are high. This may be due to deficits in the quantity or quality of resources, increased risk of morbidity or mortality, decreased time or energy, or some combination of these factors. The CFF asserts that when the stakes are high, harshness places severe constraints on elective flexibility. For example, in a modified "water jar" paradigm, Sian L. Beilock and Maci S. Decaro (2007) found that participants with higher working memory were more likely to use the suboptimal four-step solution under conditions of high stress. Similarly, Wilson and colleagues (2014) found that humans decrease their information-seeking behavior when time horizons (the maximum number of choices or task duration) are short. Additionally, Bruno B. Averbeck (2015) reviews a number of decision-making tasks demonstrating that search is less beneficial when time horizons are short and the consequences of failure are high. Thus, only responsive flexibility should occur in harsh conditions and only when it is absolutely necessary.

It follows that strategy accumulation may be most beneficial in variable environments that are redundant exposed to harshness. Merideth A. Addicott and colleagues (2017) suggested that increased exploration and strategy accrual during periods of moderation, when the benefit of a possible alternative exceed the search costs, might result in useful backup strategies that could later be deployed during periods of harshness. They even expanded this idea to account for risky behavior during human adolescence, noting that it likely provides valuable information that can be used later in adulthood (Mata, Wilke, and Czienskowski 2013). For example, hunter-gatherer toolkits are significantly predicted by proxies for risk of resource failure, suggesting that the accumulation of more tools serves to buffer harshness (Collard, Kemery, and Banks 2005; Collard et al. 2011).

Thus, in alignment with the "necessity is the mother of invention" hypothesis (Laland and Reader 1999; Reader and Laland 2001), the CFF predicts that when no other options are available, responsive flexibility should kick in, even in harsh environments. However, consistent with the "spare time" hypothesis (Kummer and Goodall 1985), it notes that elective flexibility during periods of relative calm can also provide fitness benefits, even at later times. Elective flexibility is a powerful but costly tool. Next, we look at the ways in which the price of finding and flexibly adopting an alternative strategy can be mitigated.

Mitigating Costs of Flexibility

Optimizing Search Tactics

Flexibility, whether responsive or elective, often requires searching through potential alternatives. The hunt for a good alternative strategy can be guided by previous experience or the lack thereof, or it can be an unguided, random sampling. Generalizing from previous experiences decreases the risk of failure and, ideally, concentrates search efforts into most likely avenues; whereas directed exploration prioritizes the gathering of missing information with the aim of uncovering options that would be otherwise overlooked. Both generalization and directed exploration are computationally complex, requiring one to keep track of what is known and what is not known and to make step-by-step decisions about how to update and use this information each time a new path is sampled. Random exploration bypasses calculation costs but also misses the benefits of searching among alternatives that are more likely to be advantageous.

Humans' use of random versus directed exploration is strongly tied to age and context. Young children exhibit high rates of both random and directed exploration (Meder et al. 2020). Although random exploration declines sharply with age, children's directed exploration remains higher than adults' until at least 11 years of age (Schulz et al. 2019). In other words, children try to fill in missing pieces of information rather than explore options similar to their previously successful strategies. By adulthood, exploration is reduced, and generalization becomes a primary search tactic (Wu et al. 2018). One explanation for this trajectory is that random exploration is most useful when cognitive capacities are limited, such as in early childhood, but it is quickly replaced by more complex search tactics as the required calculations become cognitively feasible (Gopnik 2020; Gopnik et al. 2017). Search tactics are also tuned to context. For example, Robert C. Wilson and colleagues (2014) found that adults used directed exploration more than random exploration when they were given fewer decision opportunities, suggesting that when each decision carries more weight, adults capitalized on the less risky directed approach. Thus, selectively applying generalization and directed and random exploration to meet both endogenous and exogenous constraints can be a valuable tool for mitigating the costs of both responsive and elective flexibility.

Socially Acquired Strategies

In humans, perhaps the most prevalent means of reducing flexibility costs is by copying successful strategies from other individuals. Indeed, our ability to extract useful skills and information from social partners is thought to have shaped human life history, which seems tuned to benefit from cultural traditions (Kaplan et al. 2000; Richerson and Boyd 2020). Social learning is especially valuable when the costs of individual search are high. For example, studies find that learning from others is advantageous when the to-be-acquired technique is complex (Caldwell, Renner, and Atkinson 2017; Derex, Godelle, and Raymond 2013), and copying fidelity increases when actions are mechanistically or conceptually unclear, as is common when behaviors are guided by contextual or conventional constraints (Clegg and Legare 2016; Froese and Leavens 2014). Another consideration is the quality of the information that can be gleaned from others.

Socially acquired strategies are most beneficial when social partners' strategies are reliably successful. In other words, learning from others is only helpful when they have useful strategies to share (see Boyd and Richerson 2005, pt. 1, for an extended discussion). For example, children prefer to copy older, presumably more experienced, demonstrators when learning a novel technique (see Wood, Kendal, and Flynn 2013b whom do a review). Additionally, a recent interactive foraging study found that adults observed each other's foraging choices more when rewards were clumped and therefore predictive of other rewards in the area, rather than when they were randomly dispersed (Wu et al. 2021).

Humans often prioritize social information over individual learning, likely because the costs and benefits of individual exploration are often unknown and are therefore eclipsed by the low cost and known benefit of a socially acquired strategy. Yet, relying too heavily on social learning can be costly if the environment in which a strategy was originally formed has now changed (Boyd, Richerson, and Henrich 2011; Lehmann, Feldman, and Kaeuffer 2010) or if the learned strategy is itself suboptimal. Research on humans' proclivity for "over-imitation" finds that both children and adults incorporate even clearly irrelevant actions when copying a demonstrated technique, leading to marked inefficiencies (Haun, Rekers, and Tomasello 2014; Hoehl et al. 2019; Horner and Whiten 2005; McGuigan, Makinson, and Whiten 2011). Furthermore, Elizabeth Bonawitz and colleagues (2011) found that children who were explicitly taught how to play with a novel toy were far less likely to explore and discover untaught functions of the toy than naive peers. Thus, social learning can be an effective way to mitigate the costs of flexibility, but it is most adaptive when used in concert with individual learning (Acerbi and Parisi 2006; Boyd, Richerson, and Henrich 2011; Boyd and Richerson 1995; Legare and Nielsen 2015; Muthukrishna 2015; Muthukrishna et al. 2018; Wood et al. 2013a).

The value of social learning is also inextricable from one's social context more generally. Cultural factors and norms affect the extent to which social information is sought (Glowacki and Molleman 2017; Stengelin, Hepach, and Haun 2019, 2020; van Leeuwen et al. 2018) and how it is transmitted (Boyette and Hewlett 2017; Clegg et al. 2020; Kline 2017; Lew-Levy, Crittenden, et al. 2019; Lew-Levy, Kissler, et al. 2019). Conventions might also dictate the "best" strategy in a given circumstance, regardless of actual effectivity (Keupp et al. 2015; Nielsen, Cucchiaro, and Mohamedally 2012). Switching strategies to match a group can result in positive social outcomes (Over and Carpenter 2013), whereas defecting from a common practice might have negative consequences for relationships or status (Watson-Jones et al. 2014). Thus, acquiring strategies through social learning comes with its own motivations and complications (see Tenpas, Schweinfurth, and Call, this volume, for a discussion on how the social environment biases cultural traits).

The CFF and Mitigating Costs of Flexibility

The CFF predicts that elective flexibility should be reduced when the consequences of failure are high. However, as described above, the costs of switching can be reduced under certain circumstances, sparking the question of whether the degree of harshness (and not just variability and predictability) in an environment predicts reliance on specific search tactics and socially acquired information. In a modeling study, Jesse Fenneman and Willem E. Frankenhuis (2020) found that higher impulsivity—choosing an action without incurring the costs (or benefits) of sampling it beforehand—was most adaptive when resource quality was low and high but not moderate. One explanation is that when harshness is high, computationally difficult search tactics are too costly and therefore suppressed; and when harshness is low, the price of sampling alternatives does not outweigh, at least in their experimental design, the now-marginal benefits. But what about when sampling costs are eradicated? Social learning solves many of the problems of being flexible under harsh conditions. The risks of failure are low because a strategy's effectiveness is easily ascertained. In humans, teaching facilitates learning by reducing the time and energy required to pick up a new skill or technique (Caldwell, Renner, and Atkinson 2017; Csibra and Gergely 2011; Kline 2017). In effect, social learning may be the golden ticket to flexibly upgrading strategies, even in harsh environments—so long as the strategy itself is still useful (Nakahashi 2007; Richerson and Boyd 2020). Future research is needed to determine how and to what extent mitigating the costs of switching strategies supports the use of elective flexibility in harsh conditions.

Concluding Remarks

In reality, environments exist along a continuum—from stable to variable, predictable to unpredictable, mild to harsh. Humans occupy a range of ecological and cultural environments simultaneously, and the concepts of variability, predictability, and harshness depend on the domain and timescale being considered. This chapter has focused on cognitive flexibility as it occurs within an individual's lifetime. However, the CFF may also be useful in deciphering how the accumulation and modification of strategies and techniques evolve within populations in response to environmental pressures. Furthermore, exposure to environmental pressures is confounded by our behaviors, such as movement patterns or our ability to buffer variability—for example, niche construction (see Chevin and Hoffmann 2017 for a discussion). By differentiating between responsive flexibility (strategy changes that occur in response to failure) and elective flexibility (strategy changes that occur proactively), the CFF provides a theoretical basis for understanding how our species balances the need to rigidly maintain techniques that work with the potential risks and benefits of flexibly sampling alternatives.

Acknowledgments

Enormous thanks to Daniel Haun, Dustin Eirdosh, Susan Hanisch, Sheina Lew-Levy, Adam Howell Boyette, and Karri Neldner, and to the editor and other chapter authors, for their helpful comments and feedback throughout the writing of this chapter. Also, thanks to Bienvenue Mbongo, Ardain Dzabatou, Francy Ntamboudila, Vidrige Kandza, and Nyambe Emmanuel for sharing and bringing to life such truly amazing Congolese fables.

Note

1. Luchins (1942) referred to cognitive set as "einstellung." Another common term is conservatism, which is the continued use of a strategy without consideration for alternatives (Brosnan and Hopper 2014; Davis et al. 2019; Hrubesch, Preuschoft, and van Schaik 2009).

References

Acerbi, A., and D. Parisi. 2006. "Cultural Transmission between and within Generations." *Journal of Artificial Societies and Social Simulation* 9 (1): 1–16.

Acuna, D., and P. Schrater. 2008. "Bayesian Modeling of Human Sequential Decision-Making on the Multi-Armed Bandit Problem." In Proceedings of the 30th annual conference of the cognitive science society (Vol. 100, pp. 200-300). Washington, DC: Cognitive Science Society.

Adamson, R. E. 1952. "Functional Fixedness as Related to Problem Solving: A Repetition of Three Experiments." *Journal of Experimental Psychology* 44 (4): 288–291.

Addicott, M. A., J. M. Pearson, M. M. Sweitzer, D. L. Barack, and M. L. Platt. 2017. "A Primer on Foraging and the Explore/Exploit Trade-Off for Psychiatry Research." *Neuropsychopharmacology* 42 (10): 1931–1939.

Allman, J., T. McLaughlin, and A. Hakeem. 1993. "Brain Weight and Life-Span in Primate Species." *Proceedings of the National Academy of Sciences* 90 (1): 118–122.

Ardiale, E., and P. Lemaire. 2012. "Within-Item Strategy Switching: An Age of Comparative Study in Adults." *Psychology and Aging* 27 (4): 1138–1151.

Arrington, C. M., and G. D. Logan. 2004. "The Cost of Voluntary Task Switch." *Psychological Science* 15 (9): 610–615.

Averbeck, B. B. 2015. "Theory of Choice in Bandit, Information Sampling and Foraging Tasks." *PLOS Computational Biology* 11 (3): 1–28.

Behrens, T. E. J., M. W. Woolrich, M. E. Walton, and M. F. S. Rushworth. 2007. "Learning the Value of Information in an Uncertain World." *Nature Neuroscience* 10 (9): 1214–1221.

Beilock, S. L., and M. S. Decaro. 2007. "From Poor Performance to Success under Stress: Working Memory, Strategy Selection, and Mathematical Problem Solving under Pressure." *Journal of Experimental Psychology: Learning, Memory, and Cognition* 33 (6): 983–998.

Benson-Amram, S., B. Dantzer, G. Stricker, E. M. Swanson, and K. E. Holekamp. 2016. "Brain Size Predicts Problem-Solving Ability in Mammalian Carnivores." *Proceedings of the National Academy of Sciences* 113 (9): 2532–2537.

Bilalić, M., P. McLeod, and F. Gobet. 2008. "Why Good Thoughts Block Better Ones: The Mechanism of the Pernicious Einstellung (Set) Effect." *Cognition* 108 (3): 652–661.

Bonawitz, E., P. Shafto, H. Gweon, N. D. Goodman, E. Spelke, and L. Schulz. 2011. "The Double-Edged Sword of Pedagogy: Instruction Limits Spontaneous Exploration and Discovery." *Cognition* 120 (3): 322–330.

Boyd, R., and P. J. Richerson. 1995. "Why Does Culture Increase Human Adaptability?" *Ethology and Sociobiology* 16:125–143

Boyd, R., and P. J. Richerson. 2005. *The Origin and Evolution of Cultures.* Oxford: Oxford University Press.

Boyd, R., P. J. Richerson, and J. Henrich. 2011. "The Cultural Niche: Why Social Learning Is Essential for Human Adaptation." *Proceedings of the National Academy of Sciences* 108 (S2): 10918–10925.

Boyette, A. H., and B. S. Hewlett. 2017. "Autonomy, Equality, and Teaching among Aka Foragers and Ngandu Farmers of the Congo Basin." *Human Nature* 28 (3): 289–322.

Brosnan, S. F., and L. M. Hopper. 2014. "Psychological Limits on Animal Innovation." *Animal Behaviour* 92:325–332.

Caldwell, C. A., E. Renner, and M. Atkinson. 2017. "Human Teaching and Cumulative Cultural Evolution." *Review of Philosophy and Psychology* 9:751-770.

Cash-Padgett, T., and B. Hayden. 2019. "Overstaying in Patchy Foraging Can Be Explained by Behavioral Variability." *bioRxiv* (December): 868596.

Charnov, E. L. 1976. "Optimal Foraging, the Marginal Value Theorem." *Theoretical Population Biology* 9 (2): 129–136.

Chevin, L.-M., and A. A. Hoffmann. 2017. "Evolution of Phenotypic Plasticity in Extreme Environments." *Philosophical Transactions of the Royal Society B: Biological Sciences* 372 (1723): 20160138.

Chrysikou, E. G., and R. W. Weisberg. 2005. "Following the Wrong Footsteps: Fixation Effects of Pictorial Examples in a Design Problem-Solving Task." *Journal of Experimental Psychology: Learning, Memory, and Cognition* 31 (5): 1134–1148.

Clegg, J. M., and C. H. Legare. 2016. "Instrumental and Conventional Interpretations of Behavior Are Associated with Distinct Outcomes in Early Childhood." *Child Development* 87 (2): 527–542.

Clegg, J. M., N. J. Wen, P. H. DeBaylo, A. Alcott, E. C. Keltner, and C. H. Legare. 2020. "Teaching through Collaboration: Flexibility and Diversity in Caregiver–Child Interaction across Cultures." *Child Development* 92 (1): e56–e75.

Cohen, J. D., S. M. McClure, and A. J. Yu. 2007. "Should I Stay or Should I Go? How the Human Brain Manages the Trade-off between Exploitation and Exploration." *Philosophical Transactions of the Royal Society B: Biological Sciences* 362 (1481): 933–942.

Collard, M., B. Buchanan, A. Ruttle, and M. J. O'Brien. 2011. "Niche Construction and the Toolkits of Hunter-Gatherers and Food Producers." *Biological Theory* 6 (3): 251–259.

Collard, M., M. Kemery, and S. Banks. 2005. "Causes of Toolkit Variation among Hunter-Gatherers: A Test of Four Competing Hypotheses." *Canadian Journal of Archaeology* 29 (1): 1–19.

Colwell, R. K. 1974. "Predictability, Constancy, and Contingency of Periodic Phenomena." *Ecology* 55 (5): 1148–1153.

Crooks, N. M., and N. M. McNeil. 2009. "Increased Practice with 'Set' Problems Hinders Performance on the Water Jar Task." *Proceedings of the 31st Annual Conference of the Cognitive Science Society* 31:643–648.

Csibra, G., and G. Gergely. 2011. "Natural Pedagogy as Evolutionary Adaptation." *Philosophical Transactions of the Royal Society B: Biological Sciences* 366 (1567): 1149–1157.

Dart, R.. 1925. "Australopithecus Africanus: The Man-Ape of South Africa."*Nature* 115: 195–199.

Davis, S. J., S. J. Schapiro, S. P. Lambeth, L. A. Wood, and A. Whiten. 2019. "Behavioral Conservatism Is Linked to Complexity of Behavior in Chimpanzees (*Pan troglodytes*): Implications for Cognition and Cumulative Culture." *Journal of Comparative Psychology* 133 (1): 20.

Deaner, R. O., R. A. Barton, and C. van Schaik. 2003. "Primate Brains and Life Histories: Renewing the Connection." In *Primates Life Histories and Socioecology*, edited by P. M. Kappeler and M. E. Pereira, 233–265. Chicago: University of Chicago Press.

deMenocal, P. B. 1995. "Plio-Pleistocene African Climate." *Science* 270 (5233): 53–59.

Derex, M., B. Godelle, and M. Raymond. 2013. "Social Learners Require Process Information to Outperform Individual Learners." *Evolution* 67 (3): 688–697.

Dingemanse, N. J., and M. Wolf. 2010. "Recent Models for Adaptive Personality Differences: A Review." *Philosophical Transactions of the Royal Society B: Biological Sciences* 365 (1560): 3947–3958.

Dingemanse, N. J., and M. Wolf. 2013. "Between-Individual Differences in Behavioural Plasticity within Populations: Causes and Consequences." *Animal Behaviour* 85 (5): 1031–1039.

Doebel, S., and P. D. Zelazo. 2015. "A Meta-Analysis of the Dimensional Change Card Sort: Implications for Developmental Theories and the Measurement of Executive Function in Children." *Developmental Review* 38: 241–268.

Duckworth, R. A. 2010. "Evolution of Personality: Developmental Constraints on Behavioral Flexibility." *The Auk* 127 (4): 752–758.

Duncker, K., and L. S. Lees. 1945. "On Problem-Solving." *Psychological Monographs* 58 (5): i–113.

Fawcett, T. W., S. Hamblin, and L.-A. Giraldeau. 2013. "Exposing the Behavioral Gambit: The Evolution of Learning and Decision Rules." *Behavioral Ecology* 24 (1): 2–11.

Fenneman, J., and W. E. Frankenhuis. 2020. "Is Impulsive Behavior Adaptive in Harsh and Unpredictable Environments? A Formal Model." *Evolution and Human Behavior* 41 (4): 261–273.

Fischhoff, B., and S. B. Broomell. 2020. "Judgment and Decision Making." *Annual Review of Psychology* 71 (1): 331–355.

Floresco, S. B. 2011. "Neural Circuits Underlying Behavioral Flexibility: Multiple Brain Regions Work Together to Adapt Behavior to a Changing Environment." Science brief, *Psychological Science Agenda* 25 (4): 1–8.

Fogarty, L., N. Creanza, and M. W. Feldman. 2015. "Cultural Evolutionary Perspectives on Creativity and Human Innovation." *Trends in Ecology & Evolution* 30 (12): 736–754.

Friedman, N. P., A. Miyake, S. E. Young, J. C. Defries, R. P. Corley, and J. K. Hewitt. 2008. "Individual Differences in Executive Functions Are Almost Entirely Genetic in Origin." *Journal of Experimental Psychology: General* 137 (2): 201–225.

Froese, T., and D. A. Leavens. 2014. "The Direct Perception Hypothesis: Perceiving the Intention of Another's Action Hinders Its Precise Imitation." *Frontiers in Psychology* 5: 1–15.

German, T. P., and H. C. Barrett. 2005. "Functional Fixedness in a Technologically Sparse Culture." *American Psychological Society* 16 (1): 1–4.

German, T. P., and M. A. Defeyter. 2000. "Immunity to Functional Fixedness in Young Children." *Psychonomic Bulletin & Review* 7 (4): 707–712.

Gigerenzer, G., R. Hertwig, and T. Pachur. 2011. *Heuristics: The Foundations of Adaptive Behavior*. Oxford: Oxford University Press.

Gittins, J. C., and D. M. Jones. 1979. "A Dynamic Allocation Index for the Discounted Multiarmed Bandit Problem." *Biometrika* 66 (3): 561–565.

Glowacki, L., and L. Molleman. 2017. "Subsistence Styles Shape Human Social Learning Strategies." *Nature Human Behaviour* 1 (5): 1–5.

Gopnik, A. 2020. "Childhood as a Solution to Explore–Exploit Tensions." *Philosophical Transactions of the Royal Society B: Biological Sciences* 375 (1803): 20190502.

Gopnik, A., S. O'Grady, C. G. Lucas, T. L. Griffiths, A. Wente, S. Bridgers, R. Aboody, et al. 2017. "Changes in Cognitive Flexibility and Hypothesis Search across Human Life History from Childhood to Adolescence to Adulthood." *Proceedings of the National Academy of Sciences* 114 (30): 7892–7899.

Gopnik, A., T. L. Griffiths, and C. G. Lucas. 2015. "When Younger Learners Can Be Better (or at Least More Open-Minded) than Older Ones." *Current Directions in Psychological Science* 24 (2): 87–92.

Grove, M. 2011. "Speciation, Diversity, and Mode 1 Technologies: The Impact of Variability Selection." *Journal of Human Evolution* 61 (3): 306–319.

Hanus, D., N. Mendes, C. Tennie, and J. Call. 2011. "Comparing the Performances of Apes (*Gorilla gorilla, Pan troglodytes, Pongo pygmaeus*) and Human Children (*Homo sapiens*) in the Floating Peanut Task." *PLOS One* 6 (6): e19555–e19555.

Haun, D. B. M., Y. Rekers, and M. Tomasello. 2014. "Children Conform to the Behavior of Peers; Other Great Apes Stick with What They Know." *Psychological Science* 25 (12): 2160–2167.

Henrich, J., and R. McElreath. 2003. "The Evolution of Cultural Evolution." *Evolutionary Anthropology* 12 (3): 123–135.

Hill, K., M. Barton, and A. Magdalena Hurtado. 2009. "The Emergence of Human Uniqueness: Characters Underlying Behavioral Modernity." *Evolutionary Anthropology* 18 (5): 187–200.

Hoehl, S., S. Keupp, H. Schleihauf, N. McGuigan, D. Buttelmann, and A. Whiten. 2019. "'Over-Imitation': A Review and Appraisal of a Decade of Research." *Developmental Review* 51:90–108.

Hopper, L. M., S. J. Schapiro, S. P. Lambeth, and S. F. Brosnan. 2011. "Chimpanzees Socially Maintained Food Preferences Indicate Both Conservatism and Conformity." *Animal Behaviour* 81 (6): 1195–1202.

Horner, V., and A. Whiten. 2005. "Causal Knowledge and Imitation/Emulation Switching in Chimpanzees (*Pan troglodytes*) and Children (*Homo sapiens*)." *Animal Cognition* 8 (3): 164–181.

Hrubesch, C., S. Preuschoft, and C. van Schaik. 2009. "Skill Mastery Inhibits Adoption of Observed Alternative Solutions among Chimpanzees (*Pan troglodytes*)." *Animal Cognition* 12 (2): 209–216.

Ionescu, T. 2017. "The Variability-Stability-Flexibility Pattern: A Possible Key to Understanding the Flexibility of the Human Mind." *Review of General Psychology* 21 (2): 123–131.

Jacobson, S. L., and L. M. Hopper. 2019. "Hardly Habitual: Chimpanzees and Gorillas Show Flexibility in Their Motor Responses When Presented with a Causally-Clear Task." *PeerJ* 7:e6195–e6195.

Jansson, D. G., and S. M. Smith. 1991. "Design Fixation." *Design Studies* 12 (1): 3–11.

Kahneman, D., S. P. Slovic, P. Slovic, and A. Tversky. 1982. *Judgment under Uncertainty: Heuristics and Biases.* Cambridge: Cambridge University Press.

Kaplan, H., K. Hill, J. Lancaster, and A. Magdalena Hurtado. 2000. "A Theory of Human Life History Evolution: Diet, Intelligence, and Longevity." *Evolutionary Anthropology* 9 (4): 156–185.

Keupp, S., T. Behne, J. Zachow, A. Kasbohm, and H. Rakoczy. 2015. "Over-Imitation Is Not Automatic: Context Sensitivity in Children's Overimitation and Action Interpretation of Causally Irrelevant Actions." *Journal of Experimental Child Psychology* 130:163–175.

Kline, M. A. 2017. "Teach: An Ethogram-Based Method to Observe and Record Teaching Behavior." *Field Methods* 29 (3): 205–220.

Kummer, H., and J. Goodall. 1985. "Conditions of Innovative Behavior in Primates." *Philosophical Transactions of the Royal Society of London. Series B: Biological Sciences* 308 (1135): 203–214.

Laland, K., B. Matthews, and M. W. Feldman. 2016. "An Introduction to Niche Construction Theory." *Evolutionary Ecology* 30: 191–202.

Laland, K. N., and S. M. Reader. 1999. "Foraging Innovation in the Guppy." *Animal Behaviour* 57:331–340.

Larcom, S., F. Rauch, and T. Willems. 2017. "The Benefits of Forced Experimentation: Striking Evidence from the London Underground Network." *Quarterly Journal of Economics* 132 (4): 2019–2055.

Laughlin, S. B., and T. J. Sejnowski. 2003. "Communication in Neuronal Networks." *Science* 301 (5641): 1870–1874.

Laureiro-Martínez, D., and S. Brusoni. 2018. "Cognitive Flexibility and Adaptive Decision-Making: Evidence from a Laboratory Study of Expert Decision Makers." *Strategic Management Journal* 39 (4): 1031–1058.

Lefebvre, L., S. M. Reader, and D. Sol. 2004. "Brains, Innovations and Evolution in Birds and Primates." *Brain, Behavior and Evolution* 63 (4): 233–246.

Legare, C. H., and M. Nielsen. 2015. "Imitation and Innovation: The Dual Engines of Cultural Learning." *Trends in Cognitive Sciences* 19 (11): 688–699.

Lehmann, L., M. W. Feldman, and R. Kaeuffer. 2010. "Cumulative Cultural Dynamics and the Coevolution of Cultural Innovation and Transmission: An ESS Model for Panmictic and Structured Populations: Evolution of Cumulative Culture." *Journal of Evolutionary Biology* 23 (11): 2356–2369.

Lemaire, P., and M. Leclère. 2014. "Strategy Repetition in Young and Older Adults: A Study in Arithmetic." *Developmental Psychology* 50 (2): 460–468.

Lew-Levy, S., A. N. Crittenden, A. H. Boyette, I. A. Mabulla, B. S. Hewlett, and M. E. Lamb. 2019. "Inter- and Intra-Cultural Variation in Learning-through-Participation among Hadza and BaYaka Forager Children and Adolescents from Tanzania and the Republic of Congo." *Journal of Psychology in Africa* 29 (4): 309–318.

Lew-Levy, S., S. M. Kissler, A. H. Boyette, A. N. Crittenden, I. A. Mabulla, and B. S. Hewlett. 2019. "Who Teaches Children to Forage? Exploring the Primacy of Child-to-Child Teaching among Hadza and BaYaka Hunter-Gatherers of Tanzania and Congo." *Evolution and Human Behavior* 41 (1): 12–22.

Luchins, A. S. 1942. "Mechanization of Problem Solving: The Effect of Einstellung." *Psychological Monographs* 54 (6): 1–95.

Luchins, A. S., and E. H. Luchins. 1950. "New Experimental Attempts at Preventing Mechanization in Problem Solving." *Journal of General Psychology* 42:279–297.

Mata, R., A. Wilke, and U. Czienskowski. 2013. "Foraging across the Life Span: Is There a Reduction in Exploration with Aging?" *Frontiers in Neuroscience* 7:1–7.

McFadden, D., M. J. Machina, and J. Baron. 1999. "Rationality for Economists?" In *Elicitation of Preferences*, edited by B. Fischhoff and C. F. R. Manski, 73–110. Dordrecht: Springer.

McGuigan, N., J. Makinson, and A. Whiten. 2011. "From Over-Imitation to Super-Copying: Adults Imitate Causally Irrelevant Aspects of Tool Use with Higher Fidelity than Young Children." *British Journal of Psychology* 102 (1): 1–18.

Meder, B., C. M. Wu, E. Schulz, and A. Ruggeri. 2020. "Development of Directed and Random Exploration in Children." *Developmental Science* 24 (4): e13095.

Meiran, N. 2010. "Task Switching: Mechanisms Underlying Rigid vs. Flexible Self-Control." In *Self Control in Society, Mind, and Brain*, edited by R. Hassin, K. Ochsner, and Y. Trope, 202–220. Oxford: Oxford University Press.

Milton, K. 1981. "Distribution Patterns of Tropical Plant Foods as an Evolutionary Stimulus to Primate Mental Development." *American Anthropologist* 83 (3): 534–548.

Miyake, A., N. P. Friedman, M. J. Emerson, A. H. Witzki, A. Howerter, and T. D. Wager. 2000. "The Unity and Diversity of Executive Functions and Their Contributions to Complex 'Frontal Lobe' Tasks: A Latent Variable Analysis." *Cognitive Psychology* 41 (1): 49–100.

Monsell, S. 2003. "Task Switching." *Trends in Cognitive Sciences* 7 (3): 134–140.

Muthukrishna, M. 2015. "The Cultural Brain Hypothesis and the Transmission and Evolution of Culture." PhD diss., University of British Columbia.

Muthukrishna, M., M. Doebeli, M. Chudek, and J. Henrich. 2018. "The Cultural Brain Hypothesis: How Culture Drives Brain Expansion, Sociality, and Life History." *PLOS Computational Biology* 14 (11): 1–37.

Nakahashi, W. 2007. "The Evolution of Conformist Transmission in Social Learning When the Environment Changes Periodically." *Theoretical Population Biology* 72 (1): 52–56

Navarro, D. J., B. R. Newell, and C. Schulze. 2016. "Learning and Choosing in an Uncertain World: An Investigation of the Explore–Exploit Dilemma in Static and Dynamic Environments." *Cognitive Psychology* 85 (March): 43–77.

Nielsen, M., J. Cucchiaro, and J. Mohamedally. 2012. "When the Transmission of Culture Is Child's Play." *PLOS One* 7 (3): e34066, 1–6.

Ntamboudila, F. K. 2020. *The King of the Jungle.* Brazzaville, Republic of the Congo.

Odling-Smee, F. J. 1988. "Niche Constructing Phenotypes." In *The Role of Behavior in Evolution*, edited by H. C. Plotkin, 73–132. Cambridge, MA: MIT Press.

O'Farrell, S., J. N. Sanchirico, O. Spiegel, M. Depalle, A. C. Haynie, S. A. Murawski, L. Perruso, and A. Strelcheck. 2019. "Disturbance Modifies Payoffs in the Explore–Exploit Trade-Off." *Nature Communications* 10 (1): 3363.

Öllinger, M., G. Jones, and G. Knoblich. 2008. "Investigating the Effect of Mental Set on Insight Problem Solving." *Experimental Psychology* 55 (4): 269–282.

Over, H., and M. Carpenter. 2013. "The Social Side of Imitation." *Child Development Perspectives* 7 (1): 6–11.

Payne, J. W., J. R. Bettman, and E. J. Johnson. 1993. *The Adaptive Decision Maker.* Cambridge University Press.

Peterson, E., and T. Verstynen. 2019. "A Way around the Exploration-Exploitation Dilemma." *bioRxiv* (November): 671362.

Pope, S. M. 2018. "Differences in Cognitive Flexibility within the Primate Lineage and across Human Cultures: When Learned Strategies Block Better Alternatives." PhD diss., Georgia State University.

Pope, S. M., J. Fagot, A. Meguerditchian, D. A. Washburn, and W. D. Hopkins. 2019. "Enhanced Cognitive Flexibility in the Seminomadic Himba." *Journal of Cross-Cultural Psychology* 50 (1): 47–62.

Pope, S. M., J. Fagot, A. Meguerditchian, J. Watzek, S. Lew-Levy, M. M. Autrey, and W. D. Hopkins. 2020. "Optional-Switch Cognitive Flexibility in Primates: Chimpanzees' (*Pan troglodytes*) Intermediate Susceptibility to Cognitive Set." *Journal of Comparative Psychology* 134 (1): 98–109.

Pope, S. M., A. Meguerditchian, W. D. Hopkins, and J. Fagot. 2015. "Baboons (*Papio papio*), but Not Humans, Break Cognitive Set in a Visuomotor Task." *Animal Cognition* 18 (6): 1339–1346.

Pope-Caldwell, S. M., and D. A. Washburn. 2022. "Overcoming Cognitive Set Bias Requires More than Seeing an Alternative Strategy." *Scientific Reports* 12 (1): 2179.

Potts, R. 1996. "Evolution and Climate Variability." *Science* 273 (5277): 922–923.

Potts, R. 2012. "Environmental and Behavioral Evidence Pertaining to the Evolution of Early Homo." *Current Anthropology* 53 (S6): S299–S317.

Potts, R., and J. T. Faith. 2015. "Alternating High and Low Climate Variability: The Context of Natural Selection and Speciation in Plio-Pleistocene Hominin Evolution." *Journal of Human Evolution* 87:5–20.

Price, E. E., S. P. Lambeth, S. J. Schapiro, and A. Whiten. 2009. "A Potent Effect of Observational Learning on Chimpanzee Tool Construction." *Proceedings of the Royal Society B: Biological Sciences* 276 (1671): 3377–3383.

Reader, S. M., and K. N. Laland. 2001. "Primate Innovation: Sex, Age and Social Rank Differences." *International Journal of Primatology* 22 (5): 787–805.

Reader, S. M., and K. N. Laland. 2002. "Social Intelligence, Innovation, and Enhanced Brain Size in Primates." *Proceedings of the National Academy of Sciences of the United States of America* 99 (7): 4436–4441.

Reader, S. M., and K. MacDonald. 2003. "Environmental Variability and Primate Behavioural Flexibility." In *Animal Innovation*, edited by S. M. Reader and K. N. Laland, 83–116. Oxford: Oxford University Press.

Richerson, P. J., and R. Boyd. 2013. "Rethinking Paleoanthropology: A World Queerer than We Supposed." In *Evolution of Mind, Brain, and Culture*, edited by Gary Hatfield and Holly Pittman, 263–302. Philadelphia: University of Pennsylvania Press.

Richerson, P. J., and R. Boyd. 2020. "The Human Life History Is Adapted to Exploit the Adaptive Advantages of Culture." *Philosophical Transactions of the Royal Society B: Biological Sciences* 375 (1803): 20190498.

Riotte-Lambert, L., and J. Matthiopoulos. 2020. "Environmental Predictability as a Cause and Consequence of Animal Movement." *Trends in Ecology & Evolution* 35 (2): 163–174.

Sakai, K. 2008. "Task Set and Prefrontal Cortex." *Annual Review of Neuroscience* 31:219–245.

Schaik, C. P. van. 2013. "The Costs and Benefits of Flexibility as an Expression of Behavioural Plasticity: A Primate Perspective." *Philosophical Transactions of the Royal Society B: Biological Sciences* 368 (1618): 20120339.

Schillemans, V. 2011. "The Perseveration Effect in Individuals' Strategy Choices." PhD diss., KU Leuven.

Schulz, E., C. M. Wu, A. Ruggeri, and B. Meder. 2019. "Searching for Rewards Like a Child Means Less Generalization and More Directed Exploration." *Psychological Science* 30 (11): 1561–1572.

Snell-Rood, E. C. 2013. "An Overview of the Evolutionary Causes and Consequences of Behavioural Plasticity." *Animal Behaviour* 85 (5): 1004–1011.

Sol, D., R. P. Duncan, T. M. Blackburn, P. Cassey, and L. Lefebvre. 2005. "Big Brains, Enhanced Cognition, and Response of Birds to Novel Environments." *Proceedings of the National Academy of Sciences* 102 (15): 5460–5465.

Sol, D., F. Sayol, S. Ducatez, and L. Lefebvre. 2016. "The Life-History Basis of Behavioural Innovations." *Philosophical Transactions of the Royal Society B: Biological Sciences* 371 (1690): 20150187, 1–8.

Speekenbrink, M., and E. Konstantinidis. 2015. "Uncertainty and Exploration in a Restless Bandit Problem." *Topics in Cognitive Science* 7 (2): 351–367.

Stengelin, R., R. Hepach, and D. B. M. Haun. 2019. "Being Observed Increases Overimitation in Three Diverse Cultures." *Developmental Psychology* 55 (12): 2630–2636.

Stengelin, R., R. Hepach, and D. B. M. Haun. 2020. "Cross-Cultural Variation in How Much, but Not Whether, Children Overimitate." *Journal of Experimental Child Psychology* 193:104796.

Stephens, D. W., and J. R. Krebs. 1986. *Foraging Theory*. Princeton, NJ: Princeton University Press.

Todd, P. M., and G. Gigerenzer. 2007. "Environments That Make Us Smart: Ecological Rationality." *Current Directions in Psychological Science* 16 (3): 167–171.

Tversky, A., and D. Kahneman. 1974. "Judgment under Uncertainty: Heuristics and Biases." *Science* 185 (4157): 1124–1131.

Ueltzhöffer, K., D. J. N. Armbruster-Genç, and C. J. Fiebach. 2015. "Stochastic Dynamics Underlying Cognitive Stability and Flexibility." *PLOS Computational Biology* 11 (6): e1004331.

van de Pol, M., S. Jenouvrier, J. H. C. Cornelissen, and M. E. Visser. 2017. "Behavioural, Ecological and Evolutionary Responses to Extreme Climatic Events: Challenges and Directions." *Philosophical Transactions of the Royal Society B: Biological Sciences* 372 (1723): 20160134

van Leeuwen, E. J. C., and J. Call. 2017. "Conservatism and 'Copy-If-Better' in Chimpanzees (*Pan troglodytes*)." *Animal Cognition* 20 (3): 575–579.

van Leeuwen, E. J. C., E. Cohen, E. Collier-Baker, C. J. Rapold, M. Schäfer, S. Schütte, and D. B. M. Haun. 2018. "The Development of Human Social Learning across Seven Societies." *Nature Communications* 9 (1): 1–7.

van Leeuwen, E. J. C., K. A. Cronin, S. Schütte, J. Call, and D. B. M. Haun. 2013. "Chimpanzees (*Pan troglodytes*) Flexibly Adjust Their Behaviour in Order to Maximize Payoffs, Not to Conform to Majorities." *PLOS One* 8 (11): 1–10.

Vicente, K. J., and J. H. Wang. 1998. "An Ecological Theory of Expertise Effects in Memory Recall." *Psychological Review* 105 (1): 33–57.

Vitousek, P. M., H. A. Mooney, J. Lubchenco, and J. M Melillo. 1997. "Human Domination of Earth's Ecosystems." *Science* 277 (5325): 494–499.

Vrba, E. S. 1992. "Mammals as a Key to Evolutionary Theory." *Journal of Mammalogy* 73 (1): 1–28.

Vrba, E. S. 1995. *Paleoclimate and Evolution, with Emphasis on Human Origins.* New Haven, CT: Yale University Press.

Watson-Jones, R. E., C. H. Legare, H. Whitehouse, and J. M. Clegg. 2014. "Task-Specific Effects of Ostracism on Imitative Fidelity in Early Childhood." *Evolution and Human Behavior* 35 (3): 204–210.

Watzek, J., S. M. Pope, and S. F. Brosnan. 2019. "Capuchin and Rhesus Monkeys but Not Humans Show Cognitive Flexibility in an Optional-Switch Task." *Scientific Reports* 9 (1): 1–10.

Wertheimer, M. 1945. *Productive Thinking.* New York: Harper.

Wilson, R. C., A. Geana, J. M. White, E. A. Ludvig, and J. D. Cohen. 2014. "Humans Use Directed and Random Exploration to Solve the Explore–Exploit Dilemma." *Journal of Experimental Psychology: General* 143 (6): 2074.

Wilson, R. C., Y. K. Takahashi, G. Schoenbaum, and Y. Niv. 2014. "Orbitofrontal Cortex as a Cognitive Map of Task Space." *Neuron* 81 (2): 267–279.

Wood, L. A., R. L. Kendal, and E. G. Flynn. 2013a. "Copy Me or Copy You? The Effect of Prior Experience on Social Learning." *Cognition* 127 (2): 203–213.

Wood, L. A., R. L. Kendal, and E. G. Flynn. 2013b. "Whom Do Children Copy? Model-Based Biases in Social Learning." *Developmental Review* 33 (4): 341–356.

Wright, T. F., J. R. Eberhard, E. A. Hobson, M. L. Avery, and M. A. Russello. 2010. "Behavioral Flexibility and Species Invasions: The Adaptive Flexibility Hypothesis." *Ethology Ecology & Evolution* 22 (4): 393–404.

Wu, C. M., M. K. Ho, B. Kahl, C. Leuker, B. Meder, and R. H. J. M. Kurvers. 2021. "Specialization and Selective Social Attention Establishes the Balance between Individual and Social Learning." *bioRxiv* (2021): 2021–02.

Wu, C. M., E. Schulz, M. Speekenbrink, J. D. Nelson, and B. Meder. 2018. "Generalization Guides Human Exploration in Vast Decision Spaces." *Nature Human Behaviour* 2 (12): 915–924.

Wyles, J. S., J. G. Kunkel, and A. C. Wilson. 1983. "Birds, Behavior, and Anatomical Evolution." *Proceedings of the National Academy of Sciences* 80 (14): 4394–4397.

Young, E. S., W. E. Frankenhuis, and B. J. Ellis. 2020. "Theory and Measurement of Environmental Unpredictability." *Evolution and Human Behavior* 41 (6): 550–556.

Zelazo, P. D. 2004. "The Development of Conscious Control in Childhood." *Trends in Cognitive Sciences* 8 (1): 12–17.

10 The Cultural Identity of Techniques: Rigidity and Flexibility among the Akha of Highland Laos

Giulio Ongaro

Introduction

This chapter focuses ethnographically on two distinct sets of techniques that I documented among the Akha, a group of swidden farmers living in highland Laos and neighboring borderlands. It shows that these two sets of techniques have diametrically opposite characteristics in terms of their flexibility and rigidity, and it aims to explain the reason for this difference. The first set that I discuss is the technical repertoire of herbal medicine mastered by the village herbalists. The second set is what the Akha call *ghanrsanrkhovq*, which can be roughly translated as "customs," a range of practices that, along with rituals and social norms, comprise everyday techniques such as cooking, dressmaking, hunting, house building, and farming. Drawing a comparison between the two, I show that while the first is typified by high flexibility in learning, the second is typified by high rigidity. While the flexibility in the learning process leads to high variability of herbal medical techniques—that is, each herbalist treasures their own distinctive arsenal of herbal pharmacopoeia and related techniques—the rigidity in the transmission of customs leads to high stability—that is, one finds high homogeneity of customs-related techniques across Akha communities in space and time. My focus here will be customs-related techniques regarding house construction and design. I argue that the key factor explaining this difference is cultural identity. Not being a marker of cultural identity, Akha herbal medical lore varies freely depending on the personal experimentation of its practitioners and independent of community-wide conventions. By contrast, Akha customs are cultural markers: their rigid transmission over time enables the perpetuation of Akha identity. Exploring the social conditions in highland Laos that make the perpetuation of cultural identity such an important value among the Akha, I argue that a study of human techniques must always be embedded within a study of the broader geopolitical context in which these are practiced and transmitted. The chapter ends with a reflection on the relation between cultural transmission and efficacy of techniques. It notes that, in different ways, the Akha-specific means of transmitting both herbal and customs-related techniques are not necessarily conducive to higher efficacy because the value attached to their mode of transmission overshadows concerns about instrumentality.

Brief Ethnographic Background

Migratory farmers of Tibeto-Burman language origin, the Akha[1] crossed the Mekong to settle villages on the Lao hills sometime in the nineteenth century, after a long southward journey from China. Their migratory trajectory led them into the confines of five nation-states. Beside Laos, they currently live in northern Thailand, in the Yunnan Province in China, in the Shan state of Myanmar, and in the northwestern tip of Vietnam—numbering some 750,000 in total, of which 113,000 live in Laos (Wang 2013, 20). The vast swath of highlands they have occupied has the geopolitical peculiarity of having been, historically, out of the reach of governments. Its remoteness impeded major lowland power centers from exerting full control over the highlanders, effectively allowing the creation of politically autonomous zones and the proliferation of ethnic identities. With a few exceptions, highland societies stand out for being very different, both in terms of culture and modes of subsistence, from the politically dominant lowlanders. To this day, driving from a lowland Lao town toward the highlands means entering a strikingly different cultural universe. People of the plains like the Lao have had a state, permanent agriculture, a writing system, and Buddhism. Highlanders have lived in a condition of statelessness (until recently, at least), shifting cultivation, orality, and animism: they practice a mixture of spirit cults and ancestors worship. As James Scott (2009) argued in his "anarchist history of Southeast Asia," people like the Akha consciously decided to live in such remote places to escape the burdens of the lowland state—bureaucracy, slavery, warfare—and its people, who regarded them as uncivilized.

Parallel to this process of rejection, over the centuries the Akha have consolidated an elaborate complex of customs—rituals, material techniques, social norms, dressing codes, and artifacts—that has served as the basis for their identity in such a context of power inequality (Geusau 2000; Tooker 2012). Throughout their history, they have placed high value on the conservation of this tradition, carefully passing down oral codes and texts from generation to generation. One can witness very similar practices—for example, house building and hunting techniques—in Akha communities far away from one another that have been separated for centuries. Many anthropologists who have studied the Akha consider this as a great example of rigid and stable cultural transmission (Lewis 1969; Kammerer 1986; Tooker 2012). In recent decades, Akha living in Thailand and China have largely abandoned their traditional customs. In Thailand, this was mostly due to the work of foreign missionaries in the 1990s. In China, many customs were eradicated even earlier by the Cultural Revolution. In socialist Laos, however, change has been occurring at a much slower pace because of remoteness and a ban on religious proselytization. Here, customs are still held firmly in place by the power accorded to the ancestors. Where I conducted fieldwork, Akha still see themselves as the contemporary bearers of an important, identity-defining tradition, handed down through the centuries by a long line of forefathers. When external influences introduce benefits or constraints that require change to this tradition, this change (the adoption or rejection of novel practices) tends to happen at a collective level, in a way that preserves a sense of identity.

My fieldwork was conducted in the village of MawPae between December 2014 and April 2016, with three follow-up visits until 2019, for a total of 19 months. My original research focused on the Akha healing tradition, which included herbal pharmacopoeia, animal sacrifice, shamanism, and a variety of other ritual and medical techniques whose proper understanding meant uncovering many other aspects of Akha society as a whole. In what follows, I first describe herbal medical techniques and, second, Akha customs.

Herbal Medicine and Related Techniques

The following section zeroes in on Akha herbal medical techniques, which, I will point out, are typified by high flexibility in learning. I will show that there is no rigid "traditional" way of teaching herbal medical skills and that, accordingly, these techniques reveal a high degree of idiosyncrasies and individual variation. This, I will argue, is due to the esoteric nature of medicinal plant knowledge. Given the secrecy that surrounds it, medicinal plant knowledge is impervious to the exogenous and stabilizing factors that, by contrast, characterize the learning of techniques related to Akha customs that I will discuss later. Moreover, with a handful of exceptions, the actual biochemical effectiveness of herbal remedies (i.e., the efficacy of the remedies beyond the "placebo effects" produced by their administration) is seemingly low and, counterintuitively, it is not a factor that is conducive to the stability of knowledge.

By the term *yavghaq*, Akha refer to any substance that, through appropriate techniques, is used for the treatment or prevention of disease, whether in humans, other animals, or crops. It has been translated as "medicine" (Lewis 1989, 213), even though the term only refers to medicine as substance, rather than medicine as a body of knowledge. *Yavghaq* comprises herbal medicine and other remedies that Akha gather from their environment, as well as modern pharmaceuticals (the use of which I will not discuss here; see Ongaro 2019, 139ff.). These treatments form what George Foster (1976) has characterized as a naturalistic medical system: they treat what for the Akha are "naturally caused" conditions— that is, ailments and illnesses that are not spiritually caused.

Along with lumber, timber, bamboo, fruits, greens, fibers, and dyes, Akha also rely on the nearby forest for their remedies. Herbs and plants provide relief for the aches and pains that ensue from the hazards of living in what Akha themselves describe as an "uneasy" environment, especially the practical activity done in fields and in the forest itself. I have seen it applied most frequently to treat ailments such as fractures, burns, cuts, stings, bruises, or animal bites, but also for skin rashes (habitually caused by caterpillars), stomach pains, and as last resort, for emergencies like cramps or seizures, often in haphazard combination with ritual. As the primary method to cure a large number of relatively minor ailments, herbal medicine is widely used. Although people might at times describe herbal medicine as "Akha medicine" (*Aqkaq yavghaq*), they do not conceptually set it apart from the pharmacopoeia of other highland groups, nor do they refer to a specifically Akha, community-wide sphere of knowledge. Rather, if they use the term "Akha medicine" in this context, they are simply referring to "herbal medicine" as opposed to the lowland pharmaceutical medicine of the Lao or the Chinese.

This is important because, as I shall describe below, there is not any specifically Akha, community-wide sphere of medical knowledge when it comes to plant medicine, which is revealing about the flexibility involved in the learning of herbal medical techniques. Certainly, general botanical knowledge is very widespread and uniform in the community. By interacting with the forest on a daily basis, Akha become acquainted with the individualities of plants from a very early age. As they go from village to field and back, or as they venture into the woods to hunt for birds, young children follow their older relatives in what are true educational paths of knowledge: they point at plants, they ask their name, and they are taught back (often favorably, with a sense of pedagogical duty). Children quickly build up a huge knowledge of arboreal species, whose vastness usually surprises anthropologists brought up in urban places. And yet, how to pick these plants and mix them in a way to produce powerful medicine is knowledge belonging to only a very few members of the community.

Herbal medicine among the Akha is a domain of practice that is shrouded in secrecy but whose effectiveness is said to depend precisely on this quality. Although many individuals in the village might possess a smattering of knowledge, only a handful of people "know a lot about medicine," and their expertise is sought after when there are ailments to be cured. Herbalists do not divulge their knowledge except for when they teach it to select apprentices. Dispersal of knowledge is said to decrease the potency of herbal medicine. "The medicines would turn bad," they say. Typically, they keep their medicinal herbs and plants in the attic (a place only homeowners can access) and make sure not to show them to anybody else. Covertness is also involved in the preparation of herbal remedies. Before applying a remedy to the sick person, herbalists prepare it at their own house and only take the final product to the house of the sick. The final product is a mixture of herbs whose individual ingredients are visually unidentifiable by others.

There are some general ways of applying medicine that are adopted by all herbalists. These modalities of treatment vary with the nature of the ailment. For instance, herbal compresses are made for fractures or blows. These are wrapped around the limb and kept for several days, refilling the herbs on a regular basis. Cuts of all kinds are treated with vulnerary leaves pounded to mush and then applied to stop the bleeding and to clean the wound. The herbalist will often spit on both the wound and the herbs, accompanying the treatment with mumbled spells. For skin conditions like spots or rashes, herbs are poulticed and smeared on the site and smeared again when they wash off. Leaves can also be chewed beforehand or soaked in rice whisky and then applied. Sometimes, as a treatment for skin conditions, a special vine is burned at one end and held a few inches from the person; the herbalist will then blow a gentle blast of smoke over the affected area. Herbal decoctions or infusions (which tend to be very bitter) are drunk for cramps or stomach pains. Occasionally, they are also drunk as tonics, to revitalize the body even in absence of any ailment. Although I have never seen it myself, in some cases of severe fatigue, medicines are burned and their aroma inhaled to reinvigorate the sick person. There is a discernible sensorial quality to all these treatments that, through their soothingness, bitterness, pungency, and overall tactile application, may trigger the senses in ways Akha have learned to associate with healing.

Outside these general rules for how herbal remedies are applied, knowledge about ingredients, mixing, and posology is kept secret. Herbalists only disclose it to daughters and sons, provided they have an interest in acquiring the knowledge and under the promise that they won't divulge it to anyone else. For any other aspirant apprentice, there is a price to pay in exchange of knowledge, depending on the degree of kinship with the herbalist and on what they want to learn, for some herbs are more expensive than others. I myself had a short apprenticeship with four herbalists, three elderly women and an elderly man reputed to be among the most knowledgeable in the district. I paid them the necessary amount to have a solid baseline for comparing their methods and herbal pharmacopoeias. (While obtaining knowledge comes with a charge, the treatment is usually free but might be recompensed later through a chicken offering if healing is successful.)

I took several walks with each of the four tutor herbalists. We walked through teak forest, across groves of fern into thick bamboo areas, and down to swampy marshes, in search of medicines. Every time a medicinal plant was spotted, the elder would tell me its name and what part is used for treatment—root, leaf, wood, or bark—and other traits of the plant, such as whether it is planted by people by seeding or whether "it was planted by the ancestors" (i.e., it grows on its own without human intervention). I would further ask when it flowers and if it is rare or common, and I learned that the rarity of the plant, especially when it is grown in distant lands, ups its value. When disclosing medical knowledge about an important plant, my teachers would explain its features in a ceremonious way, with a grave voice, conveying importance. They would add stories of how long or far one must walk to obtain this or the other plants, or of how one time they had to travel to Myanmar or other distant lands to collect a plant that they knew was only growing there. Similar to the handling of medical knowledge in other ethnic groups in the region (Sprenger 2011), the foreign and the distant connote higher potency.

Each herbalist told me that they were taught about herbal medical techniques by their parents, but that their learning also involved a good deal of experimentation. Repeated failures usually lead to discarding the remedy and trying another one or changing the posology or combination of herbs. I was able to attest that their path to knowledge is highly personal and idiosyncratic. This is also because an important way of sourcing healing plants comes from dreams (though this oneiric ability usually sets in as one has already apprehended a good deal about herbal medicine from one's teacher). All four herbalists I worked with reported that they occasionally receive inspiration about plants from the spirit-owner of people they meet in dreams. Three of them said that they meet their former teachers (their parents, mostly) who are now dead. The fourth herbalist told me that she has established a special relationship with an old couple that visits her in dreams on a regular basis. The spirit-owner of the wise old man and woman appear in her dream, whispering the name of the plant or the method of preparation into her ear. Upon waking up, she rushes into the forest to collect the plant, without talking to anyone on the way. The plant recommended in a dream is not immediately included in the herbalist's inventory, for she will test it after gathering it and keep it only if it works reliably.

It thus comes as no surprise that, when I compared the pharmacopoeia of all four herbalists, great differences emerged among them. The overlap of shared medicinal plants among herbalists is low. Often, the same herb is used for very different ailments, and the same ailment is cured with different herbs. In one instance, the same plant used by one herbalist as a soother was used by another as a stimulant. There was, in sum, great variability. This is not simply a variability in types of plants used but also in the process involved in the production of medicine such that the sequence of acts, place, and tools employed in the making of the concoctions or mixtures varied from individual to individual. The low degree of transmission of medicinal knowledge and practices, both within and between groups, appears to be a general trait of the region. Catherine Pake (1987), working on the neighboring Hmong people, found that herbalists do not share their knowledge of plants with one another (see also Dubost et al. 2019). Jean Marc Dubost (2014) surveyed the medicinal plants used by Lue, Hmong, and Lamet communities living in close contact with each other to find that only a meager 23 percent were commonly shared, most of which, it turned out, were used for completely different ailments.

These findings should prompt some observations on the question of effectiveness. Discussions about ethnopharmacology often revolve around the possibility that some plants, unbeknownst to biomedicine, might contain potent active ingredients (Anderson 1993; Heinrich and Jäger 2015). With the obvious but important exceptions of opium, betel, and ginger, all widely used among Akha, not much is known about the potential medical properties of the herbs and plants employed by expert herbalists. A lesson learned in ethnopharmacology over the last 50 years is that the failure of bioassays usually turns out to be high—not many traditional herbs and plants are discovered to contain pharmacologically active substances for the ailments they are purported to treat. Arguably, this is likely to be even more the case where there is minimal information-sharing and a lack of a collective endeavor in testing the efficacy of plants, as it happens among the Akha and neighboring groups. We also know, however, that any treatment can produce what is misleadingly called the "placebo effect," and that this is heightened by features specific to the therapeutic act—for instance, the aura of the treatment and the patient-practitioner interaction (see Moerman 2002).[2] Certainly, the secrecy that shrouds Akha herbalism generates an aura of potency around herbal treatment. It is likely that this perceived potency, coupled with the intense physical sensations of their application, might afford potent healing effects, in light of what we know from the science of placebo responses.

To recapitulate, herbal medicine and related techniques among the Akha are characterized by high flexibility in learning and high variability. The secretive nature of this knowledge entails paths of learning that are not subjected to exogenous social factors, which, consequently, prevents such techniques from achieving stability. Moreover, these features do not appear to be conducive to higher selection and effectiveness of medicinal plants. I will now discuss another set of techniques—Akha customs—that have almost the opposite characteristics to herbal lore. I suggest that the reason for this difference lies in the fact that these techniques have been chosen as important identity markers.

Customs-Related Techniques and the Role of Identity

Contrary to medicinal plants, what Akha call "customs" (*ghanrsanrkhovq*) are typified by high rigidity and stability in their transmission. I will show that for the Akha, the techniques embedded in customs are important social markers: they define Akha identity in a multiethnic social context where the manifest belonging to a specific ethnic group carries significant practical and political value. In the following discussion, I expand on this point by arguing that the study of flexibility and rigidity of techniques should always be embedded in a study of the cultural and historical context in which their learning takes place because this context often modulates their practical utility.

Previous anthropologists working on the Akha have variably defined 'customs' as 'religion', 'way of life', 'etiquette', 'ceremonies', noting that it is hard to say where ceremony begins and 'etiquette' ends, since they influence social life all the way down to mundane, everyday practices (Lewis, 1970; Geusau, 1983; Kammerer, 1986). Akha customs constitute a total social fact that encompasses at once the spheres of religion, law, kinship, economy, and healthcare. They comprise prescriptions and proscriptions that regulate much of ordinary social life (kinship rules, behavioral codes, etc.) as well as the complex non-calendric and calendric rituals that punctuate the Akha yearly agricultural cycle. They also comprise a rich body of oral stories, myths, proverbs, and shamanic texts, comparable in breath and complexity to codexes such as the Vedas or the Old Testament, themselves taking oral forms before having been written down. These customs also regulate a range of practical techniques as varied as house building (e.g., how many rungs to fit on a ladder), hunting (e.g., how to kill game), eating (e.g., how to hold a bamboo teacup in ceremonies), sleeping (e.g., the direction one should sleep), and working in the fields (e.g., how to hold a sickle). I shall now describe the general nature of these customs before focusing on techniques specifically. Indeed, my argument is that the capacity of customs in defining Akha ethnic identity is important to the extent that it permeates even very practical daily techniques whose nature, as a result, cannot be viewed solely from an instrumental lens.

"Akha is as Akha does," as one anthropologist working on the Akha once put it (Kammerer 1990, 281). Akha identity is premised on living in an Akha community, speaking the language, and adhering to Akha customs, which are considered to have been handed down by the ancestors. Importantly, being Akha entails being affiliated to a patrilineage with a long pedigree of ancestors that can be traced to the first man, SmMirOr. Every adult male in a traditional Akha village is supposed to memorize his genealogy up to this apical figure, whose recitation is required at funerals, when the deceased person joins the ranks of the ancestors (see Hanks 1974). These genealogies can go back 60 generations, spanning about 1,500 years. As linguists and anthropologists realized, Akha communities separated for centuries share the initial nodes of these genealogies, a testimony to the striking stability of Akha customs through time. Linguist Pascal Bouchery (1993, 1) remarked:

The fact that all Akha subgroups are bound by genealogical links is absolutely remarkable, if one considers the geographical distance that separates different groups of population. For instance, Akha of Thailand and Piyo subgroups, though they are for the most part ignorant of their reciprocal existence and have no contact at all, use the same common list of 20 initial ancestors names in their genealogies, with the exception of minor phonological differences. In two communities separated by more than 500km of mountainous country, we have recorded a list of initial thirty-odd nodes at a genealogical distance of 25 generations. It has been frequently argued that among oral societies, genealogical lists of ancestors are too easily manipulated for them to carry any significant historical meaning. The great similarity of Akha initial nodes of ancestors demonstrates on the contrary that a very ancient memory can remain unchanged through centuries despite migration, geographical isolation and linguistic changes.

Genealogies also keep a sense of unity among members scattered across mountain ranges and national boundaries. I often happened to see that when two Akha strangers meet and begin to converse, they reel off each other's genealogy to trace how far back they are related. The conversations that follow tend to be spirited and mutually enjoyed. Out of the recognition of common ancestry arises a sense of shared identity and connection and warmth on the part of the host.

The discussion of customs more generally is a staple of public events. The elderly revel in talking about myths, stories, pieces of customs, and how to exactly perform a ritual or organize a festival. This is especially salient when it comes to the practical organization of rituals or festivals, where the disquisition of their technical aspects of preparation to their minute details takes on a lively tone, often branching off into a discussion about the ancestral origin of such customs. Everyone respects what the elders have to say. Unlike herbal medicine, knowledge and practice of customs is public and exoteric, and it is supposed to be so: it is a marker of social identity, of what makes a person Akha.

The learning of customs and related techniques is very different from the esoteric learning of plant medicine. In the previous section, I discussed why herbal medicine is a highly idiosyncratic form of knowledge: it involves a high degree of individual experimentation, which means that the path of knowledge ends up being highly personal. The demonstrated variability of herbal pharmacopoeia across Akha herbalists (and across ethnic groups) is a consequence of this form of flexible learning.

Unlike the learning of herbal medicine, the learning of Akha customs is highly standardized. One is not supposed to deviate from the teaching of the ancestors but to follow their words precisely (with some exceptions, Ongaro, forthcoming). This principle falls in line with the overall gerontocratic character of social relations among the Akha, according to which knowledge and status grows with age. This arrangement among the Akha makes for a fairly unidirectional type of learning of customs (Strachan, Curioni, and McEllin, this volume). During communal activities as varied as hunting, house building, pig killing, meat cutting, rice threshing, coffin making, fishing, fish-trap making, ritual sacrificing, and so on, children and young teenagers are not supposed to engage with or ask questions of their elders. It is considered improper and embarrassing to do so. Elders, for their part, disregard the presence of children, who congregate a few meters from the scene and watch attentively without interacting.

To be sure, children and teenagers are not mere spectators. As they attend the scene, they do not just watch but actively engage with each other. There is undertone chatter among themselves. There is ostensive pointing to different aspects of the scene. There is whispering in the ear of a close-by peer. There is the occasional exclamation. There is observation of another peer's observing. There is, in short, what goes under the name of "perspective sharing" (Tomasello 2019). And when it comes to performing a technique for the first time (say, hunting, or the ritual sacrifice of an animal), the young individual is guided by the elders and receives a lot of "peer correction" to ensure the technique is performed correctly. The mastering of techniques thus takes place through a multitude of means and, as such, the *process* of learning can be defined as flexible (Strachan, Curioni, and McEllin, this volume). However, the *content* of what is learned is not: the overall vertical, top-down character of teaching, as far as customs and related techniques are concerned, is prominent, and it implies little possibility for personal experimentation. Customs simply need to be learned and reproduced, as they once were, unless there is a collective, community-wide decision to change them.[3]

The learning of customs in such a rigid way is facilitated by a set of general and simple principles that underlie them. I am referring here to what Claudia Strauss and Naomi Quinn (1997) termed "cultural schemas," which are defined as a socially shared network of strongly connected cognitive elements that motivate action and interpretation of the world. Overall, these cultural schemas, typified as they are by intuitiveness and simplicity, facilitate the rigid transmission of customs and techniques. When there is flexibility in the learning of a technique, this plays out within the rigidity they impose.

Such schemas among the Akha (and many other people) take the form of binary oppositions, the most important of which is the opposition between "inside" and "outside" (see table 10.1). This principle is most saliently instantiated in the construction of the Akha village. Every Akha village in highland Laos is made of a close cluster of houses, spread on a slope, encircled by a belt of forest that sets it apart from fields, other villages, and other types of forest, namely the "outside" world. Even with growing deforestation, at least a thin rim of bush is kept around the village. The village, thus structured, guarantees some protection from external forces like wild animals, foreigners, and evil spirits. Its "inside" is perceived as a safeguarding, positive domain, seen as the fount of "blessing" that, if rightly channeled by way of ritual, counters the negative forces impinging from the outside. Much of Akha ritual life reenacts the separation between these two domains, closing off the intimate haven of the village from the threats of the outside world.

The opposition between "inside" and "outside" is reproduced within the village at the level of the household (every house must have a boundary that is ritually reinforced periodically) where it further intersects with the opposition between "men" and "women." Akha houses are windowless, hence quite dark, rectangular chambers, internally arranged in a quadripartite order. Width wise, they are divided by the main floor beam, which marks the separating point between the "living" side, where eating, cooking, and working take place, and a slightly raised "sleeping side." Lengthwise, they are divided by a partitioning wall that evenly separates a "male side" from a "female side," with elders sleeping closer to the partition. The hearth, where women cook rice and other food, is typically located at the far end of the female side. The ancestral altar, the place where Akha perform 12 sacrifices per year, is also located on the female side. In such a strongly patrilineal society, one might expect to find the ancestral section in the male side of the house. Instead, it is located on the female side because the male/female binary is juxtaposed to the outside/inside binary: being the major source of blessing and protection, the ancestral section represents quintessential "insideness." Women, by virtue of cooking rice and their association with fertility, are also associated with the "inside," in opposition to men, whose distinctive activity is hunting, an "outside" occupation. Unsurprisingly, hunting rituals are performed on the male side.

House building is also coordinated by the opposition between "above" and "below." Initially built on the ground by newly married couples, houses are elevated on stilts after a period of time, keeping with a gerontocratic principle central to Akha culture that associates age and importance with "aboveness." Accordingly, ancestors are "above" their descendants (the ancestral altar hangs from the top rafter) as elders are "above" younger people, both in terms of importance and spatial coding. For instance, it is forbidden for a young person to reach for something that lies on a shelf above the head of an elder, let alone drop it. Like the inside/outside dichotomy, the above/below dichotomy is very salient and takes many forms. It guides gender relations, to the effect that man must be above and woman below, a rule that (reportedly) is most prominently realized in sexual intercourse. It is also instantiated at the household level—for example, the attic is off-limits to everyone except the house owners—where it intersects with the other salient distinction of humans versus animals: humans must be above, animals below. In the same way that outside animals cannot enter the village, inside animals cannot climb on top of roofs. Seeing a pig, dog, chicken, or goat climbing the roof is an ominous event that signals a deficiency of blessing for the household. Customs command that the reckless animal be killed and its meat distributed evenly among all villagers, except for the animal's owner.[4]

Let me make a brief example with reference to house building. After gathering the necessary amount of timber from the forest, a family asks the village chief to call a "sacrificial

Table 10.1
Akha cultural schema: a set of major internalized symbolic binary oppositions that underpins the rigidity of the learning and transmission of Akha customs

Outside	:	Inside
Man	:	Woman
Above	:	Below
Hunting	:	Agriculture
Steep	:	Level
Night	:	Day
Wild animals	:	Domesticated animals
Wet season	:	Dry season
Spirits	:	Humans
Physical deficiency	:	Physical integrity

day," which is when one male member per household in the village helps out with building a new house in exchange for a meal (usually the family kills a pig) and for future reciprocal help whenever anyone else in the village builds a house of their own. The construction takes place in one day. The fact that the construction calls up the whole village over a day already implies a degree level of rigidity and stability since it is possible to build it speedily only if everyone involved has a shared idea of the main structure and process of construction. Effective coordination depends on a baseline of commonly shared assumptions. Most importantly, rigidity in building practices is dictated by customs and the cultural schemas outlined above. For instance, before setting up the stilts, it is "custom" to level the ground because the house—the sphere of the "inside"—must be distinct from the sloped area of the "outside"; it is "custom" to divide the house in a quadripartite fashion and to build a wall that separates male and female quadrants; it is also "custom" to hoist the main house post—where a major ancestral spirit is said to reside and to which a chicken is sacrificed before construction—before any other poles, and so on. Outside these rigid customs-related rules, there is some room for flexibility and, accordingly, variability. By way of illustration, here are three different house layouts that I have seen in my village that reveal the extent of design variability (see figure 10.1).

Depending on family wealth and preferences, house design can vary. Some houses can have two hearths, for example, or an extra uncovered porch, but one will never find a house with two front ladders, or without the female/male separation, or where the female quadrant is higher than the male quadrant. Such configurations would not be permitted by customs. *Maq janr khmq*, people would say; it is a sentence that means "it is not allowed," where the particle *khmq* is exclusively used in the Akha language to denote behavior that is permissible or not permissible by Akha customs (if a person were to doggedly refuse to adhere to customs, they will have to be fined and ultimately ostracized from the community).

To be sure, not all rigidity is the result of explicit rules. There is no rule forbidding people to, say, build their house twice the size of the others or paint it pink. The reason people do not do this is mostly due to "shame" (*xavq dawr baw*). Deviating from certain kinds of community-wide conventions is derided and scorned upon; a person would be chided for "having no shame" (*xavq dawr maq baw*). Significantly, the verb "to have" in this expression

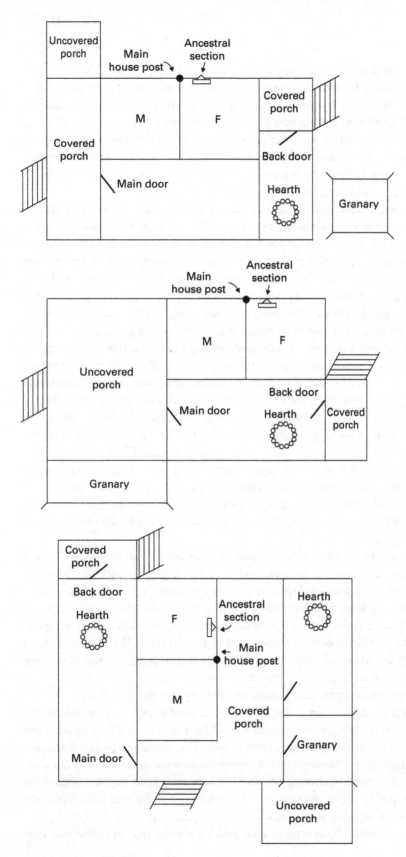

Figure 10.1
Examples of floorplan variation. Room proportions are approximate.

(*baw-e*) is usually employed for the possession of valued culturally central items, such as a house or children. Among the Akha, shame (or embarrassment or shyness) is a positive rather than negative trait: it is by exhibiting it in an appropriate context that a person maintains their reputation. Arguably, the homogenizing effect of shame is a consequence and reflection of the homogenizing and stabilizing effect of customs. Cornelia Kammerer goes as far as saying that "shame is part of customs" (1996a, 83) and that it is central to the Akha egalitarian ethos. All this suggests that there is a considerable social cost associated with flexibility and variability on the select range of techniques that act as social markers.

This brief account should give an idea of the pervasiveness of the above-mentioned cultural schemas, which are internalized by every Akha individual who grows up in the community. Importantly, these schemas run deep enough in the culture to seep into people's bodily *habitus* and even the most banausic techniques. Importantly, the fact that customs imbue techniques sometimes counters effectiveness and instrumental concerns. As I have mentioned, in building a house on stilts, Akha spend a lot of time leveling the ground when they could more easily dig holes for the main posts because the "inside" and "life-affirming spiritual forces" are associated with "levelness"; in the same way, it is important for ritual and other significant social activities to take place on level ground and not on a slope. In planting rice, men must make holes into the ground with a wooden stick while standing, while women, stooped closer to the ground, sow the seeds. The symbolic and the utilitarian merge in hunting, too, where Akha have several rules regulating the appropriate killing of wild animals (e.g., a boar that is killed while munching leaves must not be brought home, as it is seen as blurring the boundaries between the living and the dead). Similarly, in animal sacrifices, which Akha perform on a regular basis, the animal cannot be decapitated while alive: it needs to be bled to death before being chopped apart, in keeping with the parallel between life and physical integrity. Also, Akha do not perform manual work close to the ground when their back faces the sun because they see the shadow as an extension of one's soul—disturbing or piercing it with objects would thus amount to self-affliction. Doing all this is part of being Akha.

Perhaps the most salient signature of the value Akha associate with identity is found in clothing and related techniques of weaving and embroidery. Traditionally, all Akha men wear indigo culottes, an embroidered shirt, and a wide red turban, and they keep their head shaved with a long top-knot dangling from the top; whereas all women wear thigh leggings, miniskirt, bodice, a similarly embroidered indigo shirt, and heavy silver headgear. As Kammerer put it, Akha appear to be "dressed in flags" (1986, 26). The value Akha attach to their cultural identity is important to the point that it shapes the standardization of even very utilitarian techniques, like hunting, house building, or dressmaking.[5]

Of course, despite their remarkable rigidity and stability, we should not think of Akha customs as timeless. There is evidence that these customs have always evolved and diversified historically. Nevertheless, the recent expansion of the Lao state into the highlands and the introduction of new technologies have brought some new challenges to their traditional practice. For example, it was "custom" to fell trees in the forest using an axe. The introduction of the chainsaw, which makes cutting much easier, stimulated debate within Akha communities. Resisted for a few years, its use was eventually adopted after collective deliberation (except for cutting trees at funerals). The same happened to methods of storing rice in field huts during harvest. Traditionally, Akha used a particular type of handwoven container that features in many Akha myths and legends. When Chinese sacks that are used to carry sacha inchi became widely available in the area, nobody adopted them straight away to contain their harvested rice. But when a couple of households first started using the sacks in such a way, the Akha community deliberated on their adoption, and before long everyone followed suit. Overall, the arrival of new technologies and policies from the national government has slightly attenuated the salience that customs used to have (see Ongaro 2019 for a fuller account; see Kammerer 1986 and Tooker 2004 for an account of these changes in Thailand). Akha say that customs were stronger before. What is significant, however, is that the decision to abandon a practice or to take up a new one is usually decided collectively, in line with the collective identity-defining character of customs.[6]

Discussion and Conclusions

I have illustrated two sets of techniques among the Akha that exhibit diametrically opposite characteristics. On the one hand, herbal medicine is typified by high flexibility in learning. After acquiring knowledge from an expert herbalist, apprentices are relatively free to experiment with their own techniques and to expand their repertoire in idiosyncratic ways (especially through dreaming), so that the ultimate knowledge and practice one develops usually ends up being substantially different from that of the teacher. Besides, the secrecy that surrounds the practice of herbal medicine preempts sharing or "peer correction." This, in turn, accords with the amply demonstrated variability of herbal medicine know-how among the Akha and in other neighboring ethnic groups.

On the other hand, Akha ancestral customs, which are pervasive to the point of affecting a wide range of practical techniques, are typified by rigidity. Although there is a level of flexibility in the process of learning, the transmission of what is learned is remarkably rigid. The underlying ethos is to faithfully reproduce the ancestors' practices and thereby perpetuate Akha cultural identity. The rigidity in the learning of customs—facilitated by "cultural schemas" and vertical teaching—is consistent with their stability: that is, to the striking degree of homogeneity that one finds across the Akha world in space and in time. Nevertheless, as Sadie Tenpas, Manon Schweinfurth, and Josep Call (this volume) point out, the tension between flexibility and rigidity can play out at different levels and scales. And as I have shown, though rigid in their transmission, customs-related techniques display some flexibility at the community level—for example, through the collective discarding of old customs or the incorporation of novelty. These changes tend to take place after collective deliberation and in the face of socioeconomic changes affecting the Akha area in recent years.

Since the key variable determining this difference is cultural identity—its unrelatedness to herbal techniques allows technical flexibility; its salience to customs implies technical rigidity—it is worth considering once again the Akha geopolitical context, for the ultimate explanation of these patterns (particularly of the rigidity of customs) lies in the multiethnic and unequal scenario that typifies highland Southeast Asia. As Deborah Tooker (2012) argued with reference to the Akha of Thailand in the early 1980s, the shared practice of customs that has been fundamental in forging an Akha sense of collective identity has also been highly empowering in a context of unequal power relations with the lowlands. She argued that the very character of Akha customs—especially the emphasis on the inside/

outside distinction and the attachment to ancestral tradition—played a key role in strength-ening this sense of empowerment. The "outside" is the realm of lowlanders and evil spirits and of the dangers that they both represent in similar ways. The correct adherence to ancestral rules allows Akha to tap into the ancestors blessing radiating from the inside and to prevent these threatening outside forces from draining it. It is a system of customs, in short, that maps on the unequal political context in which they find themselves.

The rigidity and stability of customs and associated techniques should be seen as a product of such interethnic relations, for it helps in perpetuating the existence of the Akha as an ethnic group. The same applies to the stable *rejection* of select foreign techniques. Chief among these is writing. As Scott (2009) argued, many upland peoples of Southeast Asia have historically preferred to adhere to their oral tradition and reject writing because of its associa-tion with bureaucracy and the state, which they consciously wanted to avoid. As I mentioned earlier, the benefits afforded by recent technologies have sparked debates within Akha com-munities on how to negotiate ethnic identity, but what is significant is that the decision to incorporate or resist them has typically been a collective one.

This analysis of Akha customs arguably supports Marcel Mauss's anti-diffusionist sug-gestion in *Techniques, Technologies and Civilisation* (2006) that the most interesting aspect to study about cultural transmission is not so much the borrowing of techniques across societies (borrowing is a normal state of affairs) but the rejection thereof. Mauss surveyed a number of dramatic cases of non-transmission of even very practical technologies across the ethnographic record. He found, for instance, that Athabaskans in Alaska refuse to adopt Inuit kayaks, despite their being self-evidently more suited to the environment than their own boats; Inuits, similarly, refuse to adopt Athabaskan snowshoes. To Mauss, these exam-ples of non-transmission were eloquent about the nature and role of culture (and civilization). The fact that even very instrumental and effective techniques familiar to anyone are not adopted by a certain people allow us to see "culture" as a collective act of conscious refusal (Graeber 2013).[7]

Given that in the cognitive study of techniques, "culture" and "cultural transmission" take on a much broader meanings, I have preferred keeping the term "cultural identity" to the same effect. I have suggested that, in similar ways, the rigidity and stability of a select range of techniques should also be seen as a result of the power of cultural identity in modulating transmission. Techniques that lie outside the domain of identity markers—for example, herbal pharmacopoeia—are clearly unaffected. In fact, in the context of herbal medicine, it is the plants that one obtains from the outside (especially from faraway places) that are most cherished because they are assumed to be the most powerful. This, at least, is the case among the Akha. I should note that the ways in which cultural identity intersects with cultural transmission shows a variety of configurations across the globe. Rita Astuti (this volume), for instance, shows that among the Vezo of Madagascar, ancestral customs and cultural identity do not overlap in the same way they do among the Akha, so some areas of technical know-how that might be identity-defining (sailing techniques) are not permeated by ancestral rigidity. The role of cultural identity in inflecting the tension between flexibility and rigidity can itself be culturally diverse.

Let me end the chapter by highlighting one feature that both the mode of transmission of flexible herbal medical techniques and that of rigid customs-related techniques have in common, albeit for different reasons. This is their apparent restraint on increasing efficacy. In the case of herbal medicine, I showed that the Akha way of sourcing and preparing medicine is linked to secrecy and personal paths to knowledge (e.g., through dreams). This has the effect of precluding the sharing of knowledge and collective experimentation, which is usually key in discovering effective medical treatments. In the second case, techniques do not achieve maximal efficacy because they are infused with a nonutilitarian dimension dictated by customs. Why do people level the ground before building a house while they could more easily dig holes for the main posts? Why must one not decapitate the animal when it is alive even though it slows down food preparation? Because it is "custom" to do so. Customs steer techniques away from their purely instrumental purpose. In both cases, the value attached to the specific mode of transmission of a technique eclipses, to a certain extent, concerns about its practical utility.

Notes

1. I use "Akha" both in singular and plural form, as other anthropologists have done since the 1980s.

2. According to a popular definition (Miller et al. 2013, ix), the placebo effect "is generally understood as consisting of individuals' responses to the psychosocial context of medical treatments, 'inert' interventions, or clinical encounters, as distinct from the inherent or characteristic physiological effects of medical interventions." See Ongaro 2019, 40ff., for a critical discussion.

3. This does not mean that we should view the Akha as a people burdened by doctrine and devoid of critical thought. Quite the contrary. Because customs focus on the correctness of *practice*, they leave ample freedom to speculate about theoretical and metaphysical matters (Tooker 1992). However, given the focus on the specifically technical aspects of Akha knowledge, these matters will not be discussed further here.

4. These two dichotomies also intersect with the dichotomy between "steep" and "level." Levelness is associated with stability and "insideness," and slopes with danger and "outsideness." This matrix of binary opposites is further interwoven with the distinctions between day and night, wet and dry seasons (some activities can only take place at specific time of the day and year), and many less important distinctions.

5. See Zuckerman and Enfield 2022 for an account of a similar effect played by cultural identity on techniques of house construction.

6. This applies to many other aspects of material culture. For instance, while in 2010 every Akha man used to go to the fields wearing a red turban, by 2015, no man would wear the turban anymore. Change occurred rapidly and was a product of conscious adhering to collective ethnic identity. Working among the Akha in Thailand, Kammerer (1996b) noticed a similar sudden pattern of change regarding religion. Akha had been extremely resistant to Christianity for decades. But when conversion happened, it took place rapidly across virtually all Akha villages in northern Thailand.

7. It might be intuitive to think, for instance, that given their vital instrumental value, the adoption of techniques like weaponry should be virtually unaffected by factors other than their effective use. Even in this extreme case, we find ethnographic examples to the contrary, such as that of the Alu warriors of the Solomon Islands, who knowingly deprived themselves of an effective device that could save lives. George Brown writes: "The natives of Bougainville and Bouka [sic] Islands do not use shields in fighting and they ridicule those who use them. If Alu and Ruviana canoes are fighting, the Alu natives contend that the advantage is on their side, as only one or two men stand up and use the bow while the others paddle. In the Ruviana canoe, on the contrary, each man uses his shield, and so, having only one hand to paddle the canoe he is in the opinion of the Alu men at great disadvantage" (1910, 161). Pierre Lemonnier's aptly titled book *Technological Choices* (1993) offers us a compendium of similar cases.

References

Anderson, E. F. 1993. *Plants and People of the Golden Triangle: Ethnobotany of the Hill Tribes of Northern Thailand*. Portland: Timber Press.

Bouchery, P. 1993. "The Genealogical Patronymic Linkage System of Akha and Hani." Première Conférence Internationale sur la culture Hani/Akha, Kunming, 26 Février-3 Mars 1993.Communication : « The Genealogical Patronymic Linkage System of Hani and Akha ». Publication : Aséanie 23 (2009), cf. supra .

Brown, G. 1910. *Melanesians and Polynesians: Their Life-Histories Described and Compared*. London: Andesite Press.

Dubost, J. M. 2014. *Medicinal Plants Used in Nam Kan National Park: An Ethnobotanic Survey in Hmong, Lue and Lamet Communities*. Laos: Gibbon Experience Project.

Dubost, J. M., C. Phakcovilay, C. Her, A. Bochaton, E. Elliott, E. Deharo, M. Xayvue, et al. 2019. "Hmong Herbal Medicine and Herbalists in Lao PDR: Pharmacopeia and Knowledge Transmission." *Journal of Ethnobiology and Ethnomedicine* 15 (1): 1–15.

Foster, G. M. 1976. "Disease Etiologies in Non-Western Medical Systems." *American Anthropologist* 78 (4): 773–782.

Geusau, L. A. von. 1983. "Dialectics of Akhazan: The Interiorizations of a Perennial Minority Group." In *Highlanders of Thailand*, edited by J. McKinnon and W. Bhruksasri, 241–279. Oxford: Oxford University Press.

Geusau, L. A. von . 2000. "Akha Internal History: Marginalization and the Ethnic Alliance System." In *Civility and Savagery: Social Identity in Tai States*, edited by Andrew Turton, 122–158. London: Curzon Press.

Graeber, D. 2013. "Culture as Creative Refusal." *Cambridge Journal of Anthropology* 31 (2): 1–19.

Hanks, J. R. 1974. "Recitation of Patrilineages among the Akha." In *Social Organization and the Applications of Anthropology: Essays in Honor of Lauriston Sharp*, edited by R. J. Smith. Ithaca, 114–127. NY: Cornell University Press.

Heinrich, M., and A. K Jäger. 2015. *Ethnopharmacology*. London: Wiley & Sons.

Kammerer, C. 1986. "Gateway to the Akha World: Kinship, Ritual, and Community among Highlanders of Thailand." PhD diss., University of Chicago.

Kammerer, C. 1990. "Customs and Christian Conversion among Akha Highlanders of Burma and Thailand." *American Ethnologist* 17 (2): 277–291.

Kammerer, C. 1996a. "Begging for Blessing among Akha Highlanders of Northern Thailand." In *Merit and Blessing in Mainland Southeast Asia in Comparative Perspective*, edited by C. Kammerer and N. B. Tannenbaum. New Haven, 79–97. CT: Yale University Press.

Kammerer, C. 1996b. "Discarding the Basket: The Reinterpretation of Tradition by Akha Christians of Northern Thailand." *Journal of Southeast Asian Studies* 27 (2): 320–333.

Lemonnier, P., ed. 1993. *Technological Choices: Transformation in Material Cultures since the Neolithic*. London: Routledge.

Lewis, P. W. 1969. *Ethnographic Notes on the Akhas, Vol I*. New Haven, CT: Human Relations Area Files, Inc.

Lewis, P. W. 1989. *Akha-English-Thai Dictionary*. Chiang Mai, Thailand: Development and Agricultural Project for Akha (DAPA).

Mauss, M. 2006. *Techniques, Technology and Civilisation*. New York: Durkheim Press.

Miller, F. G., L. Colloca, R. A. Crouch, and T. J. Kaptchuk. 2013. "Preface." In *The Placebo: A Reader*, edited by F. G. Miller, L. Colloca, R. A. Crouch, and T. J. Kaptchuk. Baltimore: John Hopkins University Press.

Moerman, D. E. 2002. *Meaning, Medicine, and the "Placebo Effect."* Cambridge: Cambridge University Press.

Ongaro, G. 2019. "The 'Placebo Effect' in Highland Laos: Insights from Akha Medicine and Shamanism into the Problem of Ritual Efficacy." PhD diss., London School of Economics.

Ongaro, G. (forthcoming). "When new shamans enter the stage: traditional customs and ecstatic healing among the Akha of northwestern Laos." In *Emerging Ethnographic Perspectives on Laos*, edited by P. D. Lutz and R. Stolz. Copenhagen: NIAS Press.

Pake, C. V. 1987. "Medicinal Ethnobotany of Hmong Refugees in Thailand." *Journal of Ethnobiology* 7 (1): 13–26.

Scott, J. C. 2009. *The Art of Not Being Governed: An Anarchist History of Upland Southeast Asia*. New Haven, CT: Yale University Press.

Sprenger, G. 2011. "Differentiated Origins: Trajectories of Transcultural Knowledge in Laos and Beyond." *Sojourn: Journal of Social Issues in Southeast Asia* 26 (2): 224–247.

Strauss, C., and N. Quinn. 1997. *A Cognitive Theory of Cultural Meaning*. Cambridge: Cambridge University Press.

Tomasello, M. 2019. *Becoming Human: A Theory of Ontogeny*. Cambridge, MA: Harvard University Press.

Tooker, D. E. 1992. "Identity Systems of Highland Burma: 'Belief,' Akha Zan, and a Critique of Interiorized Notions of Ethno-Religious Identity." *Man* 27 (4): 799–819.

Tooker, D. E. 2004. "Modular Modern: Shifting Forms of Collective Identity among the Akha of Northern Thailand." *Anthropological Quarterly* 77 (2): 243–288.

Tooker, D. E. 2012. *Space and the Production of Cultural Difference among the Akha Prior to Globalization: Channeling the Flow of Life*. Amsterdam: Amsterdam University Press.

Wang, J. 2013. "Sacred and Contested Landscapes: Dynamics of Natural Resource Management by Akha People in Xishuangbanna, Southwest China." PhD diss., University of California, Riverside.

Zuckerman, C. H., and N. J. Enfield (2022). "The unbearable heaviness of being Kri: house construction and ethnolinguistic transformation in upland Laos." *Journal of the Royal Anthropological Institute* 28 (1): 178-203.

11 Why Do Children Lack Flexibility When Making Tools? The Role of Social Learning in Innovation

Nicola Cutting

The capacity of humans to manufacture and use tools has evolved far beyond that seen in any other species (Boyd, Richerson, and Henrich 2011). This is thought to be due to our propensity for cumulative culture (Boyd and Richerson 1996), where new techniques are copied throughout the social group and then improved on in a ratchet-like effect (Tomasello 1999; Tomasello, Kruger, and Ratner 1993). Cumulative culture involves two factors termed "dual engines" (Legare and Nielsen 2015)—imitation and innovation. The cumulative culture literature has predominantly focused on imitation, how techniques are transmitted through groups through social learning, with children demonstrating faithful replication of techniques from a young age. Only more recently have researchers focused efforts toward measuring propensity for the second engine—innovation. This chapter focuses on tool-innovation, which is defined as the design and manufacture of new tools or the modification of existing tools to solve a problem (Cutting et al. 2014).

In contrast to human disposition for social learning, research with children has demonstrated that capacity for innovation is somewhat weaker. However, studies into children's innovation have predominantly used paradigms based on Ramsey, Bastian, and van Schaik's (2007) where innovation is defined as an individual, asocial process devoid of external social influences. The difficulty children demonstrate in these individual problem-solving tasks may mask their ability for more socially mediated innovations (Rawlings and Legare 2020). This chapter will discuss how tool-innovation would be better categorized as a socially embedded process, where socially acquired information affects the innovation of new tools and then these innovations are transmitted back to the social group. In such a socially interconnected society, I would argue that all tool-innovation contains some degree of social influence and can never be truly asocial. Social influence could be by direct communication and teaching or by indirect social influences such as materials clearly having been made by other individuals. To date, most tool-innovation paradigms that have been used are too close to the asocial end of an asocial-to-social continuum.

Cumulative culture demonstrates the need for both technical rigidity and technical flexibility for different processes. Rigidity is required in social learning to ensure faithful transmission of information. In contrast, innovations require technical flexibility. Individual innovations can occur through flexibility in behavioral repertoires (i.e., the ability to add or remove behaviors) or flexibility in the behaviors themselves, evidenced by modifications within those behaviors. Following an overview of cumulative culture, how innovations occur, and what studying children can tell us about the development of techniques, this chapter will predominantly focus on technical flexibility in children's innovations. An overview of children's capacity for asocial and socially mediated tool-innovation will assess flexibility in children's behavioral repertoires. This will be followed by analysis of flexibility in children's tool behaviors once learned. Throughout this chapter, I will argue that the relatively new field of children's tool-innovation has focused too heavily on children's ability to innovate in asocial contexts, which may be masking capacity for technical flexibility.

How Do Tool Techniques Develop? The Role of Cumulative Culture

There is no doubt that the human species has been exceptionally successful in creating and developing an abundance of tools to aid our needs in nearly all aspects of our lives. It is difficult to imagine how our lives are not enhanced by the presence of simple tools such as cutlery and pencils to much more complex tools such as smartphones and computers. That is not to mention the hugely complex tools outside most of our daily lives such as spacecraft and hadron colliders. While we are not the only species to use tools (see Shumaker, Walkup, and Beck 2011 for a catalog of animal tool behavior), humans are unique in the sheer number and complexity of tools that we use. Our nearest relative, the chimpanzee, with which we shared a common ancestor roughly 7 million years ago, uses an array of tools including sticks for termite-fishing (Koops 2020) and nut-cracking (Boesch et al. 2019), yet they have failed to progress beyond this comparatively rudimentary set of tools. So, how have humans managed to achieve such a sophisticated technological toolkit?

Human technological success, it is suggested, is due to our capacity for cumulative technological culture: the gradual progression of techniques and tools, building on what has come before by enhancing its complexity and efficiency (Dean et al. 2014; Legare and Nielsen 2015). Over generations, these progressions accumulate to create a technique or tool that is too complex to have been invented independently by one individual (Boyd and Richerson 1996; Tennie, Call, and Tomasello 2009; Tomasello, Kruger, and Ratner 1993). Cumulative technological culture is proposed to be driven by the dual engines (Legare and Nielsen 2015) of imitation and innovation. For cumulative evolution to occur, a group must produce modifications to existing techniques, and these modifications must then be transmitted throughout the group (Charbonneau 2015). Most theories suggest that human uniqueness for cumulative technological culture is the result of our capacity for high-fidelity social learning—active teaching and faithful imitation. As such, most research has focused on the social transmission aspect of culture, with a wealth of research demonstrating humans to be faithful social learners, copying to a high fidelity from a young age. By age 2, children will over-imitate, copying causally irrelevant actions, a phenomenon that appears to be unique to humans and that clearly showcases our propensity for faithful transmission (see Hoehl et al. 2019 for a review).

While these lines of research have focused on important questions surrounding the uniqueness of human cumulative culture and its evolutionary origins, despite being recognized as essential, the second component of cumulative culture—innovation—has been somewhat neglected. Faithful transmission of techniques may be the strong driving force behind cumulative evolution, but without deviations from faithful copying, new behaviors and products would not emerge (Charbonneau 2015; Kendal, Giraldeau, and Laland 2009). The lack of focus on innovations has led to the presumption that innovation rates and ability are comparable between human and nonhuman species (Dean et al. 2014; Tennie, Call, and Tomasello 2009; Tomasello, Kruger, and Ratner 1993); however, there is little evidence to support this claim.

How Do Innovations Occur?

There are three ways that new innovations can occur: incremental improvements, serendipity, and recombination (see Muthukrishna and Henrich [2016] for a more in-depth overview, including real-world examples for each type of innovation). Innovations are commonly the product of incremental improvements of what has come before. These innovations are also referred to as modifications and must retain some aspects of their ancestor but also differ in some respect (Charbonneau 2015). These are the type of innovation that fits most neatly within the cumulative culture narrative and explains gradual advances in culture that occur over generations but often ultimately lead to a product that is unrecognizable from its beginnings.

Serendipitous innovations are nondirected and can occur because of mistakes that are made by individuals. These mistakes lead to the discovery of new techniques or products and can be made during individual learning or when attempting to replicate a technique through social learning (Henrich 2004; Powell, Shennan, and Thomas 2009). This type of innovation that is the result of good fortune rather than brought about by design often creates larger, more step-change innovations.

Finally, innovations can be the result of recombining existing knowledge or techniques in new ways. This can often give the impression of impressive step-change innovations but more accurately involves the innovator being exposed to several individual ideas that happen to come together in time (Charbonneau 2016). As such, recombination innovations in history have often been proposed by several individuals at around the same time because of similar exposure to information and ideas; an example is Darwin and Wallace's theory of evolution (Muthukrishna and Henrich 2016).

A number of disciplines have tracked and documented innovations throughout history. Developmental psychology is one approach that has been taken to investigate the potential origins of innovative ability in humans.

Why Study Innovation in Children?

The purpose of this book is to bring together ideas from different disciplines to approach the topic of flexibility and rigidity in the use and transmission of techniques. This brings the question of how children fit into this narrative. Do children produce long-lasting traditions that are transmitted within and across generations? Is there an active role for children in cumulative cultural evolution? And, as is the focus of this chapter, what is children's role in the innovation of new techniques? In this section, I will discuss ways that research with children can fit into this narrative and explain why my work on tool-innovation has taken a developmental perspective.

There are some traditions that spread and evolve that are ubiquitous with children. Playing games is one such example. Games are not restricted to any one age group; however, when we think of playing games, children are usually the first group of people that come to mind. Although this topic is difficult to study because of gaps in the historical record, Olivier Morin (2015) has systematically explored 103 French games documented by Rabelais that appeared between the Middle Ages and the early 1900s. Although the origins of some games are difficult to ascertain (i.e., they may have been designed and introduced to children by adults), Morin concludes that children ensure the longevity of games and play an active role in their development. Morin points out that what is most impressive about the transmission of children's games is that diffusion between children is horizontal. Horizontal transmission should be weaker than the more traditional vertical transmission from generation to generation, yet children's games appear to be just as long-lived as adult and cross-generational games. Thus, in the domain of games, children are strong players with the ability to produce and maintain long-lasting traditions.

Moving back to the focus of this chapter, technological culture, and thinking about recent advances in this domain, children are not likely to be the first people that come to mind. For instance, advances in smartphone technology are most likely attributed to large companies such as Apple or Samsung and the adult designers who work for them. When thinking about advances in space technology, you most likely think of them as coming from highly skilled adults with years of expertise. This is not surprising as research has shown expertise to be associated with inventions (Roux et al. 2018), and as expertise takes time to acquire it is therefore more appropriately associated with adults rather than children (see Roux et al., this volume, for an extended discussion on the role of expertise).

Although innovations in technology are not completely absent in children (see Rawlings and Legare 2020 for examples of innovations by children during the COVID-19 pandemic), the advances seen in modern technology that spread widely across the globe are rarely the product of innovations produced by children. So why have developmental psychologists, including myself, become interested in the development of innovation?

Children present researchers with the opportunity to examine the mechanisms involved in technological culture. Taking a comparative approach allows researchers to compare the abilities of humans with our nearest nonhuman primate relatives. By comparing performance on tasks requiring social learning and innovation, we can tease apart the differences and similarities between species, with the ultimate aim of discovering what is unique about humans that has allowed us and our technology to evolve beyond that of our nearest relatives. Children, rather than adults, are more appropriate to use for this research because of their relative lack of experience with the world. Of course, this does not imply that human children and chimpanzees have been raised with equivalent experiences, but children's relative naiveness to the world compared with adults gives us the best available opportunity to make comparisons. It is also assumed that the underlying cognitive and social mechanisms involved in innovation remain relatively unchanged over time, so our understanding of the development of these mechanisms in modern society can therefore inform our knowledge of these processes in our ancestors (Dean et al. 2012)

While children clearly contribute toward cumulative technological culture due to their vast capacity for learning new technologies and skills through social learning (see Hoehl et al. 2019), as stated previously, it is the second engine of cumulative culture that this chapter will focus on—innovation. Studying innovation in children gives us an opportunity to speculate when this capacity may begin in development and to explore the underlying cognitive abilities and mechanisms that are needed for innovation to occur.

The Role of Flexibility and Rigidity

Human propensity for cumulative culture demonstrates both technical rigidity and flexibility. The successful social transmission of useful tool behaviors requires rigidity. Human success in tool transmission has been credited to faithful (rigid) imitation of techniques (Dean et al. 2014; Tennie, Call, and Tomasello 2009; Tomasello, Kruger, and Ratner 1993). In contrast, the second component of cumulative culture—innovation—requires innovators to display flexibility. We must use our knowledge and skills in a more flexible manner to create new solutions, either by recombining our knowledge in new ways or by making new incremental improvements (Muthukrishna and Henrich 2016). Success in the technological domain that has allowed humans to occupy and thrive in varied environments worldwide clearly illustrates the species-level flexibility we have. What is less known is the flexibility we have as individuals in the technological domain (see Pope-Caldwell, this volume, for further discussion of this point). Behavioral flexibility can be demonstrated and measured at two levels (Ramsey, Bastian, and van Schaik 2007): flexibility in the behavioral repertoire (i.e., adding or removing behaviors) and flexibility in the behavior itself (i.e., modifying an existing behavior). Relating this to tool-innovation research, successful innovation of new tools represents repertoire flexibility. This is the first type of flexibility that I will explore. Then I will explore how the second level of flexibility differs depending on whether tool behaviors are the result of individual innovation or social learning.In the next section, I will give an overview of children's innovative abilities in the domain of tools by outlining research into independent, asocial tool-innovation that has been explored over the last decade, before going on to talk about more recent work investigating the social influences that affect children's ability to innovate.

Children's Toolmaking: Independent (Asocial) Tool-Innovation

As detailed above, children can learn new skills, including how to use tools, by watching and imitating those around them. This skill is evident from infancy (Nielsen 2006), and imitation becomes more faithful through development (Hoehl et al. 2019). In contrast to children's ability to faithfully replicate tool behaviors that they observe in others, children have great difficulty innovating tools for themselves. Initial investigations into children's ability to innovate novel tools measured innovative ability based on the notion that innovation is a predominantly asocial process that does not involve social learning (Ramsey, Bastian, and van Schaik 2007). It is also important to point out here that these studies do not expect individuals to come up with innovations that have never been generated by other individuals before—so-called historical (H) creativity (Boden 1996)—but instead measure innovative success as a new idea that is unique to the individual, or psychological (P) creativity. Under the asocial definition of innovation, behavior needs only to be new to the individual (Ramsey, Bastian, and van Schaik 2007).

The most common tool-innovation paradigm studied requires children to generate the idea of and manufacture a hooked tool from a pipe cleaner to fish a bucket out of a tall narrow tube (Beck et al. 2011) (see figure 11.1). This task was based on a study conducted with New Caledonian crows (Weir et al. 2001) in which a female crow "Betty" spontaneously manufactured a wire hook to solve the task when her mate "Abel" flew away with the tool needed to solve the problem. Children ages 3 to 10 were tested on the task, along with a "mature" sample of 16-year-olds. When asked to retrieve the bucket from the tube, children had remarkable difficulty producing the required hooked tool to complete the task. Very few children age 5 or younger were successful, with success gradually increasing with age, with just over half of children successful at age 8 and 80 percent successful at age 10 (see figure 11.2).

Children's difficulty with the "hooks task" has been shown to be robust across a number of studies conducted by multiple research groups (Cutting, Apperly, and Beck 2011; Gönül et al. 2018; Neldner, Mushin, and Nielsen 2017; Voigt, Pauen, and Bechtel-Kuehne 2019) and across cultures (Nielsen et al. 2014). Similar levels of success were observed in a sample of African Bushmen children whose culture necessitates manufacturing tools for themselves and consists of fewer premade tools than seen in Western society, where most tool-innovation research has been conducted.

Figure 11.1
Apparatus and materials used in the "hooks task." *Source*: Photo by Author.

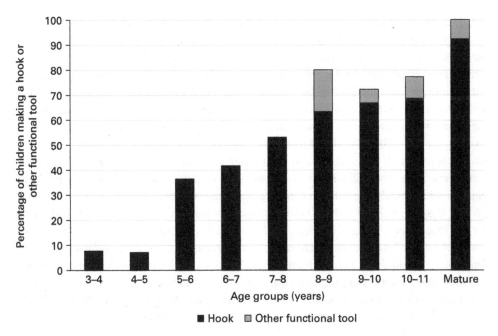

Figure 11.2
Percentage of children innovating a hook tool. *Source*: Recreated using data from Beck et al. 2011.

Most studies investigating children's tool-innovation ability have been based on the "hooks task" paradigm described above. However, a small number of other paradigms have been used and have generated similar findings. Similar levels of success were found on a task requiring children to innovate a long straight tool needed to push a reward from a horizontal tube (Cutting, Apperly, and Beck 2011). Children's difficulties innovating pipe-cleaner tools on tasks using a vertical tube (requiring a hook) and a horizontal tube (requiring a long straight tool) have also been shown to extend to studies requiring tools to be made from other materials (Cutting 2013; Neldner et al. 2019; Voigt, Pauen, and Bechtel-Kuehne 2019). The "floating peanut" task, requiring children to use water as a tool to retrieve a reward from a vertical tube by floating it to the top, is a different paradigm used to test tool-innovation ability. This task has generated similar success levels to the vertical- and horizontal-tube tasks (Hanus et al. 2011). Additionally, children aged 4 to 9 were found to have difficulty constructing variously shaped tools from LEGO sets to push a cube from one location to another inside a puzzle box (Mounoud 1996). Together, these studies suggest children's asocial innovation difficulty to be a robust phenomenon.

Why Is Independent Tool-Innovation So Difficult?

To date, research has focused on how children's tool-innovation ability may be constrained by cognitive capacity. It has been suggested that children's poor performance is due to the ill-structured nature of tool-innovation (Chappell et al. 2015; Cutting et al. 2014). Most problems we encounter in daily life are well-structured; they have clear start point and goal, and we simply choose between different options available to us. For example, in a tool-choice paradigm, we have a start-state of an apparatus containing a reward and two available tools;

the goal-state is to retrieve the reward, and the transformation to get from the start-state to the goal-state involves selecting the optimal tool to complete the task. In contrast, tool-innovation is ill-structured. The start-state and goal-state are the same as in the well-structured example, but there is little information of how to get from one to the other. The solver must generate and execute the solution for themselves (Jonassen, Beissner, and Yacci 1993; Reitman 1965). To do this involves executive ability. One must inhibit actions that are incorrect, switch between different strategies, and hold information about the problem in working memory. The difficulty of ill-structured problems is that they encompass all executive components in conjunction with each other and cannot simply be reduced to their component parts. The difficulty of ill-structured problems is demonstrated in studies with patients with frontal lobe damage (Shallice and Burgess 1991) and children with autism (White, Burgess, and Hill 2009). These participants were shown to perform at typical levels for lab-based executive tasks that tap individual executive functions but performed at comparatively lower levels in ill-structured tasks that required the use of multiple executive functions in conjunction with each other (Goel, Pullara, and Grafman 2001; Shallice and Burgess 1991; White, Burgess, and Hill 2009). It is therefore likely that the protracted development of children's executive abilities (Dumontheil, Burgess, and Blakemore 2008) may be a factor in their difficulty with ill-structured tool-innovation tasks.

Learning to Make Tools from Others: Imitation and Emulation

The above studies suggest that asocial innovation is very difficult for children. The design of these innovation studies also allows us to observe children's capacity to manufacture tools following social learning. In these studies, if children were not successful at innovating the required tool for themselves, then they next received a demonstration of how to manufacture the required tool, termed the "tool-creation demonstration." This provided children with the opportunity to imitate the correct toolmaking method. In some instances, unsuccessful children were provided with a demonstration in which the experimenter held their own pipe cleaner horizontally and manipulated one end to form the required hook tool (Beck et al. 2011; Cutting, Apperly, and Beck 2011). Importantly, in these demonstrations, the experimenter did not show the correct orientation the tool needed to be in or enter it into the apparatus. Despite not being a full demonstration of how to complete the task, the vast majority children (at least 80%) quickly modified their own pipe cleaner into the required hook tool and successfully retrieved the bucket from the tube.

These findings are in line with a wealth of research demonstrating that children easily learn how to manufacture their own tools by watching others. For example, from around 30 months, infants can manufacture a rattle toy consisting of three parts after watching a model (Barr and Wyss 2008; Hayne, Herbert, and Simcock 2003; Herbert and Hayne 2000).

Later tool-innovation studies included an additional demonstration phase for children. If children were unsuccessful at innovating a hook tool for themselves, they received what has been termed a "target-tool demonstration." In this demonstration, the experimenter showed children an example of the end-state tool, giving children an opportunity to emulate making the tool needed for the task rather than the opportunity to imitate the whole toolmaking process. As with the tool-creation demonstration described above, the end-state target-tool was presented in a horizontal orientation, and no demonstration of how to use the tool on the task was offered. Target-tool demonstrations were included in several tool-innovation studies (Beck et al. 2014; Chappell et al. 2013; Cutting et al. 2014, 2019), yielding modest improvements in children's ability to succeed on the task that did not reach the success levels seen after the tool-creation demonstrations. Children aged 6 to 7 were better able to emulate making a successful tool after seeing a target-tool example than younger children (Chappell et al. 2013), with most younger children requiring the full tool-creation demonstration to successfully complete the task.

Summary of Independent (Asocial) Innovation

So far, the presented research has shown that children find it extremely difficult to innovate simple novel tools for themselves, and this contrasts with their aptitude to learn how to manufacture tools from others. While tool-innovation undoubtedly involves cognitive skills, which likely makes it difficult, we must consider whether the tool-innovation paradigms discussed so far truly capture the nature of innovations that occur in real life. These paradigms require children to work independently with little social influence to create a novel solution they will not have encountered before. These environments stand in contrast to the highly social worlds that children inhabit.

There are two key factors of innovation studies that need to be addressed. First, real-world innovations are likely to involve more social influence. Rather than being isolated, innovators are more likely to be surrounded by other individuals and other forms of social information. Second is the type of innovation that these tasks require. Although children will likely have some experience with the properties of pipe cleaners because they are a common craft material in schools and nurseries, and children are likely to have knowledge about hooks, the requirement to create this novel tool is very much a step-change. While such step-change innovations do occur, they are likely to be a rare form of innovation and would be characterized as an innovation by recombination. The next section explores research that has looked at the scaffolding of innovations and how children use social information to help them construct novel tools without the need for full demonstrations of tool manufacture.

Socially Mediated Tool-Innovation: Can Information from Others Help Children to Innovate Tools?

Many studies have aimed to investigate the mechanisms underlying tool-innovation to try to establish where children encounter difficulty. Although not the stated purpose of these studies, their design of providing information to children within a social context allows us to draw some conclusions about socially mediated innovations more akin to those likely to occur in real life. This section outlines these studies and discusses the contribution they make to our understanding of children's ability for socially mediated innovations.

Affordances

Although children are presumed to have knowledge of the pliable properties of pipe cleaners, this was confirmed by a study in which one group of children took part in a warm-up exercise manipulating pipe cleaners by winding them around a pen and creating spiral shapes (Beck et al. 2011). Highlighting the affordance of the pipe cleaners in this social manner did not improve innovation and was therefore taken as evidence that children already possessed knowledge of pipe-cleaner properties.

As stated previously, one of the main difficulties of the current recombination-style tool-innovation paradigms is the high cognitive load placed on children. They must first generate the idea of a hook tool and then recognize the utility of the pipe cleaner in allowing them to achieve this. The tool-choice paradigm presented by Sarah Beck and colleagues (2011) demonstrated that children could easily recognize the utility of a hooked tool, quickly and effectively using it to solve the task; however, this task did not have an innovative component. Karri Neldner, Ilana Mushin, and Mark Nielsen (2017) sought to reduce cognitive load while maintaining the need for innovation. In this study, children were presented with a hook tool that had the non-hooked end curled over, preventing it from entering the apparatus. The provision of the focal affordance (hook shape) as visual information reduced cognitive load as children were only required to recognize rather than generate the appropriate affordance of the material. I would argue that the design of this task would put it in an innovation category closer to incremental improvement. Children aged 3 to 5 were nine times more likely to innovate a functional tool when the focal affordance was visible. However, successful innovation was still only seen in 45 percent of children (compared to 14% in the affordance nonvisible condition), showing that although children were helped by the reduction in cognitive load and social information, innovation was still a difficult feat for young children.

Nonfunctional Tool Examples

Another study that used social information in an attempt to scaffold children's tool-innovation presented children with a correctly shaped but nonfunctional tool (Cutting et al. 2019). The researchers presented children with oversized pipe-cleaner hooks with which to solve the vertical-tube problem. However, rather than scaffolding innovation and acting as a prompt for creating the required tool, the presence of the oversized hook appeared to hinder children's ability. In line with other studies (Beck et al. 2011; Neldner et al. 2019), children easily recognized the affordance of the hook tool, choosing to use the oversized hooked pipe cleaner significantly more than the straight pipe cleaner. However, children were poor at modifying the nonfunctional hook into a functional tool or manufacturing their own correctly sized hook from the straight pipe cleaner provided. In fact, compared to a baseline condition where children received two straight pipe cleaners, children who received an oversized hook and a straight pipe cleaner were less likely to create a functional tool to solve the task. This therefore suggests that the presence of a correct but nonfunctional tool actually hindered children's ability to solve the problem at hand.

In comparison to the original innovation by recombination hooks task, this paradigm (like Neldner et al. 2019) required a modification or innovation by incremental improvement. Children were presented with the right sort of tool, but it was nonfunctional. Considering the types of innovation discussed, it could be expected that this task should be easier for children to complete, as we would expect innovation by incremental improvement to be more common and easier to achieve than innovation by recombination. A number of explanations for children's difficulty and lower success rates were proposed in the paper (Cutting et al. 2019).

Building on the cognitive load theory of Neldner and colleagues (2019), one possibility proposed was that instead of acting as a clue to help children, the presence of the nonfunctional hook actually increased cognitive load. In contrast to the Neldner study where children simply needed to recognize the affordance of the hooked end of the material, in this study children needed to not only recognize the affordance of the hook but also realize that it was too big for the task and execute a plan of successful modification. These added requirements may have been more cognitively demanding than simply needing to recognize the solution and executing it for oneself.

Another proposed explanation was that the children's behavior was due to them being preprogrammed to learn from others, especially adults. Adults teach children and provide them with useful information; it therefore seems likely that children expect to receive useful information and help. They may interpret the testing paradigm as a situation in which the adult present is likely to provide useful and relevant information and products. They may expect that the materials they are given are ones that will be needed and will work to solve the task they are presented with. This disposition for social learning may hinder children in the context of innovation because they are not expecting to innovate; they are expecting to be taught how to solve the problem rather than figuring it out for themselves.

Building on these suggestions, I will now propose a third factor that is likely to have contributed to children's poor performance on this task—lack of expertise. Socially transmitting relevant information about aspects of the task to children is unlikely to be sufficient to help them innovate a novel tool. Valentine Roux and colleagues (this volume) show that the ability of skilled potters to create new shapes demonstrates that only the most skilled potters with the highest expertise had the flexibility to achieve these new designs. Similarly, the ill-structured problem-solving literature suggests that well-integrated structural knowledge of a concept is needed to solve ill-structured problems (Jonassen, Beissner, and Yacci 1993). It may therefore be unsurprising that providing children with small pieces of information relevant to the task might not be sufficient to induce innovations. I expand on this further in the section below, where I present a study that provided children with multiple pieces of relevant task-related information.

Multiple Scaffolds

The study (Cutting et al. 2014) explored children's ability to use information from others to innovate a hook tool. Half of children participated in an exercise that socially demonstrated pipe-cleaner bending before the innovation task; the exercise was meant to highlight the affordances of the materials. If unsuccessful on the tool-innovation task, children then received a target-tool demonstration. The design of this study allowed researchers to assess children's ability to use this social information regarding different aspects of the task. It was concluded that children's main difficulty was with generating necessary information for themselves (i.e., that a hook is needed, that pipe cleaners are pliable, etc.). When given this information, children age 5 and older were able to use it to create a successful solution to the task. However, children younger than age 5 lacked this flexibility and had great difficulty combining the different pieces of information even when presented by the researcher.

Let's revisit and attempt to apply the three explanations given for children's poor performance on innovation tasks. First is the cognitive load theory of Neldner and colleagues (2019). The current paradigm provides children with a reduction in cognitive load by presenting the various elements of the task that then need to be combined—pipecleaners bend, plus hook-shape. When provided with both elements, children over age 5 increased their chances of success, providing clear support for this theory. It is likely that younger children are lacking in the required baseline cognitive capacity to combine the information. This is supported by evidence that children's executive functions show protracted development across childhood with greater gains once children begin formal education (Hughes et al. 2010), which is around age 4 to 5 in the United Kingdom, where this study took place.

Second, the expectation that adults provide knowledge may have aided children with this task. In contrast to being given an oversized hook by an adult (Cutting et al. 2019) that may have inadvertently been interpreted as the solution, this task communicates numerous pieces of information to children, which are potentially correctly interpreted as being useful for the task.

Third, although some improvement in performance was seen with increasing social information, children did not reach high performance on the task. This could again point to the need for expertise before children can flexibly use information. It is possible that older children who were more successful have more experience with the materials, but this is difficult to disentangle from their advanced cognitive capacity. Future studies should introduce novel materials to help disentangle the roles of these factors.

Transfer of Toolmaking Knowledge

Asocial and socially mediated innovation has been shown to be difficult for young children, demonstrating that in the technical domain, children lack flexibility in their behavioral repertoires (Ramsey, Bastian, and van Schaik 2007). The next question that arises is whether children display behavioral flexibility at the second level—that is, for the tool behavior itself. For techniques to prosper, it is important that they are retained for future use (von Hippel and Suddendorf 2018). The tool-rich world we live in today could not exist if we did not retain information we learn about how to make and use tools. At the first level, it is important that once a new technique (e.g., making and using a hook tool) has been learned, this technique can then be replicated for the same task on future occasions. Second, there also needs to be some degree of flexibility in how the technique is used because it is unlikely that all tasks will require the exact same solution (i.e., there will be slight variations in the tools required).

Replicating a Learned Technique

Children demonstrate an excellent ability to manufacture identical tools on the exact same task following their own initial innovation. In a study that I participated in (Whalley, Cutting, and Beck 2017), children were presented with three trials of the hooks task, and their success on the task was stable across trials. Children who innovated a hook tool replicated their successful solution on subsequent trials. While this ability to retain useful innovative information is reassuring, this task is limited in that the trials were presented in quick succession, and we are only able to assess whether spontaneous innovations are retained for future use.

Beck and colleagues (2014) provide more substantial evidence for children's ability to retain toolmaking knowledge. Children were tested on the hooks task twice with a three-month gap in between. In the first presentation of the task (time 1), children were recorded as successfully innovating a hook tool or successfully manufacturing a hook tool following either the target-tool or tool-creation demonstration. Successful innovation was then measured at time 2. Children retained knowledge of toolmaking over the three-month period, with the ability to manufacture the tool pre-demonstration in each session rising from 0 to 71 percent in four- to five-year-olds and 16 to 68 percent in six- to seven-year-olds. There was no difference in success at time 2 depending on whether children spontaneously innovated at time 1 or received either of the demonstrations. However, low initial success rates and small sample size may be masking differences in how these factors affect retention rates.

Making Flexible Use of Learned Techniques

While exact replication of a technique is important for the retention of that technique, to drive cumulative cultural evolution, techniques need to be transferred to new tasks. The distance between the original task and the new situation will determine how much the technique evolves. Near or close transfer refers to the ability to apply learning in very similar contexts involving little flexibility in a technique, whereas far transfer refers to the ability to adapt learning to more dissimilar contexts showing greater flexibility in techniques (Sala et al. 2019). In close transfer tasks, children demonstrate some ability to flexibly transfer knowledge of making hook tools to tasks that vary only in surface characteristics. Beck and colleagues (2014) presented children with three versions of the hooks task one after the other: the original clear tube and bucket, a shorter green tube containing a blue bucket with closed-loop handle, and a cuboid clear transparent box with a square yellow bucket. Each task was presented with its own different-colored pipe cleaners and string distractors. On each version of the task, children were given an opportunity to innovate a tool for themselves and then received a tool-creation demonstration if necessary. Performance on the first task was low for all children, with five- to six-year-olds demonstrating a better ability to flexibly transfer knowledge to the new tasks (rise in success rate from 5 to 86%) than younger three- to five-year-olds (rise in success rate from 4 to 50%).

Children's ability for far transfer was tested on a task requiring them to retrieve rewards from the same apparatus using different materials. Children were unable to transfer their knowledge of hook-making with one type of material (pipe cleaners) to a second task using the same apparatus requiring them to create a hook tool using different materials (wooden dowels added together) and vice versa (Beck et al. 2014). Despite either independently solving the task by making a hooked tool or being shown how to make a tool, children were unable to use their knowledge of the tool required to make a successful tool from a new material.

Children's lack of flexibility for far transfer is confirmed by studies requiring children to make two different tools on two different tasks. Knowledge of the affordances of the pipe-cleaner materials available did not help children to make their second tool after success was achieved (either independently or with social learning) on the first task (Cutting, Apperly, and Beck 2011).

Together, these studies show that once children have learned a new tool technique, either independently or through social learning, then this knowledge is robust over time and can be readily deployed on tasks with the exact same parameters. Children also display some level of flexibility in their behavior. By age 6, children can use this knowledge on tasks with the same underlying task requirements but differing surface characteristics (close transfer or small modification). As the task requirements diverge further away from the original, children display a lack of flexibility and struggle to transfer their knowledge about the required tool shape to new materials (far transfer or large modification). This could be taken as evidence that innovation is a domain-specific rather than domain-general skill (Rawlings and Legare 2020). Additionally, this could be taken as further support for the role of expertise. As discussed previously, expertise and the ability to innovate have been closely linked (Roux, Bril, and Karasik 2018; see also Roux et al., this volume). It is therefore likely that for children to flexibly use the knowledge that they have gained, they must reach some level of expertise. Given that hook-innovation tasks have only given children a small amount of experience in making a hook tool (in most cases only one attempt), it is therefore unsurprising that children do not have enough experience to be able to flexibly use their new knowledge and modify it to new situations.

Bringing It All Together

Children demonstrate a lack of flexibility in the domain of tools. Despite having remarkable aptitude to learn how to make and use tools by imitating or being taught by others, children's ability to innovate simple tools has consistently been shown to be difficult. Initial studies into children's tool-innovation focused on independent, asocial innovation. These tasks yielded very low levels of success and were an important starting point in our understanding of children's ability to innovate. However, it seems likely these studies do not give us true insight into how innovative abilities develop. Innovations are likely to be much more socially mediated, and it is important that paradigms capture this. Some of the more recent studies described in this chapter give us an indication as to how children use social information to make innovations by modification, and they suggest a complex picture.

Social Influences in Tool-Innovation

There are two factors that we need to consider about the tool-innovation studies that have been presented in this chapter—type of innovation required and amount of social influence. These factors overlap and intertwine. I would argue that all of these studies, whether categorized as asocial or socially mediated innovation, contain some degree of social influence because of the nature of the world we live. At one end, we have social models who can demonstrate behaviors and techniques that we are able to imitate; at the other end, where there is an absence of direct modeling or instruction, information is still available to us through the context of the scenario we find ourselves in and because the materials provided for us have been manufactured and therefore must have some designed purpose (Dennett 1987). Tasks requiring individual innovation by recombination have been termed here as "asocial" and would be placed toward the asocial end of an asocial–social spectrum. However, these tasks still involve some degree of social influence. The tasks are presented to children by an experimenter who clearly has a motive for presenting the child with the apparatus and materials. The materials themselves have clearly been manufactured for some purpose (see Boyette, this volume, for a discussion on how tools give us insight into those that make them).

Studies that have investigated innovation by incremental improvement or modification contain a greater degree of social influence as innovations build on the social outputs of others. The modification studies in this chapter vary in the amount and type of social information transmitted to participants. There is some evidence to suggest that greater social input has a positive effect (Cutting et al. 2014; Neldner, Mushin, and Nielsen 2017), but these independent studies do not yet create a systematic narrative to help draw firm conclusions about the role of social influence on children's innovations or the degree of modification required and the impact they have on children's innovative ability.

Linking with the earlier discussion surrounding the requirement of expertise for innovations to occur, a logical starting point to investigate children's capacity for incremental innovations would be to train and give children expertise in making a tool of a particular configuration before then introducing new apparatus requiring the modification of the size or shape of the tool (see Pope-Caldwell, this volume, for a framework to test this notion). Once a baseline of the expertise required for flexibility for modification has been established, studies can then investigate along the social–asocial continuum in a more systematic way.

Revisiting and Expanding on the Idea of Expertise

Valentine Roux and colleagues (this volume) provide an excellent discussion and evidence for the role of expertise in the flexibility of techniques. Expertise is a factor that has been thus far overlooked in the developmental tool-innovation literature, but it is a factor that needs to be explored. In the previous section on flexible use of learned techniques, I speculate on the role expertise may play in children's ability to adapt and modify their learned tool behaviors. The topic of expertise fits with my discussion of tool-innovation as an ill-structured problem-solving task (see Cutting, Apperly, and Beck 2011; Chappell et al. 2015). In various hook-innovation studies, children have been poor at piecing together the individual components of the task to create the solution (see Cutting et al. 2014). The ill-structured problem-solving literature suggests that only experts have well-integrated knowledge that can be used flexibly to solve problems (Jonassen, Beissner, and Yacci 1993). Therefore, providing children with the pieces of information is not enough to allow them to use this information flexibly to solve the problem. The question that arises, then, and that I previously posed, is just how much of an expert do you need to be to innovate? Studies with children offer us one way to approach this question, because children's limited experience with the world means we can more easily manipulate their expertise with techniques, which will hopefully allow us to address this question.

Why Innovate If You Can Imitate?

This idea of expertise, which is very much domain-specific, is linked to a domain-general factor that has not been particularly addressed by the literature. It is the fact that children (and adults) do not have much, if any, need in their lives to innovate new tools. Society has reached a point in our cumulative technological evolution that we are surrounded by an abundance of tools for almost any purpose we can think of. Of course, there will always be advances in technology, mostly innovations by modification that will be conducted by experts who will design the new generations of smartphones and televisions with better-quality screens and more efficient batteries. There will also be some step-change innovations by invention that will change the ways that we conduct our lives in ways that we cannot possibly imagine today. But for a five-year-old child living in Western society, I would argue that they are currently surrounded by all the tools they need, and so innovation is just not something that needs to be exploited. The tools they need to use throughout their daily lives are readily available to them (financial means permitting), and rather than working out how these function for themselves, children are surrounded by people they can model their use on—parents, siblings, peers, and teachers.

If a young child wants to use or create new technology but does not know how, the most efficient way to learn is to watch other people engaging in the same activity or asking someone to model the method. Our culture is so socially connected that even if we do not have someone in our immediate network who can demonstrate something for us, then we can find a solution to our problem simply by typing prompts in ChatGPT or watching a video on YouTube. There are very few scenarios where someone is unsure of how to do something that cannot be solved by asking someone in your network or reaching out to the depths of the internet, and these are much quicker and efficient ways to reach a solution than by trying to innovate one independently (see Pope-Caldwell, this volume, for an extended discussion on how we only behave flexibly at an individual level when we are forced to).

My main argument is that most people in the modern Western world, children but also adults, just do not need to innovate especially in the domain of tools throughout their every-day lives. Building on this, I would argue that historically, for cumulative technological culture to exist, not everyone needed to be an innovator. It only takes one good idea, or one successful variation in what has gone before, for an innovation to occur. The important part, as evidenced in the literature, is for that innovation to then spread throughout the group by social transmission so that all group members adopt the new more efficient outcome (Dean et al. 2014; Tennie, Call, and Tomasello 2009; Tomasello, Kruger, and Ratner 1993). If everyone was an innovator, then the whole system would become somewhat messy. If every-one were constantly trying to improve on a tool, it would make it difficult to see which innovations were the successful ones. There would be too much data to process and no clear indication of which tool or method was the right one to use. Focusing on innovation may also take the focus away from social transmission, and without the new and more efficient ideas spreading throughout the group, cumulative evolution could not occur (see Tenpas, Schweinfurth, and Call, this volume, for a discussion surrounding how overly flexible groups can lead to cultural breakdown). We therefore need to balance a high number of social learn-ers and a much smaller number of innovators. What needs to be determined is whether the most successful groups leading to high levels of cumulative cultural evolution consist of individuals who are either innovators or social learners and what frequencies of each are needed, or whether individuals can be both innovators and social learners at different times and what the optimum frequency of innovations is. If the first option is true, then we need to identify what makes someone an innovator, and studying children is one way to answer this question.

Conclusions

Research with children has the potential to inform us about the required mechanisms and flexibility that underpin the ability to innovate new tools. As a relatively new field, research to date has predominantly used the hooks task to measure capacity for innovation by recombination. While this begins to answer some questions about the cognitive mechanisms involved in tool-innovation, the relative asocial nature of this paradigm does not replicate innovations more commonly seen in real life, which I argue are more socially influenced. More recent work has begun to investigate incremental innovations or modifications more akin to those that fuel cumulative evolution. These studies suggest some flexibility in the way that children use their tool knowledge, but these studies need to be conducted and integrated in a more systematic way that can tease apart the roles of social influence, expertise, cognitive demands, and whether the skills required for innovation are domain-specific or domain-general.

Acknowledgments

Thank you to Mathieu Charbonneau for putting together this volume and for keeping all the contributing authors on track. It has been wonderful to read and comment on each chapter and to see the emerging links between disciplines on this exciting topic.

References

Barr, R., and N. Wyss. 2008. "Re-enactment of Televised Content by 2-Year-Olds: Toddlers Use Language Learned from Television to Solve a Difficult Imitation Problem." *Infant Behavior and Development* 31:696–703.

Beck, S. R., I. A. Apperly, J. Chappell, C. Guthrie, and N. Cutting. 2011. "Making Tools Isn't Child's Play." *Cognition* 119:301–306.

Beck, S. R., N. Cutting, I. A. Apperly, Z. Demery, L. Iliffe, S. Rishi, and J. Chappell. 2014. "Is Tool-Making Knowledge Robust over Time and across Problems?" *Frontiers in Psychology* 5:1395.

Boden, M. A. 1996. "Creativity." In *Handbook of Perception and Cognition, Artificial Intelligence*, 267–291. San Diego: Academic Press.

Boesch, C., D. Bombjaková, A. Meier, and R. Mundry. 2019. "Learning Curves and Teaching When Acquiring Nut-Cracking in Humans and Chimpanzees." *Scientific Reports* 9:1515.

Boyd, R., and P. J. Richerson. 1996. "Why Culture Is Common but Cultural Evolution Is Rare." *Proceedings of the British Academy* 88:73–93.

Boyd, R., P. J. Richerson, and J. Henrich. 2011. "The Cultural Niche: Why Social Learning Is Essential for Human Adaptation." *Proceedings of the National Academy of Sciences* 108:10918–10925.

Chappell, J., N. Cutting, I. A. Apperly, and S. R. Beck. 2013. "The Development of Tool Manufacture in Humans: What Helps Young Children Make Innovative Tools?" *Philosophical Transactions of the Royal Society B: Biological Sciences* 368 (1630): 20120409.

Chappell, J., N. Cutting, E. C. Tecwyn, I. A. Apperly, S. R. Beck, and S. K. Thorpe. 2015. "Minding the Gap: A Comparative Approach to Studying the Development of Innovation." In *Animal Creativity and Innovation*, edited by A. B. Kaufman and J. C. Kaufman, 287–316. San Diego: Academic Press.

Charbonneau, M. 2015. "All Innovations Are Equal, but Some More than Others: (Re)integrating Modification Processes to the Origins of Cumulative Culture." *Biological Theory* 10:322–335.

Charbonneau, M. 2016. "Modularity and Recombination in Technological Evolution." *Philosophy & Technology* 29:373–392.

Cutting, N. 2013. "Children's Tool Making: From Innovation to Manufacture." PhD diss., University of Birmingham.

Cutting, N., I. A. Apperly, and S. R. Beck. 2011. "Why Do Children Lack the Flexibility to Innovate Tools?" *Journal of Experimental Child Psychology* 109:497–511.

Cutting, N., I. A. Apperly, J. Chappell, and S. R. Beck. 2014. "Why Can't Children Piece Their Knowledge Together? The Puzzling Difficulty of Tool Innovation." *Journal of Experimental Child Psychology* 125:110–117.

Cutting, N., I. A. Apperly, J. Chappell, and S. R. Beck. 2019. "Is Tool Modification More Difficult than Innovation?" *Cognitive Development* 52:100811.

Dean, L. G., R. L. Kendal, S. J. Schapiro, B. Thierry, and K. N. Laland. 2012 "Identification of the Social and Cognitive Processes Underlying Human Cumulative Culture." *Science* 335 (6072): 1114–1118.

Dean, L. G., G. L. Vale, K. N. Laland, E. Flynn, and R. L. Kendal. 2014. "Human Cumulative Culture: A Comparative Perspective." *Biological Reviews of the Cambridge Philosophical Society* 89 (2): 284–301.

Dennett D. C. 1987. *The Intentional Stance*. Cambridge, MA: MIT Press.

Dumontheil, I., P. W. Burgess, and S. J. Blakemore. 2008. "Development of Rostral Prefrontal Cortex and Cognitive and Behavioral Disorders." *Developmental Medicine and Child Neurology* 50 (3): 168–181.

Goel, V., D. Pullara, and J. Grafman. 2001. "A Computational Model of Frontal Lobe Dysfunction: Working Memory and the Tower of Hanoi." *Cognitive Science* 25 (2): 287–313.

Gönül, G., E. K. Takmaz, A. Hohenberger, and M. Corballis. 2018. "The Cognitive Ontogeny of Tool Making in Children: The Role of Inhibition and Hierarchical Structuring." *Journal of Experimental Child Psychology* 173:222–238.

Hanus, D., N. Mendes, C. Tennie, and J. Call. 2011. "Comparing the Performances of Apes (*Gorilla gorilla, Pan troglodytes, Pongo pygmaeus*) and Human Children (*Homo sapiens*) in the Floating Peanut Task." *PLOS One* 6 (6): e19555.

Hayne, H., H. Herbert, and G. Simcock. 2003. "Imitation from Television by 24- and 30-Month-Olds." *Developmental Science* 6:254–261.

Henrich, J. 2004. "Demography and Cultural Evolution: How Adaptive Cultural Processes Can Produce Maladaptive Losses: The Tasmanian Case." *American Antiquity* 69:197–214.

Herbert, J., and H. Hayne. 2000. "Memory Retrieval by 18- to 30-Month-Olds: Age-Related Changes in Representational Flexibility." *Developmental Psychology* 36 (4): 473–484.

Hoehl, S., S. Keupp, H. Schleihauf, N. McGuigan, D. Buttelmann, and A. Whiten. 2019. "'Over-Imitation': A Review and Appraisal of a Decade of Research." *Developmental Review* 51:90–108.

Hughes, C., R. Ensor, A. Wilson, and A. Graham. 2010. "Tracking Executive Function across the Transition to School: A Latent Variable Approach." *Developmental Neuropsychology* 35 (1): 20–36.

Jonassen, D. H., K. Beissner, and M. Yacci. 1993. *Structural Knowledge: Techniques for Representing, Conveying, and Acquiring Structural Knowledge*. Hillsdale, NJ: Lawrence Erlbaum.

Kendal, J., L. A. Giraldeau, and K. Laland. 2009. "The Evolution of Social Learning Rules: Payoff-Biased and Frequency-Dependent Biased Transmission." *Journal of Theoretical Biology* 260 (2): 210–219.

Koops, K. 2020. "Chimpanzee Termite Fishing Etiquette." *Nature Human Behaviour* 4: 87–888.

Legare, C. H., and M. Nielsen. 2015. "Imitation and Innovation: The Dual Engines of Cultural Learning." *Trends in Cognitive Sciences* 19 (11): 688.

Morin, O. 2015. *How Traditions Live and Die (Foundations of Human Interaction)*. Oxford: Oxford University Press.

Mounoud, P. 1996. "A Recursive Transformation of Central Cognitive Mechanisms: The Shift from Partial to Whole Representation." In *The Five to Seven Year Shift: The Age of Reason and Responsibility*, edited by A. J. Sameroff and M. M. Haith, 85–110. Chicago: Chicago University Press.

Muthukrishna, M., and J. Henrich. 2016. "Innovation in the Collective Brain." *Philosophical Transactions of the Royal Society B* 371:2015019220150192.

Neldner, K., I. Mushin, and M. Nielsen. 2017. "Young Children's Tool Innovation across Culture: Affordance Visibility Matters." *Cognition* 168:335–343.

Neldner, K., J. Redshaw, S. Murphy, K. Tomaselli, J. Davis, B. Dixson, and M. Nielsen. 2019. "Creation across Culture: Children's Tool Innovation Is Influenced by Cultural and Developmental Factors." *Developmental Psychology* 55 (4): 877–889.

Nielsen, M. 2006. "Copying Actions and Copying Outcomes: Social Learning through the Second Year." *Developmental Psychology* 42 (3): 555–565.

Nielsen, M., K. Tomaselli, I. Mushin, and A. Whiten. 2014. "Exploring Tool Innovation: A Comparison of Western and Bushman Children." *Journal of Experimental Child Psychology* 126:384–394.

Powell, A., S. Shennan, and M. G. Thomas. 2009. "Late Pleistocene Demography and the Appearance of Modern Human Behavior." *Science* 324:1298–1301.

Ramsey, G., M. L. Bastian, and C. P. van Schaik. 2007. "Animal Innovation Defined and Operationalized." *Behavioral and Brain Sciences* 30 (4): 407–432.

Rawlings, B., and C. Legare. 2020. "The Social Side of Innovation." *Behavioral and Brain Sciences* 43:E175.

Reitman, W. 1965. *Cognition and Thought*. New York: Wiley.

Roux, V., B. Bril, and A. Karasik. 2018. "Weak Ties and Expertise: Crossing Technological Boundaries." *Journal of Archaeological Method and Theory* 25:1024–1050.

Sala, G., N. D. Aksayli, K. S. Tatlidil, T. Tatsumi, Y. Gondo, and F. Gobet. 2019. "Near and Far Transfer in Cognitive Training: A Second-Order Meta-Analysis." *Collabra: Psychology* 5:18.

Shallice, T., and P. W. Burgess. 1991. "Higher-Order Cognitive Impairments and Frontal Lobe Lesions in Man." In *Frontal Lobe Function and Dysfunction*, edited by H. S. Levin, H. M. Eisenberg, and A. L. Benton, 125–138. Oxford: Oxford University Press.

Shumaker, R. W., K. R. Walkup, and B. B. Beck. 2011. *Animal Tool Behavior: The Use and Manufacture of Tools by Animals (Revised and Updated Edition)*. Baltimore: Johns Hopkins University Press.

Tennie, C., J. Call, and M. Tomasello. 2009. "Ratcheting Up the Ratchet: On the Evolution of Cumulative Culture." *Philosophical Transactions of the Royal Society of London. Series B: Biological Sciences* 364 (1528): 2405–2415.

Tomasello, M. 1999. *The Cultural Origins of Human Cognition*. Cambridge, MA: Harvard University Press.

Tomasello, M., A. Kruger, and H. Ratner. 1993. "Cultural Learning." *Behavioral and Brain Sciences* 16:495–552.

Voigt, B., S. Pauen, and S. Bechtel-Kuehne. 2019. "Getting the Mouse Out of the Box: Tool Innovation in Preschoolers." *Journal of Experimental Child Psychology* 184:65–81.

von Hippel, W., and T. Suddendorf. 2018. "Did Humans Evolve to Innovate with a Social Rather Than Technical Orientation?" *New Ideas in Psychology* 51:34–39.

Weir, A. A., S. J. Chappell, and A. Kacelnik. 2002. "Shaping of Hooks in New Caledonian Crows." *Science* 297 (5583): 981–981..

Whalley, C. L., N. Cutting, and S. R. Beck. 2017. "The Effect of Prior Experience on Children's Tool Innovation." *Journal of Experimental Child Psychology* 161:81–94.

White, S. J., P. W. Burgess, and E. L. Hill. 2009. "Impairments on "Open-Ended" Executive Function Tests in Autism." *Autism Research* 2 (3): 138–147.

12 A Tactful Tradition: The Role of Flexibility and Rigidity in Horse Riding and Dressage

Helena Miton

Introduction

A common challenge in using most technical crafts is achieving a specified end-goal while working with materials that impose various constraints to which the craftsperson must contingently adapt. This is particularly true and crucial whenever the materials have properties that are inconstant. For example, cheese-makers' expertise lies in their ability to adjust to variation in milk—for instance, its fat content (Paxson 2011). Techniques can also follow the opposite logic when the end-goal is adjusted to a material's specificities. Woodturners must often adjust their workpiece to accommodate knots in the raw material while simultaneously ensuring that they produce a functional end-product (e.g., making a wooden bowl rather than a plate because of a knot). Techniques involving animals—from husbandry to military use to transportation—are particularly affected by such constraints, as they need to be adjusted to the specific contextual behaviors and reactions of the animals and their physical capacities. But animals are also cognitively complex creatures, often forcing their users to adapt to their mental states, from their "personality" (or "temper") to their life experiences.

The maintenance of interspecific traditions is a fascinating feature of cultural transmission that requires learning and the use of techniques between very different living systems. Horse riding offers a particularly rich case of such interspecific systems, with its unique set of possibilities, constraints, and evolutionary trajectory.[1] Horse riding also provides unique insights into the relationship between technical rigidity and flexibility, both in terms of how a technique can be *used* (riding) and how it can be *transmitted* (to novice riders and horses). While unique, horse riding also encounters and solves problems similar to those faced by the use of techniques that do not involve interactions with another species, such as working flexibly with variable and reactive physical materials.

How can we explain the stability of horse-riding traditions, especially given that they must be maintained in two populations, one human and one nonhuman? How can an interspecies communication system remain stable through time and be used to re-create technical movements precisely over centuries? More generally, what does the stability of horse-riding traditions tell us about the relation between rigidity and flexibility in technical traditions?

In this chapter, I examine how horse-riding practices—especially dressage, which is defined by the Merriam-Webster Dictionary as "the execution by a trained horse of precision movements in response to barely perceptible signals from its rider"—emerge from complex interactions between horses and humans and can stabilize culturally by exploiting both the rigidity and the flexibility of the rider's cognition and behavior. Dressage is replete with precision movements by the horses: figures such as lateral movements (e.g., half-pass, also called *appuyer*, in which the horse moves laterally while being slightly bent around the rider's leg in the direction in which it moves) or specific gaits (e.g., piaffe, in which the horse trots while remaining in the same place without moving forward). In dressage, the precise definitions of horse movements, the specific representations of behaviors by riders, and the well-defined goals of riding practices all serve as factors of rigidity, and they promote the stability of dressage practices by imposing key boundaries around what counts as acceptable riding behaviors.

However, while the goals of the riders are rigidly defined, the fact that these have to be achieved with living horses brings about substantial variation in execution and calls for flexibility in the rider's course of action. Flexibility is thus necessary for successful riding performances if only because of the *interspecific* cultural transmission of horse-riding skills. The persistence through time of these traditions depends on individual humans being both horse riders but also "horse teachers." Long-standing horse-riding traditions thus require not only rigidly defined goals for learning but also idiosyncratic adjustments at the level of the microevents of transmission, both when training the riders (humans) and the rides (horses). This usually means that the rider's actions are adjusted both to the horse's behaviors and to the state of the horse-rider system in terms of which action is selected, its amplitude or strength, and its timing. This flexibility is required to teach horses and is explicitly represented by the human members of equestrian traditions under the concept of *tact équestre* (equestrian tact); that is, the "right" way for the rider to act, which includes elements as varied as posture, tone, or when and how to react to the horse's behaviors.

Horse-riding techniques, especially dressage, have an unusually abundant written record considering that they are mainly embodied practices. This written record makes them particularly suitable to study. This is especially true of European dressage traditions, on which I will focus. This literature is used here to illustrate the stability and relative continuity of some dressage practices, representations, and definitions transmitted between horsemen. Quotes were chosen to be representative, and horsemen cited here are among the most well-known of the French tradition (from the seventeenth to the twenty-first century). Importantly, this represents a continuous tradition aiming for "gentle" training, defined in opposition to more coercive traditions that make heavier use of force and painful stimuli (e.g., Federico Grisone). I complement these sources with experiences from my own training as a horse rider, using autoethnography to illustrate some of the teaching and learning practices used in contemporary horse riding. My training now spans over two decades and has been conducted in diverse settings (France, Hungary, and the Southwest United States) and for diverse uses (jumping, dressage, vaulting, historical reenactments, trail rides). Work in contemporary veterinary science is also referred to when relevant.

In this chapter, I first introduce the interspecies communication system at the core of horse riding and how horse riders communicate with horses once on their backs. I then examine the specific set of traditions known as dressage, asking how this interspecies communication system is used to effectively perform technical movements and how this system is maintained across generations of horses and riders. Further, I develop an argument about how rigidity and flexibility contribute to the stability of dressage traditions over time. On the one hand, rigidity plays an important role in maintaining the definitions of technical movements over time, in the passing down of rules and representations that horse riders have made about flexibility, and in fixing goals for the rider. On the other hand, flexibility plays an important role in dressage by contributing to both the performance of dressage movements themselves and to their transmission from riders to horses. When performing, flexibility can be found in the ideal of tact équestre to which riders aspire, as a way to produce movements in the face of the substantial individual variation displayed by horses. When training horses, flexibility comes from riders recruiting pedagogic mechanisms similar to those used when teaching humans. These, in turn, match learning principles known to be at play when horses learn.

Anatomy of an Interspecific Communication System

In its simplest definition, horseback riding includes a human sitting on a horse's back, with the actions between the two usually (but not necessarily) mediated by some artifacts (e.g., bridle, bit, reins, saddle).

Horse-riding traditions face a particular challenge as they must be learned and preserved by populations of not just one but two species: humans and horses. This makes horse riding closer to technical skills based on joint action (i.e., actions performed with a human partner), such as using a two-man crosscut saw, than techniques enacted individually, such as using a chainsaw.

As the transmission of these interspecific traditions involves two agents (the rider and the horse), horse riding is a complex evolving system with multiple interacting parts, each having its own (evolutionary) histories (see figure 12.1). The overall system that is formed is first and foremost an interspecies communication system, with the rider indicating a desired behavior to the horse and adjusting their actions flexibly, at the moment, so as to obtain the desired behavior or movement.

Horse domestication is assumed to have started 5,000 to 6,000 years ago, most likely in the western part of the Eurasian steppe, using local feral horses to replenish domestic herds (Levine 2005; Warmuth et al. 2012). Domestication and husbandry practices are assumed to have created positive selection pressures on horses' physiology (including their cardiac system and muscular and limb development) and cognition (especially social behavior, learning capabilities, and fear response), while having simultaneously increased the number of deleterious mutations carried along (Schubert et al. 2014). Modern breeding practices and selective breeding have further led to decreased genotypic diversity (Fages et al. 2019; Petersen et al. 2013).

Horse riding also involves material artifacts (i.e., the gear used to ride horses such as bit, bridle, saddles, stirrups, or spurs). These artifacts have been invented and improved through time (Gianoli 1969). Embodied practices, including the horses' and riders' learned behaviors and the teaching methods used for horse riding also have their own histories.

Each of the elements of this system interacts with one another in a diversity of ways. For instance, humans influenced the evolution of horses through breeding and artificial selection (Fages et al. 2019; Hanot et al. 2018; Orlando 2020). Humans and horses also influence each

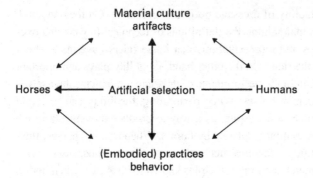

Figure 12.1
Humans use embodied practices and artifacts to ride horses. Embodied practices and artifacts interact with each other: for instance, saddles determine the rider's posture and how they can use their back, seat, and legs, while bits and bridles influence how the rider can use their hands. Humans influence horses directly through artificial selection (based on fit for particular uses, temper, and aesthetics) and less directly through their practices and interactions with horses and the artifacts used to ride and handle them.

other through their interactions and embodied practices in a range of situations (e.g., when the rider is handling the horse while on foot or when riding the horse), with some of these activities leaving osteological traces on horse remains (e.g., Taylor, Jamsranjav, and Tumur-baatar 2015).

These embodied practices also interact with the gear used to ride a horse (material culture and artifacts). Variation in saddles, bridles, and bits imply variation and constraints in their physical effects, which in turn affect how the rider and the horse will behave. Saddles determine which positions are possible and comfortable for the rider. For instance, the position of a horse rider in a Western saddle is different from the position when sitting in an English saddle. English saddles use knee rolls (padding in the front of the saddle, which is in contact with the rider's legs) to hold the rider's legs under the pelvis. In contrast, Western saddles do not have knee rolls and put the rider's legs in a more forward position. English saddles used for dressage have features that favor a vertical and seated position for the rider with large knee rolls and a deep seat, while English saddles used for jumping allow for more freedom of movement with a flatter seat and smaller knee rolls. Saddles used in different contexts or for different functions have distinctive shapes and features, which are associated with different distributions of pressure on the horse's back (Belock et al. 2012; Greve and Dyson 2013). Saddles used for horse races are made to ensure a racing position in which the jockey is not in contact with the saddle itself, and so they have a flat seat. In contrast, dressage saddles have deep seats in accordance with the rider needing to be deeply seated. Other variations in gear include the different bridles (e.g., one or two bits and pairs of reins) or bits used (some having lever effects and others not). Having two pairs of reins influences the rider's hand actions and offers the possibility of using the two bits in opposite fashion—snaffle one for actions upward, bridle one for actions downward. Other bits will allow different physical actions (Cross et al. 2017; O'Neill 2018), which in turn influence the horse's and rider's actions.

The interactions between the different elements of the horse-rider system unravel at their own timescales. Artificial selection, for instance, has effects at larger timescales than do embodied practices—across generations of horses versus within a single horse's lifetime.

Horses, through their behaviors, also provide feedback to the humans with whom they interact. How can this system, including the interactions between its different elements, constitute an interspecies communication system?

Horse and rider create a communication system. To communicate with their horses, riders use various "channels" or means of communication (aids in equestrian parlance):

- *Mouth-bridle-reins-hands*. One way to communicate between horse and rider is constituted by the horse's mouth, the bit inside the horse's mouth, and the tension applied to it through the reins. This subsystem is usually used to slow down, to stop the horse or limit its forward movement (by increasing tension on the reins), and to indicate the trajectory to follow (moving straight forward or turning by having more tension on one rein than the other).

- *Weight distribution and overall posture of the rider*. This can be mediated through a saddle and saddle pad, or without one, in the case of bareback riding. Weight distribution and posture tend to be used in combination with other signals to reinforce them, in cases such as slowing down (usually by having the rider straighten their back at the same time they increase tension on the reins). Weight distribution is often used to prepare movements by having the rider's weight anticipate other actions (e.g., when changing direction, the rider's weight changes a bit before legs and hands are used).

- *Rider's legs*. The rider's legs are usually used for impulsion, indicating balance or movement of the horse's back legs. They can be used to apply pressure (usually as a cue for the horse to speed up or put more energy into a movement) or simply through their positioning (usually as a cue for the horse about the placement of some of their limbs).

- *Voice*. This includes vocalizations such as tongue clicking. Clicks, for instance, are frequently used for upward transitions (i.e., as a cue to the horse to go faster or change gait to a faster gait).

The factors influencing even only one of these channels (such as rein tension) depend on many things, starting with the horse itself (gait, training level, head and neck position, conflict behavior, laterality), the rider (experience, handedness, position), and the equipment mediating their interaction (rein length, rein elasticity, bit type). (For a review, see Dumbell, Lemon, and Williams 2019.)

There are two specific characteristics of horse riding that are worth noting. First, in the case of horse riding, communicating recruits many of the same actions and body parts as stabilizing one's body on a horse's back does, which can complicate sensorimotor learning for both the horse and the rider (Wolframm 2013). For instance, when having a novice rider on their back, a horse is faced with a noisy signal: some of what a horse feels might be signals from the rider could actually be actions the novice rider is taking simply to maintain their balance (e.g., changing posture and straightening their back).

Second, communication with horses happens mostly through haptic signaling, which is not a modality that has been at the core of studies of communication (in contrast to language-oriented communication and mentalizing aspects such as perspective-taking). As an interspecific communication system, horse riding is peculiar in that it mostly relies on a modality that neither species uses much when interacting with their conspecifics: haptic communication. Horses groom each other as part of bonding and social interactions, but most of their social behaviors tend to occur with no direct haptic contact. As for humans, haptic communication is often recognized as understudied if important in social interactions and communication (Andersen and Guerrero 2008; Moll and Sallnäs 2009). The fact that both species use parsimonious haptic communication that mostly has an interspecific social use of its own makes a system like horse riding capable of evolving and yet still mysterious.

Dressage as Tradition

Horses have been used around the world as aids for diverse tasks, including sports and racing, as a means of transportation, and in farm work (plowing or herding cattle). Horse riding is only one among these different uses but it encompasses a diversity of practices and traditions itself, including racing, human transportation, warfare, hunting, and herding. Horse riding exists as traditions—that is, practices that have been transmitted through time and across generations. Horse-riding traditions can be traced back centuries (Ödberg and Bouissou 1999) and even millennia (Taylor, Tuvshinjargal, Bayarsaikhan 2016). Within horse-riding traditions, dressage focuses on obtaining a seemingly effortless execution of specific prescribed movements.

Although there are some general organizing principles and a "typical" curriculum for dressage horses that are bred and trained for purposes of sports and competition, horses' careers and training can show considerable variation. Training usually starts when the horse is around three years old because horses have not reached their adult size before that age and having a rider's weight on their back might compromise their growth. Years 3 to 4 and 5 to 6 are generally considered the initial "training" part of a horse's life, during which they "learn their job." This period includes breaking the horse (in French: '*débourrer*'), which refers to the initial training of a horse and specifically to getting the horse used to the gear, then the rider's weight, and finally to respond to the rider's basic actions (move forward, slow down or stop, and turn). How much refinement in training occurs after this initial training mostly depends on the use made of the horse—it can continue for several more years in the case of high-level dressage, whether it is for sports and competition purposes or as part of institutions, such as Vienna's Spanish riding school or Cadre Noir in France, aiming to maintain living traditions (Durand 2007).

Dressage can be considered an extreme case within horse-riding practices, one with very rigid and precise behavioral goals. Indeed, most uses of horses—especially when they involve more labor-oriented functions such as gathering a herd of cattle—do not ask for behaviors as well-defined and precise in their execution as those of dressage. For dressage, the horse's movements themselves are the goals, whereas the goals for most uses of horses are distinct from the horse's precise movements. This makes dressage, more than other riding practices, closely comparable to techniques involving non-animate materials (such as woodworking, pottery, or weaving) where the goal is often to produce an object conforming to a predetermined type, sometimes with substantial variation in the gestures used to adapt the production sequence to idiosyncrasies of the materials used (Bril et al. 2010; Gandon et al. 2013). With a living and dynamic organism such as a horse (here, the "material"), the technical task of producing some precisely defined figure becomes particularly challenging as the horse is reactive, has independent goals of its own, and exhibits complex behaviors.

This is not simply an effect of using horses for sports. First, dressage conflated goals and movements in military contexts before it became a sport per se. Second, some uses of horses in sports distinguish between the goal and the movements used to achieve this goal. Consider horse racing: the goal is to be the fastest to cover a given distance and to reach the finish line in a shorter amount of time than one's competitors, but the exact movements that contribute to achieving this goal do not matter. Similarly, in jumping, the goal is to jump over obstacles as fast as possible without making any of the obstacles' bars fall. In both cases, the exact movements of horses and riders matter only to the extent that they contribute to achieving this goal.

While movements are viewed as mostly instrumental for many horse-riding practices, for dressage they become the goals in themselves. Dressage is associated with an ideal of harmony and fluidity, aiming for a relative "invisibility" of the rider's actions. Producing smooth and well-defined movements with ease is a technical art that can be evaluated rigidly. But why does it matter whether goals and movements can be differentiated? In the present case, this difference means that dressage riders have to develop their ability to evaluate both their own and their horse's movements. Feedback cannot be obtained by instrumental success as it can in other horse-riding practices (e.g., racing and jumping, where feedback is provided by the time taken over a course and whether the obstacles fall down or not). Getting horses to perform such precisely defined movements uses many of the same communication channels and processes that other horse-riding practices do. Yet the importance of "feeling" and of perceptual, evaluative, and communicative abilities between rider and horse can be expected to be higher. It is in this context that both rigidity and flexibility play important roles.

Rigidity in Dressage

Rigidity can be found in dressage as practices that are precisely defined, as rules that can be faithfully passed down to new generations of riders, and finally as a kind of behavioral compass.

Rigidity contributes to the stability of horse-riding traditions by providing (1) strict definitions of practices, (2) efficient descriptions of flexible behavior, and (3) a behavioral compass and representation of goals for riders to achieve. Rigidity scaffolds stability through two main mechanisms. On the one hand, it provides fixed goals to copy, leaving little to no room for variation, as there is a well-defined and precise representation of what the horse's (and the rider's) behaviors should be. On the other hand, it provides representations that are easier to transmit, as they take the form of fixed and easy-to-remember formula (e.g., adjustment rules).

Technical Movements as Rigidly Defined Practices

Most movements found in dressage competitions today originated several centuries ago, with some even dating as far back as Xenophon's writings. For instance, both passage (a collected trot with extended times in the air) and piaffe (a trot "on the spot," in which all motion is upward and not forward) have survived entire centuries without any major change in their definitions.

Records show that the definitions of these movements have remained stable through time. One can compare, for example, the contemporary definitions proposed by the *Fédération d'Equitation Internationale* (FEI) and older definitions offered by horsemen several centuries ago. The FEI presents passage as follows in its official dressage rules: "Passage is a measured, very collected, elevated and cadenced trot. It is characterized by a pronounced engagement of the hindquarters, a more accentuated flexion of the knees and hocks, and the graceful elasticity of the movement. Each diagonal pair of legs is raised and returned to the ground alternately, with cadence and a prolonged suspension." According to the dressage rules, "Piaffe is a highly collected, elevated, rhythmical diagonal movement giving the impression of remaining in place. The horse's back is supple and elastic. The hindquarters are lowered; the haunches with active hocks are well engaged, giving great freedom, lightness and mobility to the shoulders and forehand. Each diagonal pair of legs is raised and returned to the ground alternately, with spring and regularity."[2]

The FEI definitions echo those provided by François Robichon de la Guérinière (Guérinière 1801, 61): "The passage, (from the Italian word passagio, which means a walk) is a walk, or trot, performed in measure and in time; in this motion, the horse must keep his legs long in the air, the one before, the one behind, crossed and opposite as in the trot; but the motion must be shorter, more confined, and marked, than the common walk or trot; so that each step is not to be above a foot in length, that is to say the foot that is in the air, must be put down about twelve inches before the foot which is upon the ground." De la Guérinière continues, "When a horse trots in one spot, without advancing, backing or traversing, and raises and bends his legs high with grace, he is said to piaffe." There is this also definition from François Baucher: "The piaffer is the horse's raising his legs diagonally, as in the trot, but without advancing or receding" (Baucher 1867). Both de la Guérinière's and Baucher's definitions of passage and piaffe are similar to the contemporary ones used by the FEI, although it is worth noting that the contemporary definitions are simpler than the older ones. The crucial point here is that the definitions of passage and piaffe have stayed remarkably stable across centuries.

Of course, changes have occurred: some movements have evolved, and others have simply disappeared. It is important to note that rigidity in definitions does not automatically lead to faithful transmission and even less so to stable traditions. Rigid definitions can be maintained while actual practices—the riders' and horses' behaviors—evolve and change over time. In other words, definitions and practices can become decoupled. When and how does this decoupling happen? Why is rigidity in definitions sometimes insufficient to stabilize their associated practices?

One dramatic example of this phenomenon is how *rollkur*, also known as hyperflexion, emerged and stabilized from the prescribed attitude of "bas et rond" ("low and round"). It has now been technically banned from competition, although the practice is still debated, with some practitioners and the FEI advocating for distinguishing between cases when the neck's flexion is obtained with and without force and banning the former but not the latter (e.g., van Weeren 2013). Other movements, like galloping on three legs or backward, have virtually disappeared. Both of these movements (galloping on three legs and galloping backward) are also extremely costly to acquire and produce—few riders still practice them, and those who do usually work in film and entertainment industries.

Rules: Rigid Representations of How to Be Flexible

Writings on horse riding abound with "adjustment rules." Adjustment rules are defined here as all of the rules that describe how to adjust the rider's behavior to the horse's behavior. By contrast with rules phrased as absolutes (e.g., "don't ever do X"), adjustment rules make prescriptions that are conditioned on the horse's behavior (e.g., "do X more when the horse does Y more"). They typically describe how to adjust the rider's actions and behavior depending on the horse's actions or state. De la Guérinière (1903, 28) offers such an example about the leg-spurs action: "It is recommended to always precede the contact of the iron (spur) by the application of fat-of-the legs, this is a rule that should never be lost sight of, lest it would provoke defenses."

Adjustment rules define some order or hierarchy in the rider's actions, with intensity increasing until the horse reacts. This is echoed by other contemporary ways of teaching horse riding. To make a horse move forward, legs are supposed to be used first. Then, if the horse does not react, the whip is used on the shoulder. As suggested in de la Guérinière's quote, it is assumed that disrespecting this "natural order" may jeopardize the horse's training.

These rules walk a fine line between rigidity and flexibility, as they prescribe a strict distribution of behaviors for possible states of the horse-rider system. Indeed, they order the sequence of behaviors by indicating how some actions by the rider must gradually become stronger until the desired reaction from the horse is achieved, at which point they are stopped. Flexible behavior is scaffolded through the more rigid transmission of these "adjustment rules." They represent fairly "efficient" compressed representations: in such cases, one formula can encompass a distribution of actions or action parameters over a whole range of potential situations. Moreover, these rules assume some expertise on the part of the rider— for instance, being able to recognize when the horse resists or when it yields. This echoes the use of maxims in dry stone walling, a domain in which maxims are circulated freely between craftspeople of all skill levels yet make more sense once some degree of expertise is achieved (Farrar and Trorey 2008).

Fixing Goals

As discussed above, in dressage, the horse's movements themselves are the goals to be attained. One direct implication is that dressage lacks objective feedback from easy-to-evaluate metrics of instrumental success. Indeed, dressage depends to a larger extent on the rider's ability to have the horse produce some movement and to evaluate their own behavior as well as the horse's. Rigidly defined practices and representations offer a means to stabilize dressage techniques as they explicitly fix behavioral goals. They also serve as a means for the riders to evaluate which aspects of their own and their horse's movement might depart from the prescribed goals. Rigid definitions provide a "behavioral compass" for riders to know what needs tending to: they are descriptions of movements, which can be used to evaluate whether it is the desired one and to decide among possible actions.

Flexibility in Dressage

Flexibility is understood here as a characteristic or property of (the rider's) behavior: the adjustments made while planning training, during the training session, and while riding a horse. These can take the form of choosing which communicative signal and channel to use, such as hand or leg action and whether to increase or decrease the signal's intensity or to stop signaling altogether. These can also inform choices about which training exercise is most appropriate given the horse's or the rider's previous performances. Flexibility manifests itself when the rider's behaviors vary in response to some information obtained from the horse's behavior, such as perception of speed, gait quality, or balance of the ride. Flexibility is especially salient when these informational exchanges between rider and ride respond to the changing parameters of a task (Seifert, Dovgalecs, et al. 2014), forming online dynamic loops during a course of action.

Flexibility's importance can be seen in at least three aspects of dressage traditions: (1) flexibility is at the core of the ideal of tact équestre, (2) flexibility is required to accommodate the variation that horses bring to dressage, and (3) flexibility is needed since riding horses partakes in transmitting technical and cultural information from humans to horses.

Tact Équestre: The Ideal of Flexible (Rider) Behavior

One way in which horsemen have acknowledged the centrality of flexibility in the dressage tradition has been through the concept of tact équestre (sometimes translated as "equestrian feel"). Etienne Beudant defines tact équestre as "the genius of equitation, the 'feeling of the horse'; it is the *à propos* which indicates to the rider the way and moment to act, the intensity and duration of his action. It is what makes success and helps out" (Beudant 1923, 21). In Baucher's work, the term appears around a hundred times. "This is a warning which should caution the horse rider's tact of all kinds of contingencies, and make him understand that principles, even the most exclusive of them, are subject to some variations" (Baucher 1867). Finally, from Alexis L'Hotte, "All equestrian action requires, to obtain the effect we expect, what no effect would give: *the about* and *the measure*, in other words, *the tact équestre*" (L'Hotte 1906, vi).

Tact is understood overall as an ability to act in a manner that is fit to the situation. It is an ability that is deemed essential for dressage. However, tact équestre has been considered a difficult aspect of horse riding to transmit up to this day (Zetterqvist Blokhuis et al. 2019). Yet its importance can be found fairly consistently across several centuries in written records.

Can the idea of tact équestre be further qualified with regards to its implications for the rider's cognition and properties of the rider's behavior? Tact équestre includes (at least) two types of cognitive skills. The first type concerns perceptive skills ("feeling") while the second relates more to decision-making and motor skills (i.e., when, how, and with how much intensity to act). The overarching goal is to establish a communication system that successfully accomplishes tasks from signals that are as small as possible. Signals are increased when the answer to one is not satisfying and diminished whenever possible. This principle helps prevent runaway phenomena in which increasingly "loud" signals are used (Shergill et al. 2003).

Flexibility in the Face of Variation

Early in dressage traditions, the necessity of flexible behavior—as tact équestre—is presented as emerging from the variation that horses themselves display as dynamic living beings. L'Hotte writes:

The instrument on which the musician acts is by itself inert. As a result, the conditions it presents are invariable, the same action will always produce on it the same effect. It is wholly different for the instrument that the horse-rider makes use of. Life will animate the horse and, from there, a thousand and one nuance in its manner to present itself and reply to the actions of who is riding it. There is there, for the horse rider, a whole maze of difficulties, in the middle of which he can only get lost, if he doesn't have for guide this peculiar sentiment I talked about: the equestrian tact (*tact équestre*). (L'Hotte 1906, 206–207)

In other words, flexibility is required as horses display a fair amount of variation that riders must overcome if they are to achieve rigidly defined performances. The various ways in which horse riders adjust to this variation stemming from individual differences between horses and how riders theorize this adaptation to variation echoes findings in the fields of ecological dynamics and motor control.

Expertise is often marked by the ability to dynamically adjust one's behavior and actions to variation within a task. Ecological dynamics have provided evidence and quantified experts' adaptability in a variety of sports, including climbing (Seifert, Dovgalecs, et al. 2014; Seifert et al. 2015), boxing (Hristovski, Davids, and Araujo, 2006), soccer (Chow et al. 2006), and swimming (Bideault, Herault, and Seifert 2013; Seifert, Komar et al. 2014). Ecological dynamics have also highlighted the importance of interindividual variation in explaining behavior and deciding on optimal courses of action: experts adjust their actions according to their own individual abilities (typically, body-scaled actions) and the parameters of the task. Variation is also a challenge faced by craftspeople in diverse contexts, such as dry stone walling (Farrar and Trorey 2008), cabinetmaking (Marchand 2016), cheese-making (Paxson 2011), and pottery (Roux et al., this volume).

The rider's level of expertise correlates with the rider's behaviors, some of which have been measured in studies of human movement. Expert and novice riders' positions differ in a number of ways. Most notably, expert riders are more synchronized with horses' movements than are novices (Lagarde et al. 2005), experts have a more obtuse hip-trunk angle with an overall position and orientation of the trunk that is more vertical (Schils et al. 1993), and the ways that experts use their reins differ as well (Warren-Smith et al. 2007).

So what is the variation faced by horse riders? What are the sources of such variation? How do riders adjust to it? Here, I examine the considerable individual variation among horses in terms of (1) their physical characteristics, (2) their temper, and (3) their training and past experiences.

Physical Characteristics

Horses have been selected for different characteristics, often in relation to the function for which they have been selected. Even within the same domain of horse riding—for example, horse racing—different types of racing influence some of the horses' physical characteristics. While trot races have selected for horses that are able to trot at high speeds without starting to gallop, gallop races have selected horses precisely for their gallop rather than their trot.

In some cases, a horse's physical characteristics are even presented as direct obstacles to the appropriate execution of movements. One of the horses described by Beudant is Voltigeur, "whose conformation did not allow expecting it would be successful" (Beudant 1923, 102–103). Conformation (shape or structure of a horse) and soundness (absence of lameness) of horses have also been judged, more generally, in relation to specific uses (Marks 2000). Contemporary veterinary science, for example, suggests that some breeds (in particular

German sport breeds) may be more fit for dressage than others, at least in competition, because their gaits show characteristics of regularity and amplitude of movement that are in line with what is expected in dressage competition (Barrey et al. 2002).

The role of the horsemen in the face of physical variation might be best expressed in Baucher's words. Despite being unable to change a horse's physical attributes, horsemen aim at mitigating their effects on the horse's movement: "Was I not right then in saying that if it is not in my power to change the defective formation of a horse, I can yet prevent the evil effect of his physical defects, so as to render him as fit to do everything with grace and natural ease, as the better formed horse?" (Baucher 1867, 53)

As such, the rider's role is to render the horse able to perform dressage movements to the best of their ability. Adapting to a horse's physical characteristics involves, for the rider, considering the ways that they might partially remediate a horse's defects (for horses that tend to have small steps in their gait, for instance, the rider might have them practice taking larger steps and more extended postures) or adjust the movement practiced (for instance, ask less bend from a stiff horse).

Mental Characteristics ("Personality" or "Temper")

Mental characteristics are no less important as a source of variation for the rider than horses' physical conformation. Such characteristics also affect the rider's course of action. Horses' temper has been a recurring and long-standing theme in writings on horse riding—from Xenophon's *On Horsemanship* (Xenophon 2012), where it is the main focus of part IX, to James Fillis's "Nervous Horses" (in Fillis [1902] 2005, 184–188) or the second chapter of de la Guérinière's *A Treatise upon Horsemanship*, which is aptly titled "Of the various tempers of horses" (Guérinière 1801). All of these authors discuss how individual differences in horses' temper can be substantial and must be taken into account while working with horses.

Ideally, technical movements are performed by calm and relaxed yet energetic horses. Beudant describes some horses as especially insensitive, whereas he characterizes others as more sensitive, reactive, or easily scared. An ideally sensitive or reactive horse would react promptly and easily to even faint signals from its rider but not be startled by random events in its environment (e.g., the arena's door slamming). While especially sensitive or nervous horses need to be slowed down and reassured, "colder" horses need to be made more energetic. One horse Beudant describes, Mimoun, seemingly demonstrates the worst of both worlds. Mimoun is described as both "extraordinarily fearful," getting scared by things that would scare no other animal, and at the same time lacking any impulsion or desire to move forward—to the point that Beudant gives up on teaching Mimoun the use of the whip action on the shoulder.

Training and Past Experiences

Overall, training and past experiences can influence horses' performances by facilitating or hindering some practices. In the worst cases, horses can become *rétif* (or restive); that is, having either no reactions or only defensive ones to their riders' actions and demands. Because Beudant writes about how he attempted to solve difficulties encountered with the horses that he describes, his examples are mostly negative. By contrast, more positive effects of training and past experiences can be found when he describes others' experiences after he had trained their horses. Riding well-trained horses can be particularly useful when acquiring new skills as a rider. One way that a horse's age might influence their performance is how much they lean on curves—that is, if and by how much they tend to displace their weight (putting them at risk of falling and leading to less regular and balanced gaits)—a behavior that training aims to avoid or minimize. Older horses generally lean significantly less than younger ones (Greve and Dyson 2016).

All of these differences (physical, mental, in amounts of experience) from one horse to the next affect (1) the starting point for training and (2) which path between the initial state and the goal might be available. In this context, the rider has to adjust to the variation displayed by horses themselves to generate horses that are able to reliably produce specific, rigidly defined movements—in a sense, expert horses.

Flexibility in Transmitting Dressage: Horse Riders as Horse Teachers

Training horses is, to some extent, a way of transmitting cultural content between species (Guillo and Claidière 2020). Flexibility is a common part of the pedagogic toolbox that humans use to transmit cultural contents from a teacher to a learner. I suggest that it also characterizes human practices of teaching learners of a different species—here, horses. However, as a practice, horse riding cannot be fully disentangled from training horses: cultural transmission is built into riding practices. Whenever a human rides a horse, that experience becomes part of the learned and accumulated experience of both the horse and the rider.

Horse-riding traditions are also maintained by making use of experts in both human and equine populations. One French saying in horse riding is, "A vieux cavalier, jeune cheval; à jeune cavalier, vieux cheval" (in English: To old rider, young horse; to young rider, old horse). In this case, the tradition is scaffolded by pairing trained horses with untrained riders and the inverse: the novice partner learns from the trained (expert) one. In such cases, traditions are sustained through this mutual scaffolding and "mixing" of novices and experts across species. Horses contribute to the transmission and stability of horse-riding traditions.

Riding horses also contributes to the transmission of cultural contents to the horses themselves by reinforcing their learned behaviors. The idea that horse riders teach their horses is not a modern one and can be found in most of the examined literature. For example, it can be located on several occasions in Xenophon's *On Horsemanship*: "With teaching, practice, and habit, almost any horse will come to perform all these feats beautifully, provided he be sound and free from vice"; "thus, when a horse is shy of any object and refuses to approach it, you must teach him that there is nothing to be alarmed at, particularly if he be a plucky animal"; "what we are now maintaining is that the owner ought to teach his own horse, and we will explain how this teaching is to be done" (Xenophon 2012, 16, 23, 28).

Flexibility scaffolds the horse's learning (reinforces its behavior), as suggested by some of the rules about how to increase the intensity of the rider's actions in relation to the horse's behavior. Flexibility contributes to teaching horses in two primary ways: (1) humans use flexibility as part of pedagogic strategies in the same ways that they do when teaching other humans, and (2) flexibility reflects important principles regarding how horses learn.

Horse Riders Flexibly Employ Horse-Adapted Pedagogy

One way that flexibility manifests in riders' behavior is in selecting and ordering actions following a principle of increasing intensity and also by selecting and ordering which exercises they make the horse perform. Exercises can be ordered through a form of "natural" progression, based on their difficulty and prerequisites, building from easier to more difficult versions. The selection of exercises and how difficult to make them are tools available for the rider to flexibly teach rigidly defined practices to horses.

Take, for example, the shoulder-in. In the shoulder-in, a horse moves forward while bent around the rider's inner leg toward the inside of the arena. In doing so, the horse's hind legs no longer follow the same tracks as its same-side-front counterpart, and the inner front leg follows its own track.

The shoulder-in is usually taught (both to horses and riders) by starting with the shoulder-fore, which is a less demanding version of the shoulder-in that still retains some of its main characteristics (i.e., the horse moves forward in the bend). Another way to scaffold the learning and performance of a shoulder-in is to ask the horse for the shoulder-in right after a circle or after having turned from an angle of the arena. Moving in a circle or passing a corner provides a bend to the horse that can be used to pursue the full movement. Such curved trajectories also induce the horse's inner (to the curve) back leg to go further under the horse's belly and weight, helping to lower their hips. This is used to facilitate the exercise for both horses and riders. A shoulder-in itself can also be used to help horses transition to canter, as it places the horse in the right balance to facilitate this transition. Similarly, it is possible to prepare horses for some exercises by practicing them with the rider on foot beside the horse before performing them while riding.

Transitioning from a trot (or walk) to a gallop can also be scaffolded in a similar manner, typically during or after turning around one of the arena's corners. Going past the corner "curves" the horse and puts it in a favorable balance to execute the transition.

Putting the horse in a position that is difficult to maintain can also be used to canalize and influence their behavior. For a horse that accelerates and becomes hard to control in the canter, for instance, it is possible to use a circle (and to reduce the circle's size) so that it becomes harder for the horse to speed up and maintain its balance at the same time. Put in this position, a horse will usually slow down by itself so that it can restore its balance. This approach is remarkably similar to ways in which nonlinear pedagogic strategies push learners in and out of more or less stable situations and motor patterns in order to make them learn different aspects of a task (Chow et al. 2015).

All of these aspects of flexible behavior on the rider's part reflect and mimic pedagogic interventions that are used between human teachers and learners. In other words, horse-riding traditions were made possible because humans exapted their own pedagogic abilities and strategies to horses: the rider's behavior has key characteristics frequently seen in contexts of cultural transmission to human novices.

Choosing which exercises to use and scaffolding exercises in particular ways are strategies that can be found in many situations where humans are teaching other humans. For instance, in his work on capoeira, Greg Downey identifies several types of pedagogic interventions, including scaffolding, reducing degrees of freedom, and the parsing of complex tasks (Downey 2008). These ways of scaffolding learning can be found in horse riding to some extent, too, even though they are used to scaffold the horse's learning of specific action-reaction couplings rather than imitate per se (as in Downey's example about capoeira). I focus here only on the parsing of complex tasks into simpler units.

Complex tasks employed in dressage are also usually parsed into simpler units. For example, it is common during jumping lessons to practice a whole jumping course (usually around 12 to 14 jumps) by jumping smaller sets of obstacles. In dressage, while the expected final performance might be a succession of technical movements, each movement is usually learned and practiced on its own before being executed together. Complex tasks are also parsed when they are taught and learned in parts before existing as "full" movements. Horses are taught principles like mobilizing their front or hind legs, independently and on demand, before being taught full movements. To teach them, the rider gives the adequate signal, then releases when the horse sketches the adequate response (for instance, moving only their hind legs). Only once this response is obtained reliably is it combined with other signals into a more complex movement (e.g., sidepass or half-pass).

However, some teaching behaviors that are common between humans are not used by or available to horse riders when riding and transmitting traditions to horses. This is the case for explicit verbal instructions. Although voice is used with horses, it is usually limited to a small set of signals—far from the full-length descriptions of actions found in verbal instructions used by humans when teaching other humans. Ostensive communication and demonstration are also forms of teaching that are commonly used between humans but cannot be used—or at least seem to be used much less—with horses.[3]

In the case of horse riding, interspecies communication and cultural transmission proceed through and exploit a subset of mechanisms that are usually at play when humans are teaching other humans. This is not a trivial point. It predicts that the rider must represent a number of aspects of the tasks and skills which they are trying to transmit to the horse, including the relative difficulty of variants and so in relation to the horse's specific abilities and characteristics. Flexibly using this repertoire of pedagogic interventions (e.g., choosing appropriate exercises) is at the core of the rider's expertise.

Flexibility Corresponds to Principles of How Horses Learn

Contemporary research in equitation science and learning theory have suggested that actions used by horse riders to train horses correspond to known principles in learning theory, even if horse riders often use mistaken labels. For example, removing rein tension is a form of negative reinforcement because it removes something from the horse's environment and reinforces a behavior, but it is often described as a positive reinforcement by coaches (Warren-Smith and McGreevy 2008). Here, I want to suggest that the flexibility afforded by adjustment rules in dressage reflects the riders' understanding of how horses learn. Rules for adjusting the rider's behavior to that of the horse encode useful information for riders with regard to how horses learn—that is, they implicitly inform the rider about horse cognition and, more specifically, about horses' learning capabilities.

Flexibility and equestrian feel (or tact équestre) involve principles that are in line with what is known about horses' learning, mostly based on learning theory. The adequation of practices used in training horses to learning theory has become an increasingly important topic (e.g., Cooper 1998; McCall 1990; McGreevy et al. 2018; McLean and Christensen

2017; Mills 1998), in part because of its implication for animal welfare. Practices like punishment, for instance, are considered both inefficient and nefarious for horses' learning and stress levels.

Writings by horsemen, and especially how they place flexibility at the core of their practice, echo contemporary writings in equitation science in several ways. For example, both equitation science and horsemen order exercises and movements in a similar, quasi-combinatorial way (McGreevy et al. 2018). Both start with simple responses of the horse to various aids from the rider and then combine these until they reach the most complex and difficult movements (see previous section for examples).

Other principles and examples described by horsemen also match current applications of learning theory to horse training. This is the case, for instance, with recommendations such as gradually increasing the signal asking the horse for a given action and stopping the signal at the correct behavior's onset. These recommendations highlight the importance of timing and intensity, and the coupling of signal and response facilitates horses' learning.

More general principles elaborated by both horsemen and learning theory researchers include the avoidance of conflict behavior and the need for consistency. As a final example, consider the condemnation (and perils) of (mis)using punishment, which is a recurring motif in both literatures. In applications of learning theory to training horses, "it may be difficult for the horse to associate the punishment with a specific part of its behaviour, a situation which is exaggerated if the punishment is applied inconsistently or overenthusiastically" (Cooper 1998, 41).

Granting that horses have some limited cognitive abilities, the literature also warns against the misuse of punishment as part of horses' training, focusing on its timing and the difficulties for horses of making proper associations with punishment:

Above all things, the rider of a difficult horse should never lose his temper. When a horse deserves punishment, he should get it with an amount of severity which might be regarded as the outcome of anger, but which should be proportionate to the offence. In fact, we should treat horses as we do children. We all know that nothing is worse than to punish a child when we are in a rage. A horse can in no case understand the feeling which prompts a man to punish him, and he will remember only the pain he has suffered and the occasion on which it was inflicted. His intelligence enables him to connect his action with the punishment it provoked; but it does not allow him to go further than that. On this account, if punishment is not administered at the precise moment the fault is committed, it will lose all its good effect, and will be an element of confusion in the memory of the animal. (Fillis [1902] 2005, 9)

Conclusion

As far as riding horses is concerned, flexibility can be the prerequisite to achieving (more) rigidly defined performances or outcomes. Both flexibility and rigidity participate in ensuring the cultural stability of horse-riding traditions across human and horse populations. This is by no means an exception among techniques. Several other technical domains show a similar pattern in which expert performance requires flexibility. This includes a wide range of crafts, ranging from cheese-making (Paxson 2011) to dry stone walling (Farrar and Trorey 2008). Interspecific traditions maintained among human and nonhuman populations also abound (e.g., Guillo and Claidière 2020; Manzi and Coomes 2002; Perri 2016; Ridgway 2021; Savalois, Lescureux, and Brunois 2013; Schroer 2014; Soma 2012; Stépanoff 2012).

Dressage offers a rare case study because it combines rigidly defined practices, maintained through well-documented traditions over long periods of time, with a living material—the horses—who bring variation into the practices, with sources ranging from their physical characteristics to their complex cognition and behaviors.

The stability of horse-riding traditions is ensured in part by (1) rigid definitions of practices, (2) efficient descriptions of flexible behavior that have been passed on relatively unchanged over centuries, and (3) a clear, well-defined representation of goals for riders to achieve. In contrast to these factors of rigidity, flexibility is required for performing technical gestures in the face of variation—some of which is generated by variation in horses' characteristics—but lso to transmit practices from humans to horses. This transmission is genuinely cultural, generating traditions shared by riders and horses, as it deploys pedagogic techniques that are normally used by humans in teaching other humans. Horse-riding traditions are thus maintained by this interplay of rigidity and flexibility in use and transmission, to which the horse-rider system is a necessary component.

Notes

1. There are other examples of traditions involving both humans and some nonhuman animal species, such as falconry (Schroer 2014; Soma 2012), cormorant fishing (Manzi and Coomes 2002), hunting dogs (Perri 2016; Ridgway 2021), reindeer herding and riding (Stépanoff 2012), herding dogs (Savalois, Lescureux, and Brunois 2013), or guide-dog training (Guillo and Claidière 2020), among many others.

2. A "FEI Dressage Rules," Fédération Équestre Internationale, accessed June 16, 2022, https://inside.fei.org/fei/disc/dressage/rules

3. Horses' understanding of pointing tends to be limited and is assumed to be driven mostly by stimulus or local enhancement—that is, a manipulation of their attention rather than an example of a deeper understanding of communicative interactions (Maros, Gácsi, and Miklósi 2008; Proops, Walton, and McComb 2010).

References

Andersen, P. A., and L. K. Guerrero. 2008. "Haptic Behavior in Social Interaction." In *Haptic Perception: Basics and Applications*, edited by M. Grunwald, 155–163. Birkhäuser: Springer Science and Business Media.

Barrey, E., F. Desliens, D. Poirel, S. Biau, S. Lemaire, J.-L. L. Rivero, and B. Langlois. 2002. "Early Evaluation of Dressage Ability in Different Breeds." *Equine Veterinary Journal* 34 (S34): 319–324.

Baucher, F. 1867. *Oeuvres complètes de F. Baucher: Methode d'équitation basée sur de nouveaux principes.* Paris: Dumaine.

Belock, B., L. J. Kaiser, M. Lavagnino, and H. M. Clayton. 2012. "Comparison of Pressure Distribution under a Conventional Saddle and a Treeless Saddle at Sitting Trot." *The Veterinary Journal* 193 (1): 87–91.

Beudant, E. 1923. *Extérieur et Haute École.* Imprint of Berger-Levrault.

Bideault, G., R. Herault, and L. Seifert. 2013. "Data Modelling Reveals Inter-Individual Variability of Front Crawl Swimming." *Journal of Science and Medicine in Sport* 16 (3): 281–285.

Bril, B., R. Rein, T. Nonaka, F. Wenban-Smith, and G. Dietrich. 2010. "The Role of Expertise in Tool Use: Skill Differences in Functional Action Adaptations to Task Constraints." *Journal of Experimental Psychology: Human Perception and Performance* 36 (4): 825–839.

Chow, J. Y., K. Davids, C. Button, and M. Koh. 2006. "Organization of Motor System Degrees of Freedom during the Soccer Chip: An Analysis of Skilled Performance." *International Journal of Sport Psychology* 37 (2–3): 207–229.

Chow, J. Y., K. Davids, C. Button, and I. Renshaw. 2015. *Nonlinear Pedagogy in Skill Acquisition: An Introduction.* New York: Routledge.

Cooper, J. J. 1998. "Comparative Learning Theory and Its Application in the Training of Horses." *Equine Veterinary Journal* 30 (S27): 39–43.

Cross, G. H., M. K. P. Cheung, T. J. Honey, M. K. Pau, and K.-J. Senior. 2017. "Application of a Dual Force Sensor System to Characterize the Intrinsic Operation of Horse Bridles and Bits." *Journal of Equine Veterinary Science* 48:129–135.e3.

Downey, G. 2008. "Scaffolding Imitation in Capoeira: Physical Education and Enculturation in an Afro-Brazilian Art." *American Anthropologist* 110 (2): 204–213.

Dumbell, L., C. Lemon, and J. Williams. 2019. "A Systematic Literature Review to Evaluate the Tools and Methods Used to Measure Rein Tension." *Journal of Veterinary Behavior* 29:77–87.

Durand, P. 2007. "Le Cadre noir." *Revue historique des armées* 249:6–15.

Fages, A., K. Hanghøj, N. Khan, C. Gaunitz, A. Seguin-Orlando, M. Leonardi, C. McCrory Constantz, et al. 2019. "Tracking Five Millennia of Horse Management with Extensive Ancient Genome Time Series." *Cell* 177 (6): 1419–1435.e31.

Farrar, N., and G. Trorey. 2008. "Maxims, Tacit Knowledge and Learning: Developing Expertise in Dry Stone Walling." *Journal of Vocational Education and Training* 60 (1): 35–48.

Fillis, J., Hayes. (1902) 2005. *Breaking and Riding: With Military Commentaries*. Lincoln: University of Nebraska Press.

Gandon, E., R. J. Bootsma, J. A. Endler, and L. Grosman. 2013. "How Can Ten Fingers Shape a Pot? Evidence for Equivalent Function in Culturally Distinct Motor Skills." *PLOS One* 8 (11): e81614.

Gianoli, L. 1969. *Horses and Horsemanship through the Ages*. New York: Crown Publishers.

Greve, L., and S. Dyson. 2013. "The Horse–Saddle–Rider Interaction." *The Veterinary Journal* 195 (3): 275–281.

Greve, L., and S. Dyson. 2016. "Body Lean Angle in Sound Dressage Horses In-Hand, on the Lunge and Ridden." *The Veterinary Journal* 217: 52–57.

Guérinière, F. R. de la. 1801. *A Treatise upon Horsemanship*. Translated by W. Frazer. Calcutta: Hircarrah Press.

Guérinière, F. R. de la. 1903. *A la française: pages choisies de la Guérinière*. Paris: Berger-Levrault & Cie.

Guillo, D., and N. Claidière. 2020. "Do Guide Dogs Have Culture? The Case of Indirect Social Learning." *Humanities and Social Sciences Communications* 7 (1): 1–9.

Hanot, P., A. Herrel, C. Guintard, and R. Cornette. 2018. "The Impact of Artificial Selection on Morphological Integration in the Appendicular Skeleton of Domestic Horses." *Journal of Anatomy* 232 (4): 657–673.

Hristovski, R., K. Davids, and D. Araujo. 2006. "Affordance-Controlled Bifurcations of Action Patterns in Martial Arts." *Nonlinear Dynamics, Psychology, and Life Sciences* 10 (4): 409–444.

Lagarde, J., C. Peham, T. Licka, and J. A. S. Kelso. 2005. "Coordination Dynamics of the Horse-Rider System." *Journal of Motor Behavior* 37 (6): 418–424.

Levine, M. 2005. "Domestication and Early History of the Horse." In *The Domestic Horse: The Origins, Development and Management of Its Behaviour*, edited by D. S. Mills and S. M. McDonnell, 5–22. Cambridge: Cambridge University Press.

L'Hotte, A. 1906. *Questions équestres*. Paris: Plon.

Manzi, M., and O. T. Coomes. 2002. "Cormorant Fishing in Southwestern China: A Traditional Fishery under Siege." *Geographical Review* 92 (4): 597–603.

Marchand, P. T. H. J. 2016. *Craftwork as Problem Solving: Ethnographic Studies of Design and Making*. Farnham, UK: Ashgate Publishing.

Marks, D. 2000. "Conformation and Soundness." *Proceedings of the American Association of Equine Practitioners* 46: 39–45.

Maros, K., M. Gácsi, and Á. Miklósi. 2008. "Comprehension of Human Pointing Gestures in Horses (*Equus caballus*)." *Animal Cognition* 11 (3): 457–466.

McCall, C. A. 1990. "A Review of Learning Behavior in Horses and Its Application in Horse Training." *Journal of Animal Science* 68 (1): 75–81.

McGreevy, P., J. W. Christensen, U. K. von Borstel, and A. McLean. 2018. *Equitation Science*. Hoboken, NJ: John Wiley & Sons.

McLean, A. N., and J. W. Christensen. 2017. "The Application of Learning Theory in Horse Training." *Applied Animal Behaviour Science* 190:18–27.

Mills, D. S. 1998. "Applying Learning Theory to the Management of the Horse: The Difference between Getting It Right and Getting It Wrong." *Equine Veterinary Journal* 30 (S27): 44–48.

Moll, J., and E.-L. Sallnäs. 2009. "Communicative Functions of Haptic Feedback." In *Haptic and Audio Interaction Design*, edited by M. E. Altinsoy, U. Jckosch, and S. Brewster, 1–10. Dordrecht: Springer.

Ödberg, F. O., and M.-F. Bouissou. 1999. "The Development of Equestrianism from the Baroque Period to the Present Day and Its Consequences for the Welfare of Horses." *Equine Veterinary Journal* 31 (S28): 26–30.

O'Neill, M. 2018. "The Effect of Rein Type and Bit Type on Rein Tension in the Ridden Horse." PhD diss., University of Plymouth.

Orlando, L. 2020. "Ancient Genomes Reveal Unexpected Horse Domestication and Management Dynamics." *BioEssays* 42 (1): 1900164.

Paxson, H. 2011. "The 'Art' and 'Science' of Handcrafting Cheese in the United States." *Endeavour* 35 (2): 116–124.

Perri, A. R. 2016. "Hunting Dogs as Environmental Adaptations in Jōmon Japan." *Antiquity* 90 (353): 1166–1180.

Petersen, J. L., J. R. Mickelson, E. G. Cothran, L. S. Andersson, J. Axelsson, E. Bailey, D. Bannasch, et al. 2013. "Genetic Diversity in the Modern Horse Illustrated from Genome-Wide SNP Data." *PLOS One* 8 (1): e54997.

Proops, L., M. Walton, and K. McComb. 2010. "The Use of Human-Given Cues by Domestic Horses, Equus Caballus, during an Object Choice Task." *Animal Behaviour* 79 (6): 1205–1209.

Ridgway, M. 2021. "Hunting Dogs." *Veterinary Clinics: Small Animal Practice* 51 (4): 877–890.

Savalois, N., N. Lescureux, and F. Brunois. 2013. "Teaching the Dog and Learning from the Dog: Interactivity in Herding Dog Training and Use." *Anthrozoös* 26 (1): 77–91.

Schils, S. J., N. L. Greer, L. J. Stoner, and C. N. Kobluk. 1993. "Kinematic Analysis of the Equestrian—Walk, Posting Trot and Sitting Trot." *Human Movement Science* 12 (6): 693–712.

Schroer, S. A. 2014. "On the Wing: Exploring Human-Bird Relationships in Falconry Practice." PhD diss., University of Aberdeen.

Schubert, M., H. Jónsson, D. Chang, C. Der Sarkissian, L. Ermini, A. Ginolhac, A. Albrechtsen, et al. 2014. "Prehistoric Genomes Reveal the Genetic Foundation and Cost of Horse Domestication." *Proceedings of the National Academy of Sciences* 111 (52): E5661–E5669.

Seifert, L., J. Boulanger, D. Orth, and K. Davids. 2015. "Environmental Design Shapes Perceptual-Motor Exploration, Learning, and Transfer in Climbing." *Frontiers in Psychology* 6: 1–15.

Seifert, L., V. Dovgalecs, J. Boulanger, D. Orth, R. Hérault, and K. Davids. 2014. "Full-Body Movement Pattern Recognition in Climbing." *Sports Technology* 7 (3–4): 166–173.

Seifert, L., J. Komar, T. Barbosa, H. Toussaint, G. Millet, and K. Davids. 2014. "Coordination Pattern Variability Provides Functional Adaptations to Constraints in Swimming Performance." *Sports Medicine* 44 (10): 1333–1345.

Shergill, S. S., P. M. Bays, C. D. Frith, and D. M. Wolpert. 2003. "Two Eyes for an Eye: The Neuroscience of Force Escalation." *Science* 301 (5630): 187–187.

Soma, T. 2012. "Contemporary Falconry in Altai-Kazakh in Western Mongolia." International Journal of Intangible Heritage 7:103-111

Stépanoff, C. 2012. "Human-Animal 'Joint Commitment' in a Reindeer Herding System." *HAU: Journal of Ethnographic Theory* 2 (2): 287–312.

Taylor, W. T. T., B. Jamsranjav, and T. Tumurbaatar. 2015. "Equine Cranial Morphology and the Identification of Riding and Chariotry in Late Bronze Age Mongolia." *Antiquity* 89: 854–871.

Taylor, W. T. T., T. Tuvshinjargal, and J. Bayarsaikhan. 2016. "Reconstructing Equine Bridles in the Mongolian Bronze Age." *Journal of Ethnobiology* 36 (3): 554–570.

van Weeren, P. R. 2013. "About Rollkur, or Low, Deep and Round: Why Winston Churchill and Albert Einstein Were Right." *The Veterinary Journal* 196 (3): 290–293.

Warmuth, V., A. Eriksson, M. A. Bower, G. Barker, E. Barrett, B. K. Hanks, S. Li, et al. 2012. "Reconstructing the Origin and Spread of Horse Domestication in the Eurasian Steppe." *Proceedings of the National Academy of Sciences* 109 (21): 8202–8206.

Warren-Smith, A. K., R. A. Curtis, L. Greetham, and P. D. McGreevy. 2007. "Rein Contact between Horse and Handler during Specific Equitation Movements." *Applied Animal Behaviour Science* 108 (1): 157–169.

Warren-Smith, A. K., and P. D. McGreevy. 2008. "Equestrian Coaches' Understanding and Application of Learning Theory in Horse Training." *Anthrozoös* 21 (2): 153–162.

Wolframm, I. 2013. *The Science of Equestrian Sports*. New York: Routledge.

Xenophon. 2012. *On Horsemanship*. Radford, Virginia: SMK Books/Wilder.

Zetterqvist Blokhuis, M., J. Bornemark, L. Birke, Södertörns högskola, and Institutionen för kultur och lärande. 2019. "Interaction between Rider, Horse and Equestrian Trainer a Challenging Puzzle." PhD diss., Södertörns högskola.

13 Exploring Cultural Techniques in Nonhuman Animals: How Are Flexibility and Rigidity Expressed at the Individual, Group, and Population Level?

Sadie Tenpas, Manon Schweinfurth, and Josep Call

The existence of animal culture, understood as those group-typical behavior patterns that are shared by members and that rely on socially learned and transmitted information (Laland and Hoppitt 2003), is commonly accepted today. Following over 50 years of debate, we have convincing evidence of cultural behavior in birds (Aplin 2019), cetaceans (Allen et al. 2013; Day, Kendal, and Laland 2001), and primates (Whiten and van de Waal 2017), among others (Laland and Evans 2017; Whiten 2019). Cultural traditions are distinct behavior patterns, shared by at least two members of a group, that persist over time and that new practitioners acquire, in part, through socially aided learning (Fragaszy and Perry 2003). These traditions can include techniques that produce change toward a material goal (Lamon et al. 2017). For example, chimpanzees (*Pan troglodytes*) create and use stick tools during termite-fishing (Goodall 1964). However, for the purpose of this chapter, we will extend our definition of techniques beyond cases toward material goals to include more cultural traditions, such as different manual techniques to perform a certain action. As such, we understand techniques as complex actions whose instrumental function is to produce changes in the environment, both physical and social.

The ability to ratchet up the complexity of cultural traditions, leading to both increased efficiency and productivity, is created by cumulative culture (Dean et al. 2014; Tennie, Call, and. Tomasello 2009). It is a phenomenon primarily observed, and by some exclusively described, in humans (Dean et al. 2014). Although the learning of many animal species is influenced by the observation of and/or interaction with another animal or its products (Heyes 1994), there is minimal evidence of cumulative culture in animals (but see Sasaki and Biro 2017; Schofield et al. 2018). The richest evidence of animal culture comes from one of our closest living relatives, chimpanzees (Marshall-Pescini and Whiten 2008).

Long-term research on chimpanzees has provided us with extensive documentation of cultural traditions, highlighting variation between communities and distinct cultural repertoires. Most famously, Andrew Whiten and colleagues (1999) collected candidate cultural behaviors from seven long-term field sites across Africa. Using a systematic approach, known as the method of exclusion or the ethnographic method, candidate behaviors shown only to occur within some communities were determined to be cultural when the variability between groups could not be otherwise explained by ecological or genetic differences, thus leaving social learning as the remaining source of the observed variation.

While chimpanzees show a great variety of traditions, there is little evidence that these traditions are frequently updated through additional knowledge or behaviors (e.g., Gruber et al. 2009; but see Biro et al. 2003). This tendency toward stasis is especially striking considering that chimpanzees demonstrate an impressive ability to innovate solutions to novel problems (Bandini and Harrison 2020). So, while chimpanzee cultures may not reflect the same process or complexity of human cumulative culture, chimpanzee culture still undergoes cultural evolution in which a balance must be struck between the creation, variation, or application of cultural behaviors and the high degrees of similarity between individual performances, leading to stable traditions that persist across generations despite potential disruptive influences. Thus, the questions arise: Which mechanisms and factors drive chimpanzee cultures toward stasis, which drive them toward change, and how can we understand this in respect to cultural techniques?

To address these questions, we will begin by exploring the relationship between flexibility and rigidity in culture at the individual, group, and population level. Using these concepts, we will investigate how flexibility and rigidity in chimpanzee culture are currently understood and how social learning is used to explain them. In doing so, we will highlight the limitations of what we term "social learning theory"—namely, the reliance on explaining culture through high-fidelity transmission mechanisms and the inability to meaningfully incorporate the roles of ecology and genetics into cultural transmission. We will then address an alternative theory originating in the cultural evolutionary literature, "cultural attraction theory," that sheds light on factors supporting cultural variation and stability. Further, we will discuss the limitations of cultural attraction theory. Due to its origin and use within human cultural evolution, cultural attraction theory has been critical of social learning theory, both because individual and social learning are highly intertwined in human societies and because they argue cultural evolution need not be reliant on high-fidelity transmission mechanisms. However, cultural attraction theorists still have to discuss the precise role of low-fidelity social learning mechanisms and how they might support cultural transmission. Drawing from both theoretical perspectives, we will finally propose a framework that combines these two theories to provide a more satisfactory explanation of the occurrence of flexibility and rigidity in chimpanzee techniques.

Defining Flexibility and Rigidity in Primate Techniques

To fully capture the concept of cultural evolution, we will illustrate flexibility and rigidity in primate cultural techniques at the individual, group, and population level. Beginning at the individual level, we define a spectrum with *liberalism*, describing an individual's tendency or disposition to change, on one end, and *conservatism*, describing an individual's tendency or disposition to remain the same, on the other end (table 13.1). Note that we expect to see variation both within and between individuals along this spectrum. Further, when we consider liberal (and later flexible) behavior, we do not differentiate between behavioral changes that result in the acquisition of novel behavior and changes that result in the utilization of alternative but familiar behaviors already existing within the individual's repertoire (cf. Pope-Caldwell, this volume, for a more thorough breakdown of these distinctions). On one level, behavioral change can occur as an opportunity, where an individual may alter their technique through modification, variation, or invention of behavior, often improving the technique's efficiency. On a second level, some cases of behavioral change are driven by necessity, in which access to the behavioral outcome is dependent on improvement of the technique. By distinguishing the necessity for a change in an individual's behavior, or lack thereof, we can further break down liberal and conservative behavior with respect to adaptive decisions. We will do so next by considering examples of behavioral change at the individual level on the scale of distinct decisions (i.e., each moment at which the decision to change or not is made). Over larger timescales, such as days, months, and years, individuals may oscillate between behaviors or abandon behaviors for another, and this variation is expected.

To illustrate liberalism and conservatism at the individual level and the role of necessity, we use as an example chimpanzee nut-cracking behavior in response to an environmental change. Environmental changes that influence changes in behavior can occur at the individual level, within a life span, such as seasonal changes or natural disasters. At the group and population level, environmental changes occur more slowly, across generations, such as effects of climate change or other human impacts. For our example, let us imagine a scenario in which a chimpanzee uses a wooden hammer to crack open nuts. An environmental change resulted in increased hardness of the shells of the nuts used for extractive foraging such that a wooden hammer was no longer hard enough to crack the nut. Therefore, the liberal individual selects a different hammer, such as a stone, to crack the nut. Under this condition, the individual is behaving *innovatively*, responding to the necessity for change. Alternatively, if we consider an example in which the nutshell remains the same hardness yet the liberal individual switches hammers anyway, the individual is behaving *creatively* by changing behavior, despite there being no need for change. When considering the individual behaving conservatively, the same scenarios apply. If the conservative individual maintains their behavior but there is need for change, the individual is behaving *perseveringly*. If the conservative individual maintains their behavior and there is no need for change, the individual is behaving *consistently*. While the terms we have introduced highlight distinct adaptive decisions, they do not necessarily need to be considered as binary. Likely, these decisions bookend a spectrum wherein individuals might behave somewhere in between innovatively and creatively if changing the hammer material is useful but not strictly necessary. For example, imagine that the nuts become harder, where it becomes more difficult but not impossible to use the wooden hammer. In this case, switching to a stone hammer is useful but not necessary. Conversely, in this context, if the individual chooses not to change the hammer material, they would be behaving somewhere between perseveringly and consistently.

Broadening to the group level, we introduce social factors guiding behavioral change (or lack of) using the same context and framework as before to understand how groups behave as a product of their group members. As we expect individuals to vary in their tendency toward liberal and conservative behavior, so too should we expect the composition of these individuals to vary within each group, thus affecting the overall tendency of the group. In defining a spectrum at the group level, we have *flexibility*, describing a group's tendency to change a behavior, on one end, and *rigidity*, describing a group's tendency to maintain a behavior, on the other end (table 13.1). Envisioning the same nut-cracking scenario, when a flexible group needs to change their behavior and does so, this change can be described as an *advancement*. When a flexible group does not need to change their behavior but does anyway, this change can be described as a *shift*. Alternatively, when a rigid group needs to change their behavior but does not, this lack of change can be described as *fixedness*. When a rigid group does not need to change their behavior and does not, this lack of change can be described as *stasis*. As before, these terms can be understood as the two ends of a spectrum of behavior, and we may have groups behaving at times somewhere between an advancement and a shift, or fixedness and stasis.

Expanding further, we can examine the cultural repertoire of a group to understand the population dynamics as a product of evolution. Problems arise when a group leans too far toward either end of the flexible–rigid spectrum. For example, if a group leans too far toward the flexible end, such that it is constantly creating and abandoning cultural behaviors, the group is not able to retain long-term cultural information and risks disadvantage when facing old problems again. Over evolutionary time, an overly flexible group can lead to cultural *breakdown* (table 13.1). Conversely, if a group leans too far toward the rigid end, such that it rarely acquires new cultural behaviors or modifies existing ones, then the group is unlikely to innovate new solutions or adapt to changing environments. Over evolutionary time, an overly rigid group can lead to cultural *stagnancy* (table 13.1). Therefore, there must be a balance between flexible creation, variation, and application of cultural behaviors while also maintaining relatively rigid transmission and performance, allowing cultural traditions to form and persist across generations.

One way we can understand these dynamics is through frequency-dependent evolution, which describes how the fitness of traits within a population is related to their frequency within the population (Ayala and Campbell 1974). Within social learning, traits may be subject to frequency-dependent biased transmission, specifically conformism and anti-conformism

Table 13.1
Definitions of flexibility and rigidity at the individual, group, and population level

Term	Definition	Level
Liberalism	The tendency or disposition to change	Individual
Innovation	When an individual needs to change behavior and does	
Creativity	When an individual does not need to change behavior but does anyway	
Conservatism	The tendency or disposition to remain the same	
Perseverance	When an individual needs to change behavior but does not	
Consistency	When an individual does not need to change behavior and does not	
Flexibility	The tendency to change behavior	Group
Advancement	When a group needs to change behavior and does	
Shift	When a group does not need to change behavior but does anyway	
Rigidity	The tendency to maintain behavior	
Fixedness	When a group needs to change behavior but does not	
Stasis	When a group does not need to change behavior and does not	
Breakdown	When a group leans too far toward flexibility, such that it is constantly creating and abandoning cultural behaviors and thus is not able to retain long-term cultural information and risks disadvantage when facing old problems again; over evolutionary time, an overly flexible group can lead to cultural breakdown	Population
Stagnancy	When a group leans too far toward rigidity, such that it rarely acquires new cultural behaviors or updates existing ones and thus is unlikely to innovate new solutions or adapt to changing environments; over evolutionary time, an overly rigid group can lead to cultural stagnancy	

strategies. For instance, conformity is the preferential copying of the most common behavior, and anti-conformity is the preferential copying of rare behavior (Barrett, McElreath, and Perry 2017). As the frequency of cultural traits implicitly holds important information on those different traits, learners can use frequency to select locally adaptive behaviors or avoid selecting maladaptive behaviors (Nakahashi, Wakano, and Henrich 2012). Conformist learners may use the frequency of the most common trait as a cue for which behaviors to adopt and integrate, which can be adaptive in that they learn information from their group while avoiding potential errors (Nakahashi, Wakano, and Henrich 2012). This strategy has been demonstrated to work well in spatially heterogeneous environments, but without additional flexible strategies, conformity may keep more adaptive behaviors from spreading, leading to cultural stagnancy (Barrett, McElreath, and Perry 2017; Kendal, Giraldeau, and Laland 2009; Nakahashi, Wakano, and Henrich 2012). Anti-conformity may be such a flexible strategy, offering fitness value to individuals who can more flexibly adapt to their changing environment, thus introducing new behaviors. However, in a population of only anti-conformists, no cultural behavior can become stable and will eventually disappear. This highlights the interplay of conformists and anti-conformists, which are dependent on the other strategy. If there is the right frequency of both, cultural traits can emerge and persist.

When considering primate cultural techniques, comparative researchers have historically explained the relative stability through the fidelity of social learning mechanisms. A social learning mechanism of high-fidelity would allow for rigid transmission of techniques, retaining a high degree of similarity between performances of the demonstrator and observer such that the technique remains stable as it is transmitted between individuals over a long period of time and protected from disruptive influences such as miscopying or the loss of necessary information required to sustain a tradition (Charbonneau 2020). Mechanisms of lower fidelity offer less rigidity in the transmission of a technique, such that subsequent performances may differ from that of the demonstrator, being more susceptible to said disruptive influences and thus not able to retain a high degree of similarity. In this use of fidelity, comparative researchers are specifically employing "propensity fidelity" (outlined in Charbonneau 2020), focusing on the specific mechanisms involved in the transmission of a cultural trait, where certain transmission mechanisms are more or less faithful than another in perpetuating a tradition. We will illustrate these mechanisms and their associated fidelity as described by social learning theory in the next section by addressing them in chimpanzee behavior.

Social Learning and Cultural Fidelity in Chimpanzees

To this day, social learning mechanisms have been the dominant source for understanding the spread and persistence of chimpanzee cultural behavior. It has been argued that high-fidelity social learning mechanisms can stabilize cultural traditions through faithful transmission from one individual to the next. Imitation, the copying of another's actions (Tomasello 1996), and teaching, the active facilitation of another's learning (Hoppitt and Laland 2013), are understood to result in a high degree of fidelity and play an important role in the evolution of culture (Whiten et al. 2009). Other social learning mechanisms such as emulation, copying the environmental or end-results of another's actions (Tomasello 1996), and local and stimulus enhancement, when the actions of another draw the attention of the observer to particular locations or stimuli, respectively (Hoppitt and Laland 2013), are thought to be of lower fidelity. In such cases, when low-fidelity mechanisms are used, the techniques performed by the demonstrator are not retained and performance can be dissimilar between episodes of transmission from one individual to the next (Hoppitt and Laland 2013).

Following the onset of research into the culture of wild chimpanzees, the idea that an ape had the ability to ape (i.e., imitate) and the existence of culture went hand in hand (Whiten et al. 2009). However, subsequent experimental research into the social learning abilities of chimpanzees questioned and critiqued this notion, reporting that unlike humans, chimpanzees are not natural imitators but rather emulators (Tomasello et al. 1987). In an experiment, chimpanzees were exposed to a demonstrator using a rake tool to acquire a food-reward (Tomasello et al. 1987). Individuals who observed the demonstrator were more likely to adopt the behavior than those who did not observe the demonstrator. However, they did not acquire the technique used by the demonstrator to obtain food from awkward positions. These results suggest a mechanism of higher fidelity than stimulus enhancement but lower fidelity than imitation, termed emulation (Tomasello 1990). Later studies reaffirmed chimpanzees' tendency to emulate end-results rather than imitate a demonstrator's actions (Call, Carpenter, and Tomasello 2005; Tomasello and Call 1997; Whiten et al. 2004; see also Whiten et al. 2009).

Many of the aforementioned studies relied on a single episode of transmission between one demonstrator and an observer, but cultural transmission requires multiple iterations of these episodes to sustain traditions. To address this discrepancy, Andrew Whiten, Victoria Horner, and Frans de Waal (2005) conducted a transmission chain experiment with chimpanzees to experimentally replicate spontaneous imitation through the transmission of a novel food-processing technique. In doing so, an "artificial fruit" that could be opened by either using a stick to lift a hook ("lift" method) or poking a stick into a trap ("poke" method) was presented to three groups of chimpanzees. For each group, one individual was selected and exposed to the artificial fruit. One was trained with the lift method, one was trained with the poke method, and another was not trained at all. The results revealed that no individual in the untrained group was able to access the food using either the lift or poke method. Importantly, individuals of both trained groups were able to learn the seeded technique through observation of group members. However, the initial technique introduced into the trained groups was not exclusively retained by all group members, nor were all individuals able to open the artificial fruit. This variation attracted criticism.

Nicolas Claidière and Dan Sperber (2010) highlighted two features of Whiten, Horner, and De Waal's (2005) results, critiquing their argument that imitation offers sufficient fidelity to explain the stability of the observed transmission. First, because not all individuals performed the demonstrated technique, stimulus enhancement by increased manipulation of the device could have led other individuals to engage with the device. Thereby, the naive individual could have explored the device individually and discovered their own technique, leading to several techniques in the group. Even if the individual were strictly adhering to imitation to solve the problem, increased interaction may still have resulted in the observing individual spontaneously discovering the alternative technique. Second, had the groups received more naturalistic exposure to the artificial fruits, such as unrestricted access for a longer period of time, one might expect that the individuals eventually perform the most efficient method, or both if equally efficient, rather than that technique initially propagated through imitation. This outcome is illustrated by the fact that members of the lift group more often converted to the poke method than the other way around. Thus, Claidière and Sperber suggested that the social learning mechanisms demonstrated in the experiment may act as propagation mechanisms, but that complementary stabilization mechanisms, such as ecological availability, reward-based factors (that combine an ecological and a psychological aspect), content-based psychological factors, and source-based psychological factors (Sperber and Claidière 2008), must exist to explain wild observations of cultural transmission that are not being modeled in these experiments.

This criticism highlights a more fundamental problem of social learning theory that is demonstrated by the method of exclusion, which defines cultural behavior based on social learning alone (Whiten et al. 1999). Using this method, chimpanzees have been described to demonstrate at least 39 cultural traits (Whiten et al. 1999). When only considering social learning mechanisms, these group behaviors might be difficult to explain by mechanisms of lower fidelity than imitation (such as stimulus enhancement) alone. For example, variations observed in tool-use techniques like ant-dipping, in which one group uses a long wand with one hand and the ants are wiped off with the other (McGrew 1974) and another group uses a short wand and ants are transferred directly to the mouth (Nishida 1973), could use differential copying of either technique. Here, one could imagine a mix of imitation, other social learning mechanisms, and individual learning is likely at play, as was implied by previous experimental research (Whiten et al. 2004).

However, while the method of exclusion allows cataloging and identifying cultural traits, genetic and ecological factors are often difficult to entirely exclude (Laland and Hoppitt 2003; Schuppli and van Schaik 2019). For example, subsequent studies on ant-dipping techniques in chimpanzees revealed that the performed variants are influenced by the nature of the ants' behavior (Humle and Matsuzawa 2002). Techniques differ based on the abundance of ants, their aggressiveness, and the severity of their bite, such that when each of these factors were high, the chimpanzees used the long wand and hand-swiping technique. Under the method of exclusion, this ecological explanation would lead to the exclusion of ant-dipping as a cultural trait. However, we would argue that by excluding behavioral variation understood to be influenced by ecological factors, we may be overlooking the very mechanisms that support cultural variation and stability similar to those complementary mechanisms proposed by Claidière and Sperber (2010).

Ecological factors are arguably necessary to the existence of culture. For example, the use of olive oil is a cultural feature of Mediterranean Basin cuisine, which has evolved with the historic availability of olive trees (Vossen 2007). However, we would not exclude dishes involving olive oil from being cultural traditions of these regions because olives have historically not been available in other parts of the world, as one would do if employing the method of exclusion here. Rather, we understand that these recipes are created, taught, and otherwise passed down through generations and they continue to be adapted and supported by the availability of olives. While this example is most certainly oversimplified (as the Mediterranean Basin encompasses many distinct cultures and olives have since been cultivated around the globe), one can understand that such ecological features not only make culture distinct but may additionally provide a consistent feature, which maintains these cultural traditions over generations. This idea should also be applied to animal cultures. Take again the example of the ant-dipping variations, one could imagine that a distinct technique observed in one population is propagated through the group by social learning mechanisms but is stabilized by the nature of the ant species. Unlike the transmission chain experiments, individuals interacting with more aggressive ants may be less likely to switch to or discover an alternative, potentially more efficient technique, dissuaded by the severe bites and thus maintaining the behavior over time.

Both the transmission chain experiments and the method of exclusion highlight the emphasis placed on imitation and other high-fidelity social learning mechanisms in explaining cultural variation and stability by social learning theory. In discounting the role of lower fidelity social learning mechanisms in combination with excluding influential factors such as ecology and genetics, social learning theory may be overlooking the very mechanisms supporting cultural variation and stability. In the next section, we will explore those factors outside of social learning mechanisms that may have been excluded or overlooked but that may assist in the propagation and stabilization of cultural techniques. Here, we will shift our focus to psychological and ecological factors by addressing cultural attraction theory in the realm of animal culture and later applying it to examples of chimpanzee cultural techniques.

Cultural Attraction Theory and Animal Culture

Cultural attraction theory can explain how cultural traits (such as norms, beliefs, skills) change in their distribution and form over time (Buskell 2017; Scott-Phillips, Blancke, and Heintz 2018; Sperber 1985, 1996). Initially introduced to understand human culture, cultural attraction theory was intended to reconcile two observations: (1) that at the micro-level, transmission of information between humans is generally not a copying process and typically results in modifications, and (2) that at the macro-level, cultural information is relatively stable within entire populations and often remains so across generations (Claidière and Sperber 2007). These observations suggest that the micro-process of transmission is not itself faithful enough to explain macro-stability. Thus, it contrasts social learning theory in which cultural traditions are mainly based on high-fidelity social transmission (Claidière and Sperber 2007). Instead, cultural attraction theory posits that the cognitive mechanisms producing social transmission at the micro-level create *cultural chains* of causally related events. Thereby the micro- and macro-level are interconnected. Mental representations (knowledge, beliefs, intentions) allow for public productions (artifacts, behavior, speech) that influence the mental representations of others and thus both levels act as positive feedback loops (Scott-Phillips, Blancke, and Heintz 2018). Transformations in these chains are biased because they are not totally random. Hence, over time, cultural traits emerge within a population. In a system of cultural chains, *cultural attraction* is the probabilistic favoring of specific traits. *Factors of attraction* that bias cultural traits influence an individual's mental representation and thus the production of traits, which can be applied to nonhuman cultures.

Cultural attraction theory suggests that factors of attraction can be divided into three categories: reconstructive learning, motivational factors, and ecological factors (Buskell 2017). To understand factors related to reconstructive learning, we must turn to the transformative nature of transmission assumed by cultural attraction theory. As such, an individual acquiring a new cultural trait rarely strictly copies the variant(s) but instead draws on the information transmitted in addition to personal background knowledge, inferential abilities, and interests to produce their own variant(s) (Claidière and Sperber 2007). Therefore, reconstructive learning consists of processes involved in inferential learning that are influenced by individual beliefs, emotions, judgments, and cognition (Buskell 2017). This is distinct from motivational factors that make one want to use or transmit a particular variant (Buskell 2017; Morin 2015). Finally, ecological factors encompass environmental elements influencing cultural traits. Ecological factors can range from features of the biological or physical environment, such as food and material resources, to behaviors and artifacts, including public representations used for communication (Buskell 2017; Heintz and Claidière 2015; Morin 2015; Scott-Phillips, Blancke, and Heintz 2018).

The factors of attraction outlined by cultural attraction theory contribute to a variety of items that influence cultural traits, resulting in a similar distribution of those traits between a given time step and another. This phenomenon has been termed within cultural attraction theory as hetero-impact, where one item of a population can affect the evolution of another item; for instance, item A influences the frequency of item B (Claidière, Scott-Phillips, and Sperber 2014; Sperber, pers. comm.). Conversely, homo-impact describes the impact of one item in a population on the evolution of itself; for instance, item B influences the frequency of item B (Claidière, Scott-Phillips, and Sperber 2014; Sperber, pers. comm.). Whereas "impact" describes the evolutionary relationship between items from cause to effect, "attraction" identifies attractors by viewing the same relationship from the opposite perspective, from effect to cause (Sperber, pers. comm.). Copying processes, like that of imitation, would be described as homo-attraction as propagation by high-fidelity social learning mechanisms would result in self-similar reproductions of a cultural item. Cultural attraction theory makes a point to highlight the idealization of models of cultural evolution that rely on copying processes for the reproductive success of cultural items, noting that, in the context of human cultures, it would be surprising if the success of our traditions were the product of imitation alone, given the many ways with which we can share information (Claidière, Scott-Phillips, and Sperber 2014). Rather, cultural attraction theory posits that a variety of mechanisms contribute to social learning, few of which would be of high enough fidelity to support cultural stability (Sperber, pers. comm.). Cultural attraction theory thus de-emphasizes imitation within cultural evolution, contrasting human social learning research that holds copying to be a key process, and has yet to discuss the roles of different social learning mechanisms explicitly. Therefore, while cultural attraction theory does not deny or exclude the role of social learning mechanisms contributing to cultural fidelity, its main objective has been to identify the range of processes outside of copying processes that contribute to cultural fidelity. In doing so, much of the literature on and surrounding cultural attraction theory to date has not discussed the contribution of specific low-fidelity social learning mechanisms in detail, not so dissimilar from how the literature on social learning has yet to discuss the contribution of alternative influences like ecology and genetics in more detail. Until now, these theories have developed mostly in isolation from one another; however, given their differences in focus, we believe both theories to be complementary and that when pursued jointly, they may shed light on the contributions to cultural evolution that the other has yet to explore. Therefore, in an attempt to merge the concepts of social learning theory and cultural attraction theory, we have developed a theoretical framework, which we will now detail and describe in the context of animal culture.

When considering the factors of attraction that may bias animal cultural variants, we imagine three *classes* that are separate but highly interactive: one *social*, one *individual*, and one *ecological*. These categories are similar to, but importantly distinct from, those described for cultural attraction theory, as we will attempt to incorporate the social transmission mechanisms highlighted by social learning theory into a new framework. The social class of factors of attraction are related to the social environment, including accessibility to observe conspecifics or their products and community structure, as well as social learning mechanisms. The individual class of factors of attraction are related to an individual's psychology including emotional, motivational, and sensorimotor processes, be they determined or learned. Finally, the ecological class of factors of attraction are related to ecological availability and opportunity and environmental state, ranging from climate considerations to the available vegetation and landscape features (cf. Pope-Caldwell, this volume, for a discussion on the influence of dynamic environments). As stated previously, we imagine these classes separately but recognize that they are highly interactive, influencing one another greatly. For example, when considering individual psychological mechanisms and the social environment, combinations of the two might uniquely influence cultural variants through social learning strategies such as dominance rank based social learning which occurs when chimpanzees selectively observe and copy group members of higher dominance (Kendal et al. 2015, 2018). Further, given the species, we may see that the strength of each class's ability to bias cultural traits varies, though importantly, each class will always have some influence, even if only minimal. As factors of attraction can influence across different levels, we would also expect to see variation of each class's strength both within and between individuals, which in turn guides liberal and conservative behavior. In the next section, we will illustrate how our combined approach offers new understanding toward the flexibility and rigidity of chimpanzee cultural techniques.

Applying Cultural Attraction Theory to Chimpanzee Techniques

Techniques can be complex actions that produce changes in the environment, both physical and social. This definition can be extended to cultural traditions that include actions beyond tool use, such as specific social grooming techniques. We analyze techniques by focusing on three aspects: (1) the *function* of a target behavior, (2) the *material means* necessary to produce the target behavior, and (3) the *actions* with which the target behavior is achieved. For example, in the case of ant-dipping, extracting ants is the function; selecting the type of tool is the material means; and hand wiping the tool is the action. Between groups and subgroups, we expect variation in techniques to occur at one or more of these aspects.

We will focus on three techniques to illustrate our approach. Table 13.2 presents these techniques as candidate examples for a strong influence by each class, along with potential social learning mechanisms and other factors of attraction supporting them. Through our integrated conceptualization of social learning and cultural attraction, we will illustrate how each technique is differently influenced by the three classes and how this affects the expression of flexible and rigid behavior, with sponging demonstrating strong individual influence, ant-dipping demonstrating strong ecological influence, and hand-clasping demonstrating strong social influence (table 13.2). Crucially, even if a technique is mainly affected by one of the classes, the others are still relevant.

Table 13.2
Summary of three chimpanzee techniques and the proposed social learning mechanisms and factors of attraction supporting them

Behavior	Technique	Social learning mechanisms	Factors of attraction
Sponging			
	Insert item in mouth Chew and suck on item Insert item in hole Repeat	Local + stimulus enhancement	Chew plant material repeatedly (wadging) Insert objects/fingers in hole
Ant-dipping			
	Insert stick in ant nest or near trail May swirl stick Remove stick Harvest ants by two techniques: a. Direct stick to mouth b. Sweep stick with hand and pop ants from hand into the mouth Repeat	Emulation or Imitation	Aggressive ants Avoid insect bites Insert tool in hole
Hand-clasping			
	Grab and lift arm high in the air Hold arm against own's by two techniques: a. Interlocking hand palms b. Pressing wrists or forearms Groom armpit Switch sides	Ontogenetic ritualization or Imitation	Low branch nearby absent Prevent partner from lowering her arm Increase own arm's comfort Coordinate actions

For our three classes, we conceptualize that the stronger the influence, the more conservative the individual tends to be and, in turn, the more rigid a group may be such that when novel innovations appear, they are unlikely to diffuse within the group. Under conditions where the influence lessens, the individual can behave more liberally, leading to innovations that can be adopted and demonstrated flexibly at the group level. For any given cultural behavior, there is always influence by each class, and the strength of each class's influence interacts with the others. For example, returning to ant-dipping behavior, we might find that a given ant-dipping technique is heavily influenced by the social class; however, in environments where the ecological class also has a strong influence (i.e., the ants are very aggressive), we may find that the behavior is rigid within this context but can flexibly switch between contexts in accordance with shifting influence. This relationship between the classes' strength can be highlighted by the chimpanzees of Bossou, Guinea, who use two ant-dipping techniques that varies due to ant behavior (Humle and Matsuzawa 2002). When the ants are more aggressive, the ecological influence becomes stronger, constraining the behavior within that context despite additional strong social influences. When the strong ecological influence is lessened, an alternative technique may be employed guided by another class.

To begin discussing our candidate chimpanzee techniques, let us first consider sponging behavior, which we view as strongly influenced by the individual class. Sponging is a foraging technique in which a wad of leaves and/or other vegetation is folded and/or chewed and used to collect water, then squeezed in the mouth (Goodall 1964). Considered to be a universal behavior, sponging is not recognized as a cultural trait by the method of exclusion

(Whiten et al. 1999). Further, when captive chimpanzees have been provisioned with the materials to sponge, they spontaneously invented the technique without prior experience or observation (Kitahara-Frisch and Norikoshi 1982). Such findings suggest that given the correct environment, chimpanzees can perform sponging behavior relying predominantly on their individual dispositions—namely, two seemingly innate behaviors: wadging, which involves extracting liquid by chewing and compressing matter in the mouth (Goodall 1989; Teleki 1973), and poking fingers and objects into holes and crevices (Köhler and Winter 1925). However, a recent observation of the transmission of a novel sponging variation using moss as different material means has provided evidence for social learning and therefore suggests sponging might be a cultural trait (Hobaiter et al. 2014).

The study by Catherine Hobaiter and colleagues (2014) also made observations that highlight the potential roles of the ecological and social classes. The novel moss-sponging variation occurred under an unusual ecological context of discovering a novel clay pit that had been repeatedly flooded by the nearby river. Consequently, the clay pit attracted larger groups, which may have increased competition. In addition to individual dispositions toward sponging, the social influence of competition combined with the unusual ecological context may have fostered the initial innovation of moss-sponging, allowing space for variation in material means that could diffuse through the group through lower-fidelity social learning, such as local or stimulus enhancement.

The second candidate technique we will be considering is ant-dipping, which we believe to be strongly influenced by the ecological class. Ant-dipping is a foraging technique in which an individual selects a wand tool, inserts the wand into a nest or near a trail of ants, may move or hold the wand still to collect ants, then removes the ants by either bringing the wand directly to the mouth or by using the opposite hand to remove ants and put into the mouth (Humle 2011). In contrast to sponging, ant-dipping is only present in some chimpanzee populations, and the technique differs in terms of material means and actions (Humle and Matsuzawa 2002; Humle 2011; McGrew 1992; Whiten et al. 1999; Yamakoshi 2001). Depending on the level of aggression of the ants present, chimpanzees select wands of different length and apply different ant-gathering techniques to avoid severe bites. Under conditions with high probability for bites, the ecological influence becomes more important. The behavior is predicted to be more rigid to avoid discomfort. In a setting where ants vary in aggressiveness or show low levels of aggression, behaviors are predicted to be more flexible, allowing for more variation. Indeed, chimpanzees from sites with more aggressive ants use longer tools and collect the ants by hand wiping them from the wand whereas those from sites with less aggressive ants were dipped with shorter tools associated with the direct-to-mouth method (Schöning et al. 2008). Strikingly, the ecology can probably not exclusively explain ant-dipping techniques because chimpanzees from two sites with ants that show the same level of aggression use differently sized wands (Möbius et al. 2008). This finding suggests that social learning mechanisms, such as emulation or imitation, may play an important role in maintaining these variations (Humle 2011). In combination with social learning, individual experience of developing chimpanzees may further reinforce a group-specific technique as they encounter ant bites. Through combining strong ecological influences with social learning and individual experience, we can imagine that variation of techniques between groups is maintained.

Our third candidate technique, representing strong social influence, is not material-based but a social custom that is only present in some groups and varies between groups. Hand-clasping is a social grooming technique where two individuals clasp hands or press wrists or forearms (or other combinations of two) together overhead and groom the other with their free hand (McGrew and Tutin 1978). Because of the dyadic nature of this grooming technique, it is unlikely that variations in the actions of this behavior would arise out of, or be maintained by, individual dispositions alone. It might have originated by holding arms overhead in a comfortable manner for which branches are needed (McGrew et al. 2001), highlighting an ecological influence. Still, groups kept under the same ecological condition show different techniques, suggesting that variations are socially determined (McGrew et al. 2001; Nakamura and Uehara 2004; van Leeuwen et al. 2012). In combination with these individual and ecological influences, as well as coordination, we can speculate that high-fidelity social learning mechanisms help maintain group-specific variations. For example, mechanisms such as imitation or ontogenetic ritualization, in which two individuals shape one another's behavior through repeated interaction (Tomasello and Call 1997), may support transmission. In this context, one could imagine that the strong influences of social learning and the dyadic nature of the technique maintain group rigidity in addition to the appropriate ecological context (i.e., no available branches) and limited individual influence.

Beyond our three candidate cultural techniques, we can use our framework to examine other techniques, offering new perspectives into how they might be shaped by the influences of our three classes. For example, nut-cracking, a foraging technique in which chimpanzees place a nut on an anvil and push or pound a hammer until cracked open (Boesch et al. 1994), is one of the most well-known examples of chimpanzee cultural techniques, yet less is understood about how it emerges and persists. Through the lens of our new framework, we can speculate that nut-cracking may be strongly influenced by the ecological class. Nut-cracking is a highly complex technique that is rare among chimpanzees and varies between groups. Thus far, nut-cracking has only been described in three wild communities across two geographically distinct populations: Bossou in Guinea and Taï Forest in Côte d'Ivoire (Whiten et al. 2001) and Ebo Forest in Cameroon (Morgan and Abwe 2006, through indirect evidence). Within the Taï Forest, ecological factors such as rainfall, raw material availability, fruit production patterns, and fruit availability are understood to be similar throughout the area (Luncz, Mundry, and Boesch 2012). However, in an experimental study, it was demonstrated that between three groups living in the forest, ecological factors related to nut hardness influenced each group's technique, in which they adapted to seasonal changes in group-specific patterns (Luncz, Mundry, and Boesch 2012). Further, another study found that ecological opportunity over necessity influenced the presence or absence of nut-cracking in that despite having access to the necessary material means with which to nut-crack, low density and low distribution of high-value nuts may limit the invention and transmission of nut-cracking (Koops, McGrew, and Matsuzawa 2013). Similar to ant-dipping, nut-cracking appears to be influenced by ecological constraints combined with social learning to maintain group-specific variations. Social learning mechanisms that have been suggested to support nut-cracking include high-fidelity copying mechanisms such as imitation and teaching (Biro et al. 2003; Boesch 1996; Boesch et al. 2019). However, an experimental study investigating the individual and social learning mechanisms involved in nut-cracking found that a captive group was unable to nut-crack spontaneously or after a stepwise demonstration (Neadle, Bandini, and Tennie 2020). The researchers suggest that their finding might indicate the presence of a sensitive period in which chimpanzees learn nut-cracking when both ecology and development allow for it (Neadle, Bandini, and Tennie 2020). Considering this evidence together, under our framework we can imagine that strong ecological influence combined with social learning mechanisms to main group-specific variation and individual development uniquely contribute to nut-cracking as a cultural trait.

In summary, we have described three classes of factors of attraction: individual, social, and ecological. Not only do these classes act together to produce specific behaviors, but their respective strengths can limit the variations and inventions an individual can produce and in turn limit how innovations spread through a group. Subjected to strong influence, fixedness occurs. Conversely, overall weaker influence produces more variability and thereby flexibility. Depending on the species, the potential impact each class exerts over the transmission and performance of techniques may vary. For example, an animal group that does not highly affiliate or cohabitate between group members may not see as strong an influence by the social class and may not have as many socially transmitted and sustained techniques. Conversely, we can imagine that humans are very strongly influenced by the social class, allowing for the use of mechanisms like teaching and the accumulation of techniques beyond the individual. Further, individual factors may more strongly influence humans allowing for a wide array of stylistic preferences liberally demonstrated. As such, an appropriate balance between the influences by each class can provide the opportunity for liberal invention and variation by individuals and flexible yet stable adoption through social learning by groups, resulting in cultural evolution.

Conclusion

Our three-class framework combines the insights of cultural attraction and social learning theories to produce a more complete understanding of animal cultural behavior. We do not view flexibility and rigidity as a single dimension in which individuals or groups behave one way or the other. Rather, we understand the expression of flexible and rigid behavior as the outcome of the dynamic process created by the unique influence of each class in a given context. This framework helps us to unify cultural attraction theory and social learning theory by including the role of social learning mechanisms within cultural attraction theory and integrating the role of ecology and genetics within social learning theory, while discussing the contribution of low-fidelity social learning within both theories. We do not believe that the differences between each theory should represent mutually exclusive reasoning. On the contrary, when brought together, they offer solutions for understanding one another. As such, we use this combined framework to illuminate that chimpanzee populations are not strictly flexible or rigid; rather, their flexibility and rigidity arise differentially as an outcome of dynamic attraction by factors within each of our three classes. Through this lens, we understand that the traditions and techniques demonstrated in chimpanzee cultures are a product of combined social, individual, and ecological influences that have allowed for the evolution of the distinct repertoires we observe today.

Acknowledgments

We would like to thank Mathieu Charbonneau and Dan Sperber for the invitation and opportunity to contribute to this book, as well as all the authors for their collaboration and feedback in developing this chapter, and for the support of the European Research Council (Synergy Grant 609819 SOMICS to Josep Call).

References

Allen, J., M. Weinrich, W. Hoppitt, and L. Rendell. 2013. "Network-Based Diffusion Analysis Reveals Cultural Transmission of Lobtail Feeding in Humpback Whales." *Science* 340 (6131): 485–488.

Aplin, L. M. 2019. "Culture and Cultural Evolution in Birds: A Review of the Evidence." *Animal Behaviour* 147:179–187.

Ayala, F. J., and C. A. Campbell. 1974. "Frequency-Dependent Selection." *Annual Review of Ecology, Evolution, and Systematics* 5 (1): 115–138.

Bandini, E., and R. A. Harrison. 2020. "Innovation in Chimpanzees." *Biological Reviews* 95 (5): 1167–1197.

Barrett, B. J., R. L. McElreath, and S. E. Perry. 2017. "Pay-Off-Biased Social Learning Underlies the Diffusion of Novel Extractive Foraging Traditions in a Wild Primate." *Proceedings of the Royal Society B: Biological Sciences* 284 (1856): 20170358.

Biro, D., N. Inoue-Nakamura, R. Tonooka, G. Yamakoshi, C. Sousa, and T. Matsuzawa. 2003. "Cultural Innovation and Transmission of Tool Use in Wild Chimpanzees: Evidence from Field Experiments." *Animal Cognition* 6 (4): 213–223.

Boesch, C. 1996. "Three Approaches to Investigating Chimpanzee Culture." In *Reaching into Thought: The Minds of the Great Apes*, 404–429. Cambridge: Cambridge University Press.

Boesch, C., D. Bombjaková, A. Meier, and R. Mundry. 2019. "Learning Curves and Teaching When Acquiring Nut-Cracking in Humans and Chimpanzees." *Scientific Reports* 9 (1): 1515.

Boesch, C., P. Marchesi, N. Marchesi, B. Fruth, and F. Joulian. 1994. "Is Nut Cracking in Wild Chimpanzees a Cultural Behaviour?" *Journal of Human Evolution* 26 (4): 325–338.

Buskell, A. 2017. "What Are Cultural Attractors?" *Biology & Philosophy* 32 (3): 377–394.

Call, J., M. Carpenter, and M. Tomasello. 2005. "Copying Results and Copying Actions in the Process of Social Learning: Chimpanzees (*Pan troglodytes*) and Human Children (*Homo sapiens*)." *Animal Cognition* 8:151–163.

Charbonneau, M. 2020. "Understanding Cultural Fidelity." *British Journal for the Philosophy of Science* 71 (4): 1209–1233.

Claidière, N., T. C. Scott-Phillips, and D. Sperber. 2014. "How Darwinian Is Cultural Evolution?" *Philosophical Transactions of the Royal Society B: Biological Sciences* 369 (1642): 20130368.

Claidière, N., and D. Sperber. 2007. "The Role of Attraction in Cultural Evolution." *Journal of Cognition and Culture* 7 (1–2): 89–111.

Claidière, N., and D. Sperber. 2010. "Imitation Explains the Propagation, Not the Stability of Animal Culture." *Proceedings of the Royal Society B: Biological Sciences* 277 (1681): 651–659.

Day, R. L., J. R. Kendal, and K. N. Laland. 2001. "Validating Cultural Transmission in Cetaceans." *Behavioral and Brain Sciences* 24 (2): 330.

Dean, L. G., G. L. Vale, K. N. Laland, E. Flynn, and R. L. Kendal. 2014. "Human Cumulative Culture: A Comparative Perspective." *Biological Reviews* 89 (2): 284–301.

Fragaszy, D. M., and S. Perry. 2003. "Towards a Biology of Traditions." In *The Biology of Traditions: Models and Evidence*, 1–32. Cambridge: Cambridge University Press.

Goodall, J. 1964. "Tool-Using and Aimed Throwing in a Community of Free-Living Chimpanzees." *Nature* 201 (4926): 1264–1266.

Goodall J. 1989. *Glossary of Chimpanzee Behaviors*. Tucson, AZ: Jane Goodall Institute.

Gruber, T., M. N. Muller, P. Strimling, R. Wrangham, and K. Zuberbühler. 2009. "Wild Chimpanzees Rely on Cultural Knowledge to Solve an Experimental Honey Acquisition Task." *Current Biology* 19 (21): 1806–1810.

Heintz, C., and N. Claidière. 2015. "Current Darwinism in Social Science." In *Handbook of Evolutionary Thinking in the Sciences*. Dordrecht: Springer.

Heyes, C. M. 1994. "Social Learning in Animals: Categories and Mechanisms." *Biological Reviews* 69 (2): 207–231.

Hobaiter, C., T. Poisot, K. Zuberbühler, W. Hoppitt, and T. Gruber. 2014. "Social Network Analysis Shows Direct Evidence for Social Transmission of Tool Use in Wild Chimpanzees." *PLoS Biology* 12 (9): 1001960.

Hoppitt, W., and K. N. Laland. 2013. *Social Learning: An Introduction to Mechanisms, Methods, and Models.* Princeton, NJ: Princeton University Press.

Horner, V., and A. Whiten. 2005. "Causal Knowledge and Imitation/Emulation Switching in Chimpanzees (*Pan troglodytes*) and Children (*Homo sapiens*)." *Animal Cognition* 8 (3): 164–181.

Humle, T. 2011. "Ant-Dipping: How Ants Have Shed Light on Culture." In *The Chimpanzees of Bossou and Nimba*, 97–105. Tokyo: Springer.

Humle, T., and T. Matsuzawa. 2002. "Ant-Dipping among the Chimpanzees of Bossou, Guinea, and Some Comparisons with Other Sites." *American Journal of Primatology* 58 (3): 133–148.

Kendal, J., L. A. Giraldeau, and K. Laland. 2009. "The Evolution of Social Learning Rules: Payoff-Biased and Frequency-Dependent Biased Transmission." *Journal of Theoretical Biology* 260 (2): 210–219.

Kendal, R., L. M. Hopper, A. Whiten, S. F. Brosnan, S. P. Lambeth, S. J. Schapiro, and W. Hoppitt. 2015. "Chimpanzees Copy Dominant and Knowledgeable Individuals: Implications for Cultural Diversity." *Evolution and Human Behavior* 36 (1): 65–72.

Kendal, R. L., N. J. Boogert, L. Rendell, K. N., Laland, M. Webster, and P. L. Jones. 2018. "Social Learning Strategies: Bridge-Building between Fields." *Trends in Cognitive Sciences* 22 (7): 651–665.

Kitahara-Frisch, J., and K. Norikoshi. 1982. "Spontaneous Sponge-Making in Captive Chimpanzees." *Journal of Human Evolution* 11 (1): 41–47.

Köhler, W., and E. Winter. 1925. *The Mentality of Apes*. Translated by E. Winter. London: K. Paul Trench Trubner.

Koops, K., W. C. McGrew, and T. Matsuzawa. 2013. "Ecology of Culture: Do Environmental Factors Influence Foraging Tool Use in Wild Chimpanzees, *Pan troglodytes verus*?" *Animal Behaviour* 85 (1): 175–185.

Laland, K., and C. Evans. 2017. "Animal Social Learning, Culture, and Tradition." In *APA Handbook of Comparative Psychology: Perception, Learning, and Cognition*, 441–460. Washington, DC: American Psychological Association.

Laland, K. N., and W. Hoppitt. 2003. "Do Animals Have Culture?" *Evolutionary Anthropology: Issues, News, and Reviews* 12 (3): 150–159.

Lamon, N., C. Neumann, T. Gruber, and K. Zuberbühler. 2017. "Kin-Based Cultural Transmission of Tool Use in Wild Chimpanzees." *Science Advances* 3 (4): 1602750.

Luncz, L. V., R. Mundry, and C. Boesch. 2012. "Evidence for Cultural Differences between Neighboring Chimpanzee Communities." *Current Biology* 22 (10): 922–926.

Marshall-Pescini, S., and A. Whiten. 2008. "Chimpanzees (*Pan troglodytes*) and the Question of Cumulative Culture: An Experimental Approach." *Animal Cognition* 11 (3): 449–456.

McGrew, W. C. 1974. "Tool Use by Wild Chimpanzees in Feeding upon Driver Ants." *Journal of Human Evolution* 3 (6): 501–508.

McGrew, W. C. 1992. *Chimpanzee Material Culture: Implications for Human Evolution*. Cambridge: Cambridge University Press.

McGrew, W. C., L. F. Marchant, S. E. Scott, and C. E. Tutin. 2001. "Intergroup Differences in a Social Custom of Wild Chimpanzees: The Grooming Hand-Clasp of the Mahale Mountains." *Current Anthropology* 42 (1): 148–153.

McGrew, W. C., and C. E. Tutin. 1978. "Evidence for a Social Custom in Wild Chimpanzees?" *Man* 13 (2): 234–251.

Möbius, Y., C. Boesch, K. Koops, T. Matsuzawa, and T. Humle. 2008. "Cultural Differences in Army Ant Predation by West African Chimpanzees? A Comparative Study of Microecological Variables." *Animal Behaviour* 76 (1): 37–45.

Morin, O. 2015. *How Traditions Live and Die*. Oxford: Oxford University Press.

Morgan, B. J., and E. E. Abwe. 2006. "Chimpanzees Use Stone Hammers in Cameroon." *Current Biology* 16 (16): R632–R633.

Nakahashi, W., J. Y. Wakano, and J. Henrich. 2012. "Adaptive Social Learning Strategies in Temporally and Spatially Varying Environments." *Human Nature* 23 (4): 386–418.

Nakamura, M., and S. Uehara. 2004. "Proximate Factors of Different Types of Grooming Hand-Clasp in Mahale Chimpanzees: Implications for Chimpanzee Social Customs." *Current Anthropology* 45 (1): 108–114.

Neadle, D., E. Bandini, and C. Tennie. 2020. "Testing the Individual and Social Learning Abilities of Task-Naïve Captive Chimpanzees (*Pan troglodytes sp.*) in a Nut-Cracking Task." *PeerJ* 8:e8734.

Nishida, T. 1973. "The Ant-Gathering Behaviour by the Use of Tools among Wild Chimpanzees of the Mahali Mountains." *Journal of Human Evolution* 2 (5): 357–370.

Sasaki, T., and D. Biro. 2017. "Cumulative Culture Can Emerge from Collective Intelligence in Animal Groups." *Nature Communications* 8 (1): 15049.

Schofield, D. P., W. C. McGrew, A. Takahashi, and S. Hirata. 2018. "Cumulative Culture in Nonhumans: Overlooked Findings from Japanese Monkeys?" *Primates* 59:113–122.

Schöning, C., T. Humle, Y. Möbius, and W. C. McGrew. 2008. "The Nature of Culture: Technological Variation in Chimpanzee Predation on Army Ants Revisited." *Journal of Human Evolution* 55 (1): 48–59.

Schuppli, C., and C. P. van Schaik. 2019. "Animal Cultures: How We've Only Seen the Tip of the Iceberg." *Evolutionary Human Sciences* 1:1–13.

Scott-Phillips, T., S. Blancke, and C. Heintz. 2018. "Four Misunderstandings about Cultural Attraction." *Evolutionary Anthropology: Issues, News, and Reviews* 27 (4): 162–173.

Sperber, D. 1985. *On Anthropological Knowledge*. Cambridge: Cambridge University Press.

Sperber, D. 1996. *Explaining Culture: A Naturalistic Approach*. Cambridge: Blackwell Publishers.

Sperber, D., and N. Claidière. 2008. "Defining and Explaining Culture (Comments on Richerson and Boyd, Not by Genes Alone)." *Biology & Philosophy* 23 (2): 283–292.

Teleki, G. 1973. *The Predatory Behavior of Wild Chimpanzees*. Lewisburg, PA: Bucknell University Press.

Tennie, C., J. Call, and M. Tomasello. 2009. "Ratcheting Up the Ratchet: On the Evolution of Cumulative Culture." *Philosophical Transactions of the Royal Society B: Biological Sciences* 364 (1528): 2405–2415.

Tomasello, M. 1990. "Cultural Transmission in the Tool Use and Communicatory Signaling of Chimpanzees." In *"Language" and Intelligence in Monkeys and Apes: Comparative and Developmental Perspectives*, 274–311. Cambridge: Cambridge University Press.

Tomasello, M. 1996. "Do Apes Ape?" In *Social Learning in Animals: The Roots of Culture*, 319–346. Cambridge: Academic Press.

Tomasello, M., and J. Call. 1997. *Primate Cognition*. Oxford: Oxford University Press.

Tomasello, M., M. Davis-Dasilva, L. Camak, and K. Bard. 1987. "Observational Learning of Tool-Use by Young Chimpanzees." *Human Evolution* 2 (2): 175–183.

van Leeuwen, E. J., K. A. Cronin, D. B. Haun, R. Mundry, and M. D. Bodamer. 2012. "Neighbouring Chimpanzee Communities Show Different Preferences in Social Grooming Behaviour." *Proceedings of the Royal Society B: Biological Sciences* 279 (1746): 4362–4367.

Vossen, P. 2007. "Olive Oil: History, Production, and Characteristics of the World's Classic Oils." *HortScience* 42 (5): 1093–1100.

Whiten, A. 2019. "Cultural Evolution in Animals." *Annual Review of Ecology, Evolution, and Systematics* 50:27–48.

Whiten, A., J. Goodall, W. C. McGrew, T. Nishida, V. Reynolds, Y. Sugiyama, C. E. G. Tutin, R. W. Wrangham, and C. Boesch. 1999. "Cultures in Chimpanzees." *Nature* 399 (6737): 682–685.

Whiten, A., J. Goodall, W. C. McGrew, T. Nishida, V. Reynolds, Y. Sugiyama, C. E. G. Tutin, R. W. Wrangham, and C. Boesch. 2001. "Charting Cultural Variation in Chimpanzees." *Behaviour* 138 (11): 1481–1516.

Whiten, A., V. Horner, C. A. Litchfield, and S. Marshall-Pescini. 2004. "How Do Apes Ape?" *Animal Learning & Behavior* 32 (1): 36–52.

Whiten, A., V. Horner, and F. B. De Waal. 2005. "Conformity to Cultural Norms of Tool Use in Chimpanzees." *Nature* 437 (7059): 737–740.

Whiten, A., N. McGuigan, S. Marshall-Pescini, and L. M. Hopper. 2009. "Emulation, Imitation, Over-Imitation and the Scope of Culture for Child and Chimpanzee." *Philosophical Transactions of the Royal Society B: Biological Sciences* 364 (1528): 2417–2428.

Whiten, A., and E. van de Waal. 2017. "Social Learning, Culture and the 'Socio-Cultural Brain' of Human and Non-Human Primates." *Neuroscience & Biobehavioral Reviews* 82:58–75.

Yamakoshi, G. 2001. "Ecology of Tool Use in Wild Chimpanzees: Toward Reconstruction of Early Hominid Evolution." In *Primate Origins of Human Cognition and Behavior*, 537–556. Tokyo: Springer.

Discussion: The Cumulative Culture Mosaic

Kim Sterelny

Preliminary Skirmishing

There is a consensus that the late-evolving hominins are distinguished from their ancestors and from the rest of animal life by the richness, the centrality, and the adaptive power of their cultural traditions (Boyd 2018). According to this view, more recent hominins have been able to both accumulate information culturally, building on the informational capital of their ancestors, and, increasingly over time, have been able to innovate as well, shaping their lore to local circumstances, though as Dietrich Stout (this volume) points out, the progressive character, directionality, and discreteness of this contrast has been constantly overstated. Likewise, there is a consensus that these distinctive features of hominin culture depend on distinctive powers of hominin minds (Michael Tomasello has been especially influential here; see Tomasello 1999, 2014).

This chapter develops an alternative picture of the dynamics of cumulative culture. First, I will argue that one change over deep time is a shift from the cultural transmission of specific procedures to the cultural transmission of expertise. To the extent that this transition has taken place, there is no trade-off between the reliable transfer of the community's informational capital and its improvement. The social and cognitive mechanisms that support expertise support both. If this is right, to some degree, there is a false contrast between flexibility and rigidity: the reliable reproduction of a technical procedure over many generations does not imply an inability to innovate. Second, I defend (in agreement with others) the existence of an important distinction between the skills involved in instrumental actions targeted at the physical and biological world and those involved in social signaling. Their learning mechanics and causal drivers differ. Third, I argue that there is a recent but fundamental change in the mechanisms that support stability and innovation, deriving from the very different organization of learning and childhood in preliterate and postliterate communities. The overall picture from these three arguments is a view of cumulative culture in which imitation and copying play a much less central role. Fourth, I will suggest that the features that drove the initial stage of cumulative culture were ecological and economic. The cognitive differences between humans and great apes in cultural learning mostly evolved later and, to some degree, in response to cumulative culture rather than as its precondition. I will build this picture through a series of themes that link many (but not all) of the chapters in this volume.

A False Trade-Off?

Human cognitive capacities embedded in cultural traditions support both incremental accumulation and innovation. As Dietrich Stout points out, it is difficult to demonstrate for many late-evolving cultural traits that they are *improvements* over their older prequels. But many are incremental in that earlier forms were *essential preliminaries* for the establishment of later forms. The hafted tools that began to appear about 500 thousand years ago offer many mechanical and safety advantages, so very likely they are improvements over handheld forms. But in any case, they could not have developed without the earlier stone tool traditions and the skills they established. A hafted adze could not have been the first hominin stone tool. Many chapters in this volume treat this conjoint capacity for innovation and accumulation as a puzzle—for example, the chapters by Adam Boyette; Nicola Cutting; Sadie Tenpas, Manon Schweinfurth, and Josep Call; and Sarah Pope-Caldwell. As they present the issue, the expectation is that the cognitive and social factors that make cumulative culture possible tend to suppress innovation. Accumulation depends on stable traditions that in turn depend on faithful copying, which yet again depends on suppressing temptations to diverge from local current practice. So there is a trade-off between the reliable transmission of current lore and improving on that lore. Cutting, for example, proposes to address this trade-off through the idea that innovators and innovation are rare. Most individuals most of the time, perhaps even all of the time, learn by learning from others what those others already know. In exploring the interactions between innovation and stability in structured populations, Tenpas and colleagues even suggest that managing the trade-off between flexibility and rigidity is so difficult that it requires group selection to explain the persistence of both stable traditions and innovation. For example, as they suggest in chapter 13 of this volume:

Problems arise when a group leans too far toward either end of the flexible–rigid spectrum. For example, if a group leans too far toward the flexible end, such that it is constantly creating and abandoning cultural behaviors, the group is not able to retain long-term cultural information and risks disadvantage when facing old problems again. Over evolutionary time, an overly flexible group can lead to cultural breakdown. Conversely, if a group leans too far toward the rigid end, such that it rarely acquires new cultural behaviors or modifies existing ones, then the group is unlikely to innovate new solutions or adapt to changing environments. Over evolutionary time, an overly rigid group can lead to cultural stagnancy. Therefore, there must be a balance between flexible creation, variation, and application of cultural behaviors while also maintaining relatively rigid transmission and performance, allowing cultural traditions to form and persist across generations.

In my view, these concerns (and especially those about the paralyzing effects of conformity) are driven by modeling cultural learning too closely on genetic transmission—in particular, by treating learning about current practice from others as if the *only source* of relevant information comes from the model. On this conception, stability is the result of novices aiming to, and succeeding in, copying the action sequences of models. Something like that is true of genetic transmission, for it is a template-copying process, with the local environment playing no role in modifying or editing the genes that one generation inherits from its predecessor.

In explaining the intergenerational stability of cultural traits, it is important to distinguish between traits that function to support instrumental action on the physical and biological environment ("world-facing" traits) and those functioning to mediate social interactions ("community-facing" traits). These are usually social signals.[1] The template-copying view is a misleading framework for explaining the stability of world-facing traits—traits like recognizing and exploiting the local flora. First, in itself, variation does not require special explanation. Cultural transmission is never perfect: there will always be some variation generated, deliberately or inadvertently, in the flow of information from model to novice. This will create variation in novice behavior in (for example) their attempts to ant-fish, crack nuts, or spear bush-babies. This variation is an innovation opportunity when such variants are improvements and when novices (or third parties) *notice* the improvement. That is quite likely with world-facing cultural traits because these are almost acquired by hybrid learning, as Dietrich Stout points out[2] (and see also Sterelny 2006, 2012). Techniques are learned not just by observation but also by practice and experiment. For the most part, novices do not acquire the technique of models primarily by attending to, memorizing, and replicating their action sequences. Agents learn as well from experiment and practice. Moreover, as Adam Boyette illustrates in this volume, access to the raw materials, tools, and products of models is also crucial. As a consequence, agents get success and failure signals from the natural world, not just the social world. Variation from received practice can be rewarded by positive feedback from the world. Conversely, stability is partially explained by novices being guided back to the model's technique when they get failure signals. Nut-cracking is hard, so young chimps need social input to achieve competence. But it is rewarding. Physical traces suggest that the chimp nut-cracking tradition has been stable for a couple of thousand years. Its stability depends on (i) its rewards, since nut meat is nutritious and dense in oils and protein; (ii) the fact that social information is reliably available in the developmental environments of these subpopulations of chimps; and (iii) the intrinsic task demands of nut-cracking (given chimp baseline competences, they have only one effective nut-cracking routine available). Other task domains are less constrained. In domains where multiple solutions are possible, variation and improvement can be reinforced with success signals. All forms of learning generate variation, but since hybrid learning is a version of socially guided experimentation, it improves the prospects of lucky accidents and the prospects of exploiting those accidents (as hybrid learners attend to signals from the world). On this view, and in contrast to those of Nicola Cutting (this volume), agents regularly innovate, even as they acquire skills that are common in their communities, in that they produce solutions that are *novel to them*. Innovation is rare only in the sense of solutions that are *novel to the community*. Once the stock of skills and material culture reaches a certain threshold, the space of innovation by recombination opens up. Once cordage is invented, many uses are possible, but while individuals will discover many of these by experiment, most will already have been tried by others.

As a consequence of considerations like these, I am unpersuaded that capacities for flexibility or innovation trade off against the capacities that support reliable transmission of existing capacity. The same cognitive and social factors that explain the accuracy and reliability of cultural transmission—rightly seen as the essential foundation of incremental improvement—also support innovation. The crucial idea here is that in recent human cultures, both stability and flexibility depend on the socially scaffolded acquisition of expertise.

Here my views converge with those of Stout and the experimental work of Valentine Roux and colleagues, whose writings on expertise (in this volume on pottery-making) and in their earlier work on stone bead knapping make clear this dual role of expertise in both stability and innovation (Bril, Roux, and Dietrich 2005; Biryukova and Bril 2008). Experts reliably preserve technical traditions. Indeed, both in the case of stone bead making and in potting, artisans are recognized as experts because they regularly and accurately reproduce the traditional range of artifact designs. More than others, they contribute to the stability of these designs over time. But as Roux and colleagues show experimentally, the most expert, and hence those who most reliably sustain traditional forms, are also the most successful innovators. In their work with knappers, the bead makers were asked to make beads with unusual materials and with nonstandard hammers. All were competent, but some had served long apprenticeships, and the acknowledged experts were most able to adapt to these novel task demands. Likewise, confronted with the requirement to make pots of a novel and technically demanding shape, the acknowledged experts responded most successfully.

Valentine Roux and her colleagues' work shows experts bringing their expertise to bear on novel challenges,[3] demonstrating experimentally that expertise is not an unusually large set of precise, canned responses to a domain's challenges. The cognitive science of expertise is controversial, but expertise often involves an integrated complex of perceptual capacities, motor skills, procedures, and more general declarative information, allowing experts to give effective advice to others and to guide their own improvement through focused practice and error recognition (see, for example, Christensen et al. 2015; Christensen, Sutton, and McIlwain 2016). An expert-level skill is a cognitive mosaic. For example, one central natural history skill of foragers is tracking, and expert tracking requires more than an on-board field guide to tracks. Tracks are often partial, interrupted, and overlaid with the traces of other animals. Tracks need to be interpreted, not just recognized, and ethnography shows that foragers read tracks through the lens of a very rich understanding of the target and local environment (Liebenberg 1990, 2013; Shaw-Williams 2014, 2017).

I suggest that instead of seeing capacities to preserve and to innovate as trading off against one another in the explanation of the cumulative yet dynamic character of human culture, we should instead be probing the cognitive, social, and life history requirements of the reliable reconstruction of expertise, generation by generation. On this view, one change through deep history was a shift from the social transmission of skills tied to quite specific procedures, contexts, and material substrates (as in chimp nut-cracking) to increasingly flexible, less context- and function-bound versions of that skill (perhaps exemplified in Oldowan toolmaking) to full expertise. There were no sharp boundaries in such a transition, and there is no default expectation that it was coordinated in time over different world-facing cultural traits—for example, that tracking and knapping became forms of expertise at roughly the same time in hominin history. That said, expertise very likely has deep roots in human history, extending back to Acheulian toolmaking. A mid-Pleistocene hominin cannot become a skilled knapper just by memorizing a set of specific procedures by watching one or more models. Stone is too heterogeneous and unpredictable in its precise fracture pattern for such a rigid learning strategy to succeed. Stout's account of adze making in Papua New Guinea is a wonderful depiction of craft transmission (Stout 2002). There, copying probably does play an important role. Apprentice adze-makers almost certainly attend to the action pattern of masters. But this is just one source of social information, and those acquiring these world-facing capacities are exposed to relevant information from the world, not just their social partners; they see, hear, and touch stone. Even so, high-fidelity imitation probably does play an important role in acquiring expertise in this domain. However, in others it is not at all central, and in all these world-facing capacities, it is at most part of a package. Consider an Australian Aboriginal boy learning tracking skills from the males in his community, gradually acquiring this skill. If he is lucky, they will point out tracks and traces and name them, as some African groups do (Dira and Hewlett 2016). They may even make tracks of different species to show their shapes (Morrison 1981 has a set of photographs of such Aboriginal-made tracks). Making and identifying tracks is a children's game in some Aboriginal communities (Love 2009). But much of the boy's acquisition of this craft will be socially scaffolded practice: walking with this group of experts through the bush, beginning at about 12 years of age (Meggitt 1965); listening to their views of what they see and looking where they look; seeing and smelling the dung they inspect for freshness; acquiring more-or-less explicit ecological and natural history lore, plus rich and subtle pattern recognition skills. Practice is much more than training motor skills; pattern recognition is at least as important. Australian foragers became highly skilled trackers. They did so with very considerable social support. Almost certainly, the reliable rebuilding of tracking expertise at the N + 1 generation depended on this support. But since socially supported learning did not take the form of a novice copying from a model (as perhaps a young Hadza bowyer might), socially supported learning does not suppress the capacity to innovate. After European arrival, Australian foragers could track them and their animals.[4]

No doubt some cultural transmission is done by rigidly learning specific routines, as in Giulio Ongaro's example of Akha genealogies; perhaps community-facing cultural traits are often acquired this way (see below for an explicit discussion of community-facing traits). Moreover, it is likely that some innovation is the result of play, as subadults explore new uses for established technologies (Lew-Levy et al. 2020). As noted earlier, cordage, once invented, has an enormous range of possible uses, and it is quite likely that some of these were discovered by fooling around. But the incremental improvement of an existing technology must usually depend on a full mastery of the baseline capacities and hence on expertise. Once we think of human cultural learning as rebuilding expertise across the generations, it is no longer puzzling that human cultural learning makes possible both the preservation of existing informational capital and innovations from that capital. That frame shift is further supported by a second theme in these essays: childhood in different social worlds.

Nurturing Experts

A second theme that links many of the chapters in this volume is the very deep differences between social learning in small pre-state societies (especially forager communities) and knowledge acquisition in large-scale state-structured societies. This contrast is the explicit focus of Boyette's chapter. He points out that from the perspective of deep history, these late-emerging large-scale societies—especially Western, educated, industrialized, rich, and democratic (WEIRD) societies—are an outlier. These state-structured societies came to organize education, and as a consequence, (a) children are often segregated from older and more advanced near-peers; (b) language (and even literacy) becomes the dominant vehicle of information coding; (c) children have sharply reduced opportunities of learning by experiment and practice, even at home (especially as WEIRD societies became ever more risk averse with respect to children[5]); (d) to the extent that children learn from a skilled model, those children have very little choice in model selection; (e) there is very heavy emphasis on explicit teaching; and (f) teachers typically have had little or no prior connection with learners. There is no prior establishment of trust and affection, and the personal dynamics within the classroom between the children, or between a child and teacher, often have little influence on whether the learner stays within the group or leaves it, so (g) children typically have little control over learning schedule or content.

Small pre-state communities, especially forager communities, contrast sharply on just about all these dimensions. Forager learning is self-directed, experimental learning by playing and doing, mostly in social contexts with peers and near-peers. Adult teaching is important (adults are often nearby and readily observed), but it is more occasional, and these adults are part of the child's social world. One very important adult contribution is the provision of material resources with which to experiment. As Boyette points out, even very young BaYaka forages have access to blades, from their own houses and those of other adults, and they use that access to experiment and practice. One striking feature of forager cultural learning is its efficiency. Virtually all (surviving) forager children become competent adult foragers, masters of a portfolio of challenging skills. (For further reviews of social learning in small-scale and especially forager communities, see Lew-Levy et al. 2017; Lew-Levy and Boyette 2018; Lew-Levy et al. 2018).

The chapter by James Strachan, Arianna Curioni, and Luke McEllin complements Boyette. Their example is a homely one drawn from WEIRD life—passing on a family recipe in a modern kitchen—but it too underscores the difference between informal, peer-to-peer learning and WEIRD institutional learning, while showing the efficiency and reliability of informal transmission. Strachan and colleagues' distinctive twist is to see these noninstitutionalized examples of learning as a special case of collaboration or joint action. When someone learns a recipe from a cook who is already competent, model and novice interact, and this interaction can often be seen as a special case of collaboration or joint action. Even if passing on the skill is a subsidiary aim of the cook (who is mainly concerned with producing the dish), to some extent they have a common goal of increasing the novice's competence, and there is typically a back-and-forth process of mutual adjustment ("show me that again"; "here, taste it and you can taste the oregano"). The novice is not passive, and the model adjusts to novice response. This process of interaction and mutual adjustment is a dominating feature of social learning in forager communities (and to a considerable extent in subsistence farming communities, though these are more regimented). It is a muted aspect of learning in formal educational institutions (in part because of typical model–novice ratios in such contexts).

It would be wrong to think these contrasts marked a sharp dichotomy. Lab work in science education involves learning by doing. Likewise, Bert De Munck's study of the changing character of apprenticeship in early modern Europe traces the shift in craft skill transmission from a mode that had much in common with craft learning in pre-state communities (though with much more specialization and division of labor) to a mode that was much more shaped by legal and market institutions. Even at the beginning of the early modern period in Europe, it is an intermediate case. Adult models are drawn from the social world of the apprentice; peer-peer interaction and learning by doing was central, but the whole process was much more adult-directed than the learning environments that Boyette describes. But by the end of De Munck's time frame, around the end of the eighteenth century, the system is much more formalized and regimented, and the master–apprentice relationship was largely seen as a fee-for-service contractual relationship. Helena Miton's chapter on equestrian skills is best read as a case study on such intermediate cases, combining quite significant elements of top-down direction and explicit teaching with a good deal of practice, interactivity, and learning assisted by material scaffolds. So there is a continuum of possibilities connecting the self-directed, peer-influenced experiences of forager children with the institutional and adult-dominated experiences of WEIRD children. Even so, three important considerations are suggested by the contrast.

First, informal, forager-like learning regimes have an immense amount of redundancy. The novice has access to the relevant phenomena through a range of cultural channels (observation, demonstration, advice, material props), a range of models, and on multiple occasions. BaYaka adults use blades so routinely that a child has more-or-less unlimited opportunities to observe blade work in action and, over time, many opportunities for advice or demonstration. Forager social worlds are physically very intimate, so there is a lot of public information (Hewlett et al. 2019). Likewise, there are multiple direct channels, as the child experiments with different blades, at different tasks, on different substrates, and again with no real constraint on the number of trials or the time to competence. Strachan and colleagues (in this volume) likewise point out that the would-be cook has access to many informational channels, now including YouTube videos as well as recipe books. In these informal regimes, the novice has a good deal of control over the choice of channel, model, and repetition frequency. In more institutionalized settings, the channel range shrinks, and the novice has less control over channel, model, and repetition count.

Second, it is likely that both stability and innovation have very different explanations in small-scale, traditional communities versus state society. In state society, the preservation of information (and misinformation) is supported by physical media: journals, books, and their descendants. The material culture of preliterate communities can have secondary functions as information stores. To some degree, technology can be reverse-engineered from exemplars, so some information is stored in usable form in tools themselves. The same is true of some of a community's physical effects on the environment: trails store information about safe and efficient routes, and campsites indicate safe or convenient camping places. Moreover, some preliterate communities have purpose-built memory aids. Nonetheless, in preliterate worlds, the informational capital of generation N is mostly stored in their heads and bodies, so it has to be re-created by the $N+1$ generation, albeit with much direct and indirect support from N. In a postliterate world, much of the informational capital of N is stored in the world, and those stores survive N's death. Almost always, the $N+1$ generation will modify and add to those stores, but they do not have to re-create their inheritance. Moreover, to a considerable extent (especially with utilitarian, world-facing information and technology), innovation depends on specialization and the division of informational labor and on the institutions and infrastructure that support those specialists. Especially over the last few centuries, state societies and their institutions support targeted research and development (R&D) by specialists. Targeted R&D is not the only source of innovation in these communities. Serendipitous discovery still happens, though often requiring the skills of a specialist to recognize lucky chance. But R&D is the typical source of recent innovation.

Third, it gives us an alternative perspective on Cutting's puzzling data on young children's capacity to innovate. Her chapter reviews experiments in developmental psychology that seem to show that youngish children are poor at innovation. The studies present children with standard opportunity-to-innovate tasks mostly derived from animal innovation studies. To succeed, the child must variously shape a hook to fish a reward out of a tube, add materials to a tube to float the reward to the top, and use a probe to push a reward out of a tube. Few children, especially younger children, succeed in these tasks. Children are not regularly successful until they are around age 10. These results are interesting and surprising. Cutting suggests that these failures were the result of the test situation being too unstructured and hence having a heavy cognitive load. The children had too much to think about and track. The considerations above suggest instead that it might be the result of too much structure: both the goal and the materials available to solve the goal were set exogenously. Forager ethnography suggests that children learn in self-directed and social activities and often in play and exploration. They innovate and learn by fooling around in company and in experiment. Such opportunities were not present in these studies. There is always a tension in experiments between control and ecological validity; perhaps the trade-offs here have too seriously sacrificed ecological validity.

World-Facing and Community-Facing Capacities

Culture is cumulative in both intimate, pre-state communities and the gigantic state-structured societies of recent times, but as we have just seen, the underlying mechanisms are probably very different. Previously, I distinguished between world-facing and community-facing traits. The class of world-facing traits is somewhat broader than Stout's specification of a technology. For Stout, technologies are world-facing because they support distinctively human social lives and life histories. But they do so by means of material interventions on the environment. Neither tracking nor forager natural history knowledge (and perhaps not even the control of fire) count as technologies by his lights. While I agree that in many of its usages, the concept of a technology has expanded beyond usefulness, a skill like tracking plays a role in human life comparable to a paradigm technology like trap-making. Moreover, as I argued above, they are similar in their transmission from one generation to the next.

World-facing traits contrast with culturally transmitted capacities that enable agents to navigate their local community. To flourish in a human community, an agent must master its lore, norms, customs, values, and signal systems (including but not only the local languages). Many of these cultural products have been built incrementally: the wonderfully complex Yiwara kinship system was not invented in one go (Gould 1969). This, too, is a continuum rather than a sharp dichotomy: there are intermediate and hybrid cases. Some apparently utilitarian technologies also function as social signals, as Ongaro shows in his discussion of Akha village architecture. A house is not just a house, as home organization signals identification with and adherence to the norms of that culture. Likewise, some tools signal the prestige and wealth of their owner, and expense is a design feature, not a drawback (Hayden 1998). Equestrian skill, too, seems an intermediate case. It clearly began as a world-facing utilitarian skill but over time has acquired a variety of social signal functions (in part because of the high cost of horses, thus plugging into the same signaling mechanisms that

Hayden describes), and the display of skill includes elements that are highly conventionalized, precisely specified, and ritual-like, as Miton's chapter shows. Moreover, some capacities are both world-facing and community-facing. A skilled tracker will recognize the individual tracks of every member of the community and can often reconstruct the nature of an interaction from the track pattern, making (for example) illicit liaisons more difficult to conceal.

Even so, the distinction is important because both the learning mechanisms and the selective dynamics are probably different. First, there is skill transmission. In general, as argued above, in small-scale communities, world-facing capacities are acquired through hybrid learning. The skills of a high-class tracker, a good hunter or herbalist, or a skilled maker of tools or textiles are mostly acquired by socially guided and scaffolded rediscovery, as argued by Stout and illustrated by Strachan in this volume. In the transition to expertise, novices learn by doing, practicing, and playing, often in company with somewhat more advanced near-peers and more occasionally with explicit advice or interaction with adult experts. For example, there is ethnographic evidence of such explicit advice and instruction when adolescent boys are out hunting with adults (Dira and Hewlett 2016), perhaps because a failed experiment will impose costs on the adult hunters, not just the half-skilled apprentices. Novices gain information from cues from the nonsocial world, not just signals from the social environment, quite often even when their plans miscarry. Danny Naveh's account of skill transmission in a forager community begins with a description of two boys repeatedly experimenting with trap construction and eventually succeeding after a couple of days of failure (Naveh 2016). While there was no explicit adult input at the time, success was achieved on the back of hunting expeditions with skilled adults and, on those trips, seeing and handling functional traps. Likewise, the world will provide feedback to young trackers if they misidentify tracks or try to follow stale tracks.

There are some elements of learning by doing in the acquisition of community-facing capacities. Children will sometimes guess the rules of local games, or the local norms of generosity, from older children's reactions to their violations and the violations of others. But many norms (for example) are not observable regularities made available in public information; sometimes it is because they are not observed,[6] but often it is because the categories that structure the regularities are idiosyncratically internal to the relevant culture. For example, in some Australian Aboriginal cultures, an agent's potential sexual partners are restricted to specific moieties or divisions of a moiety (known as a section, so that one can never marry in one's own section). This pattern (and its occasional violations) is only visible given the kinship-based assignment of individuals to their respective sections. Likewise, in many of these cultures, it is forbidden to interact casually with anyone who is even eligible to be your mother-in-law. With these culturally specific, highly conventional categories, often there are too many degrees of freedom for error to be informative. Unless the agent already has a near-complete social map of their environment, in social interaction with a misstep, they may well realize that something has gone wrong but that realization leaves too many possibilities open.

In sum, learning trackway reading or the local botany is usefully conceptualized as socially scaffolded rediscovery. That is not true of learning the Walbiri language or learning the kinship rules that determine the identity of one's potential mothers-in-law. This information is cultural in a double sense: it is information about one's culture and community and is learned almost entirely from members of that community. Strachan suggests that one outcome of interactive cultural learning is an increase in standardization and stereotyping. Models increase the similarity of their own demonstrations over time and likewise reduce their between-model differences. All this is to reduce signal ambiguity. This is a plausible suggestion with community-facing traits like ritual greetings. But it is much less plausible for world-facing traits, as variation is often an adaptive response to variation in the material substrate, agent phenotype, or both.

Celia Heyes has argued that while utilitarian skills can probably be transmitted without high-fidelity imitation, such imitation is probably required for an important subclass of community-facing traits: explicit and implicit rituals and, more generally, many community signal systems (Heyes 2013, 2021a, 2021b). As Heyes points out, in many pre-state cultures, shared community ritual and dance is central to community life and social cohesion (see, for example, Lewis 2013, 2015). These traits have two features that make imitation essential. First, they are conventional: the correct move sequence in a ritual dance is whatever that community considers to be correct. Second, often the novice must master a precise series of bodily movements. In such cases, precision is often important, and apparently small variations—raising your left hand rather than your right or raising your right hand a half a second later—count as serious errors, in part because they lead to coordination failures. Artisan skills are rarely error intolerant in just this way. They cannot be very stereotyped (contrary the suggestion by Strachan, Curioni, and McEllin that stereotyping improves transmission) because the artisan must respond to variation in raw materials. Moreover, the novice can and must adapt the model's routines to the novice's own specific phenotype: a left-handed novice uses her left hand for a task in which the right-handed model uses her right. The correct performance of a ritual requires agents to precisely coordinate with one another, so the right action must be produced at the right time. When an artisan is making a tool, there is no such requirement, so an artisan pressure flaking a Kimberley point can take his time, working at his own pace and in part sequencing the tasks at his discretion. He does not have to pressure flake in coordination with the knapper to his left.

So arguably an important subset of cultural traits depends on high-fidelity imitation. György Gergely and Ildikó Király (this volume) disagree, reanalyzing a well-known experimental paradigm of social learning in young children. In response to a model turning on a touch-sensitive lamp with their forehead, young children often turn it on the same way, but only if the context suggests that the use of the head matters. Children imitate, but only in causally opaque contexts. Gergely and Király supplement and revisit these experiments, arguing that children recognize the model's goals but choose their own method of implementing those goals. In cases where others see selective imitation, they see the recognition of a complex action sequence, with children recognizing subgoals nested within a complex action. In support of their analysis, they point out that the children's motor sequences, even when supposedly imitating, do not closely resemble those of the models. As I see it, Gergely and Király's analysis of the lamp experiment is consistent with Heyes's hypothesis about the role of imitation in cultural learning. Lamp operation is a world-facing trait; snake dances and similar action sequences are community-facing. In one view, the crucial difference between procedures that demand imitation and those that allow hybrid learning is causal opacity: imitate when the operation is causally opaque. If the relation between a procedure and its outcome is causally opaque—if the agents do not and cannot understand why it works—the novice adopts the strategy of imitating the exact procedure.

Gergely and Király argue against the idea that imitation is triggered by causal opacity. My more radical claim is that the notion of causal opacity is flawed because it is hopelessly heterogeneous. It lumps together many different cases. First, it considers cases where there is a genuine causal connection between procedure and result, while understanding the mechanism requires information inaccessible to the agent. Joseph Henrich's (2016) example of manioc processing is such a case (and in these cases, opacity comes in degrees). Second, there are cases where the procedure purports to have an effect, but where there is no actual effect, as in some folk medicines. Third, there are cases of supposed occult causation, as in sorcery and some rituals, where (a) there is no effect, (b) agents believe there is an effect, and (c) agents think they understand the mechanism. Fourth, there are cases where the point of the procedure is not the production of an effect. The relation between gang insignia or tribal tattoos and what they denote is not causal. Wearing a mongrel mob patch does not cause the wearer to be a member of that bikie gang. In my view, then, causal opacity is not relevant to the snake dance and kindred examples. Rather, the source of information about correct performance is entirely social; what counts as correct performance in these examples is determined by community opinion. Moreover, the lamp experimental paradigm is not one in which success depends on a precisely copied action sequence. Had it so depended, the children would almost all have failed.

So, despite the existence of intermediate cases and variation within clear cases of the two broad categories, the transmission mechanisms of world-facing and community-facing traits probably differ. Their evolutionary dynamics probably differ, too. On many views, cumulative culture is central to the ecological release of the hominin lineage, with selection acting on culturally transmitted variations producing both local adaptation and incrementally built complex adaptation. However, selection is more likely to adaptively tune world-facing traits because they tend to have more variation visible to selection. Rita Astuti's chapter provides an ideal case study. In her discussion of Vezo canoe sail rigging, she describes the community practice as it shifted from sails attached to two fixed poles to a system with a single pole and a mobile boom, which is a more flexible system, allowing fast enough adjustment of sails to tack into the wind. These advantages drove a shift from one design to the other in a generation—a shift that would have been much more difficult had mast layout been part of a large adaptive complex; if, for example, the shift had required a redesign of the canoe hull, different cordage, or different nets and gear. Selection is more efficient when a trait can and does vary independently of other aspects of the community practice—in one terminology, when traits are modular.

Community-facing traits are often in part coordination devices, making social interactions more predictable (Boyd 2018). For example, norms, customs, and values support expectations about how others will act: about what is shared, what is private, and how others will treat and educate their children; about predictable contributions to common projects and others' use of common space. This makes it possible to invest in the future, for the more opaque the future, the less rational it is to defer current reward. As one's own plans will often miscarry if others act contrary to expectation, there is significant social pressure both for common norms—all the members of the village have the same views as to what is common property and what is the preserve of specific individuals—and for conformity to these norms. So community-facing traits tend to vary less within the community. To the extent that cultural selection derives from agents in the N + 1 generation noticing and adopting successful innovation at the N generation, this makes within-community cultural selection less effective. As Astuti shows, Vezo sailors were able to see the advantages of the new rigging and were free to adopt it.

Community-facing traits tend to be less free to vary independently of one another. Moreover, variation often involves significant social costs. Ongaro's ethnography illustrates both constraints. As he shows, house design is not free to vary independently of the overall organization of space in an Akha village or the internal organization of space within the house itself. If cultural selection acts on anything, it acts on that overall organization of space and the ideology that supports it. In contrast to Akha house design, specific Vezo canoe rigging systems had not been co-opted as signals of cultural identity, so there was no cost of reduced esteem or alienation from support networks in shifting to the new rig. Moreover, their existing skill set and boating equipment were readily transferable to the new rig. Transition costs to the new rig were low. Contrastingly, as Ongaro's ethnographic descriptions show, an Akha housebuilder trying out a radical new design would pay very serious social costs. He contrasts Akha herbal lore (a world-facing trait) with the norms of village organization. Herbal lore, like Vezo canoe rigging, has not been co-opted as a signal of social identity and hence is free to vary, and does vary, individually.

Bottom line: Adaptive fine-tuning is likely to the extent that (a) a cultural trait can vary independently from the rest of an agent's repertoire; (b) it can do so without the agent incurring social costs; (c) the trait variation has its effects, positive or negative, over short time horizons, so agents can make repeated trials; (d) those effects are public, available for third-party observation.

The Vezo canoe rigging case study has all these characteristics, and world-facing traits will quite often be like this.[7] Community-facing traits can be like this, too: a griping new story, a catchy song, an engaging new game. But this is much less typical. Even in these cases, trialing the new idea requires others to respond as audience members or game participants. The pioneer of the new rig, Badiga, could and did try out the new Vezo rig entirely autonomously. Stout is skeptical of a view of cumulative culture that sees it as generating a progressive trend of improvement. In the light of this discussion, his skepticism seems most appropriate for community-facing traits. There an evolutionary model of increasing diversity may be more appropriate (Gould 1996).

The Economics and Ecology of Innovation

Over the last 20 years, there has been serious focus on the cognitive, demographic, and life history scaffolds of stability, of the reliable preservation of the cognitive capital of a community. Preservation is indeed essential to further improvement. But innovation is necessary, too. What are the cognitive, economic, and demographic scaffolds of innovation? A welcome feature of these essays is attention to this issue. I have discussed the cognitive supports of innovation above; here I take up ecological and economic issues.

Ecological considerations take center stage in Pope-Caldwell's chapter. She revisits an old question: To what extent is innovation driven by necessity; to what extent is it driven by opportunity? Pope-Caldwell suggests that in unforgiving environments, the risk costs of innovation are high enough to require the strategy of innovating only when innovation is essential (that is, when the payoff to the current strategy falls toward zero). In her terminology, this is "response flexibility." In more benign environments, under some conditions, a less risk-averse, opportunity-driven strategy ("elective flexibility") can pay. What makes environments more or less forgiving? Pope-Caldwell distinguishes three dimensions of environmental variation: variability, predictability (the extent to which change can be anticipated), and harshness or risk. As she sees it, response flexibility is adaptive when the agent can reliably predict the failure of a current strategy, independently of variation or harshness. Elective flexibility is contraindicated in harsh environments and in stable or predictably variable environments. But it is adaptive in unpredictable but not harsh environments. So innovation is necessity-driven when current methods fail. It is opportunity-driven in environments that are variable in the right ways and where the costs of failed experiments are not too high.

Aiming to identify general features of environments that reward different strategies of innovation seems to be the right approach. But I have significant reservations about the particular implementation of this approach. One reservation is that it seems to be an equilibrium model: "Specifically, in stable environments," Pope-Caldwell writes, "the usefulness of a strategy does not change over time, so responsive flexibility is not needed and elective flexibility is not pragmatic. *Once an optimal strategy is found*, there is little benefit derived from maintaining, using, or seeking alternatives. Thus, under stable conditions, both responsive and elective flexibility are maladaptive—and should be suppressed—so long as a working strategy can be maintained." And again: "However, the conditions that predict elective flexibility are more complex. In predictably variable environments, . . . elective flexibility is not useful because *even though optimal strategies rotate over time*, the set of useful strategies is unchanged" (italics added, in both cases). To which the obvious questions are: How is an optimal strategy found? And how do agents know that they have found it? The Vezo shift in canoe design had nothing to do with environmental change or declining rates of return from the old design. There is little discussion in Pope-Caldwell's chapter about how agents identify the kinds of environment they are in. That does not matter if the search process is conceived as one that involves *only* blind variation, vertical transmission, and selective retention. But it does matter if this analysis is intended to illuminate cultural evolutionary processes. For in many conceptualizations of cultural evolution, novices select models on the basis of model success. The Vezo illustrate this. In these forms of cultural evolution, agents have at least some knowledge of the kind of environment they are in. In these frameworks, the character of the environment matters, but so does the informational access of the agent to the environment. Pope-Caldwell herself notes that real environments have stochastic variation, so it is not trivial to determine that a previously successful routine is now reliably failing.

Likewise, missing is any response to niche construction ideas: this analysis treats environmental parameters as fixed externally and agents as responding to those parameters. That does not capture the relations between hominins and their environment. Harshness or risk, for example, does not seem external to agent strategies in this way. Both technological and social innovations modify risk. For example, forager sharing norms have often been interpreted as risk-reduction strategies. With these norms in place, desert environments with all their objective dangers do not present as threatening to their inhabitants (for an insightful discussion of these different perceptions of desert environments, see Hiscock and O'Connor 2005).

Finally, let's return to cumulative culture and the hominin-chimp contrast. Tenpas, Schweinfurth, and Call (this volume) make the uncontroversial empirical point that chimp material culture does not show much, if any, evidence of incremental improvement, despite the fact that in experimental contexts, chimps are quite good innovators. This has typically been explained as a reflection of relatively limited chimp cultural learning capacities (see, for example, Henrich and Tennie 2017). If the argument of section 2 holds up, that cannot be right. For at least some world-facing capacities can be transmitted reliably through a mix of cognitively undemanding cultural channels and individual, exploratory trial and error. It is also theoretically problematic: we would expect sophisticated cultural learning capacities to follow, rather than precede, cumulative culture: there is selection for such capacities only once agents live in a social environment in which important information flows culturally.[8]

Here is an alternative suggestion. Perhaps chimp material culture is not cumulative because their technology is expedient. That is, it is used on the spot and then discarded. This contrasts with so-called maintainable technologies—lightweight, multipurpose, and readily repaired, such as the San bow-arrow-quiver system—and with "reliable" technologies—robust, highly engineered, typically special-purpose systems, such as specialist Inuit harpoon systems (Bleed 1986; Bousman 1993). Foragers use expedient tools, typically when there is little time pressure, when raw materials can be expected to be readily available, and when transport costs are high. Richard Gould describes Australian Western Desert foragers using expedient stone tools, simple flakes, to work wood (Gould 1980). These are low-cost, low-investment tools with little obvious sign of cumulative improvement. The contrast with maintainable and reliable tool systems is marked. The San bow system is a multipart composite, requiring different raw materials for the bow, the bow-string, the arrow shaft, the arrow heads (plus poison and poison applicator), and the quiver. Time, effort, and skill have gone into their construction (and adept use). Improved tools are also more expensive tools (hence curated tools), not used once and abandoned. Quite independently of their social learning powers, chimp social life may not support rational investment in more expensive, curated technologies. First, the investing chimp needs secure possession of the tool and any resources made available by its use. The more effective the tool, the more valuable it is to others. In alpha-dominated social worlds, subordinates lack secure possession of tools and their products. Second, there has to be a potential, profitable, step-improvement from some item in the existing material repertoire. In hominin evolution, access to large game carcasses probably provided the economic rationale for the Acheulean. In comparison to the Oldowan, Acheulean tools require a higher skill base, and so they were an incremental advance over that earlier technology (Stout and Chaminade 2012; Stout et al. 2015). But those tools were also more expensive, using more stone and taking longer to make. There seems to be no counterpart in chimp ecology: no obvious resource unexploited or poorly exploited through an inability to shift from expedient to modestly more expensive, curated tools.

Concluding Summary

Time to finish. I suggest the following five take-home messages from these reflections. (1) Insofar as a directional conception of human cultural evolution is appropriate, it is best framed as a gradual transition from the transmission of specific, context-bound, limited-purpose skills to more flexible, more context-independent, expertise-based capacities. (2) Once human adaptive capacities were largely based on expertise, there is no trade-off between the preservation and the improvement of these repertoires of skill. (3) The transmission mechanisms of world-facing traits are importantly different from those of community-facing traits, with hybrid informational pathways playing a much larger role with world-facing traits. (4) The mechanisms of stability and innovation in larger and more complex societies differ from those of smaller, less complex ones. Transmission in small-scale, preliterate communities depends much more on autonomous self-directed learning, with much horizontal transmission, though with adult protection, material support, some explicit teaching, and skill modeling. (5) In the initial triggering of the distinctive cultural trajectory of the hominin lineage, ecological and economic differences were probably more important than cognitive ones. In *pan* worlds, there is no apparent first step on the pathway to technical and cooperative takeoff. There is no obvious resource that is exploitable only with a modest technical advance on existing capacities, but profitable enough when exploited to stabilize the practice, becoming a platform for further improvement. Scavenging bone marrow from large animal carcasses (more frequent and more visible in open habitats) may have been the equivalent hominin opportunity. Once that distinctive trajectory was established, culture-cognition coevolution very likely kicked in, driving further divergence.

Acknowledgments

It is a pleasure to acknowledge the constructive feedback from Mathieu Charbonneau, Peter Hiscock, Ross Pain, Dietrich Stout, and Dan Sperber on earlier versions of this material. Likewise, it is a pleasure to thank the Australian Research Grants Council for its support for my work on human evolution over many years and grants.

Notes

1. A similar distinction is defended in Jagiello, Heyes, and Whitehouse 2022.

2. Stout diagnoses the problem as conceiving of cultural in terms of information rather than material forms and practice. I think that is a relatively harmless oversimplification; the problem is to think of information transmission as if it were template copying. See Sterelny and Hiscock (forthcoming).

3. No doubt, there is a good deal of variation from skill to skill, however. Work on African bowyers suggests that they have fairly limited capacity to innovate on their traditional bow designs (Harris, Boyd, and Wood 2021).

4. Sadly, almost certainly it is no longer possible to do Roux and colleague's type of experiments with trackers to determine the extent to which they can reconstruct a track-maker from tracks of an unfamiliar kind and to determine whether the most expert trackers within the existing repertoire would also be the best able to extrapolate beyond that repertoire. Louis Liebenberg (1990) has argued that the tracking of true experts is guided by a good deal of theory, in addition to superb observational and pattern recognition skills, and if that is right, we would expect expertise to have this expanding effect.

5. I was a child in Australia in the 1950s and 1960s, and my parents would now risk criminal prosecution for activities permitted to me—for example, swimming unsupervised in farm dams or bushwalking by myself as a nine-year-old. None of this was unusual at the time.

6. There is a difference between behavior that is not observed and behavior that is forbidden, but this difference is behaviorally manifested only when forbidden behavior takes place and is detected.

7. However, some technology (like some net fishing) can only be used collaboratively, and some is so expensive, like large fish traps, that its cost has to be spread over several agents. These cannot be trialed without others buying in.

8. One might suggest that in a fission-fusion foraging organization, valuable but transient information (e.g., the location of a specific carcass, the state of a waterhole) might select for specifically cultural learning capacities,

prior to the cumulative improvement of long-life informational capital. But sharing such information would seem to require the kinds of communicative tools that must be built cumulatively.

References

Biryukova, E. V., and B. Bril. 2008. "Organization of Goal-Directed Action at a High Level of Motor Skill: The Case of Stone Knapping in India." *Motor Control* 12:181–209.

Bleed, P. 1986. "The Optimal Design of Hunting Weapons: Maintainability or Reliability." *American Antiquity* 51 (4): 737–747.

Bousman, C. 1993. "Hunter-Gatherer Adaptations, Economic Risk, and Tool Design." *Lithic Technology* 18 (1–2): 59–86.

Boyd, R. 2018. *A Different Kind of Animal*. Princeton, NJ: Princeton University Press.

Bril, B., V. Roux, and G. Dietrich. 2005. "Stone Knapping: Khambhat (India), a Unique Opportunity?" In *Stone Knapping: The Necessary Conditions for a Uniquely Hominin Behaviour*, edited by V. Roux and B. Bril, 53–71. Cambridge: McDonald Institute for Archaeological Research.

Christensen, W., K. Bicknell, D. McIlwain, and J. Sutton. 2015. "The Sense of Agency and Its Role in Strategic Control for Expert Mountain Bikers." *Psychology of Consciousness: Theory, Research, and Practice* 2:340–353.

Christensen, W., J. Sutton, and D. McIlwain. 2016. "Cognition in Skilled Action: Meshed Control and the Varieties of Skill Experience." *Mind & Language* 31:37–66.

Dira, S., and B. S. Hewlett. 2016. "Learning to Spear Hunt among Ethiopian Chabu Adolescent Hunter-Gatherers." In *Social Learning and Innovation in Contemporary Hunter-Gatherers*, edited by H. Terashima and B. Hewlett, 71–81. Dordrecht: Springer.

Gould, R. A. 1969. *Yiwara: Foragers of the Australian Desert*. Sydney: Collins.

Gould, R. A. 1980. *Living Archaeology*. Cambridge: Cambridge University Press.

Gould, S. J. 1996. *Full House: The Spread of Excellence from Plato to Darwin*. New York: Harmony Press.

Harris, J., R. Boyd, and B. Wood. 2021. "The Role of Causal Knowledge in the Evolution of Traditional Technology." *Current Biology* 31:1–6.

Hayden, B. 1998. "Practical and Prestige Technologies: The Evolution of Material Systems." *Journal of Archaeological Method and Theory* 5 (1): 1–55.

Henrich, J. 2016. *The secret of our success: How culture is driving human evolution, domesticating our species, and making us smarter*. New Jersey: Princeton University Press.

Henrich, J., and C. Tennie. 2017. "Cultural Evolution in Chimpanzees and Humans." In *Chimpanzees and Human Evolution*, edited by M. Muller, R. Wrangham, and D. Pilbeam, 645–688. Cambridge, MA: Harvard University Press.

Hewlett, B., J. Hudson, A. Boyette, and H. Fouts. 2019. "Intimate Living: Sharing Space among Aka and Other Hunter-Gatherers." In *Towards a Broader View of Hunter-Gatherer Sharing*, edited by N. Lavi and D. Friesom, 39–56. Cambridge: McDonald Institute for Archaeological Research.

Heyes, C. 2013. "What Can Imitation Do for Cooperation?" In *Cooperation and Its Evolution*, edited by K. Sterelny, R. Joyce, B. Calcott, and B. Fraser, 313–332. Cambridge, MA: MIT Press.

Heyes, C. 2021a. "Primer on 'Imitation.'" *Current Biology*: R215–R240.

Heyes, C. 2021b. "Imitation and Culture: What Gives?" *Mind & Language* 38 (1): 42–63.

Hiscock, P., and S. O'Connor. 2005. "Arid Paradises or Dangerous Landscapes: A Review of Explanations for Paleolithic Assemblage Change in Arid Australia and Africa." In *Desert Peoples: Archaeological Perspectives*, edited by P. Veth, M. Smith, and P. Hiscock, 58–77. London: Blackwell.

Jagiello, R., C. Heyes, and H. Whitehouse. (2022). Tradition and invention: The bifocal stance theory of cultural evolution. *Behavioral and Brain Sciences* 45: e249

Lew-Levy, S., and A. Boyette. 2018. "Evidence for the Adaptive Learning Function of Work and Work-Themed Play among Aka Forager and Ngandu Farmer Children from the Congo Basin." *Human Nature* 29:157–185.

Lew-Levy, S., N. Lavi, R. Reckin, J. Cristóbal-Azkarate, and K. Ellis-Davies. 2018. "How Do Hunter-Gatherer Children Learn Social and Gender Norms? A Meta-Ethnographic Review." *Cross-Cultural Research* 52 (2): 213–255.

Lew-Levy, S., A. Milks, N. Lavi, S. M. Pope, and D. Friesem. 2020. "Where Innovations Flourish: An Ethnographic and Archaeological Overview of Hunter-Gatherer Learning Contexts." *Evolutionary Human Sciences* 2:e31.

Lew-Levy, S., R. Reckin, N. Lavi, J. Cristóbal-Azkarate, and K. Ellis-Davies. 2017. "How Do Hunter-Gatherer Children Learn Subsistence Skills? A Meta-Ethnographic Review." *Human Nature* 28:367–394.

Lewis, J. 2013. "A Cross-Cultural Perspective on the Significance of Music and Dance on Culture and Society: Insight from BaYaka Pygmies." In *Language, Music and the Brain: A Mysterious Relationship*, edited by M. Arbib, 45–65. Cambridge, MA: MIT Press.

Lewis, J. 2015. "Where Goods Are Free but Knowledge Costs: Hunter-Gatherer Ritual Economics in Western Central Africa." *Hunter Gatherer Research* 1 (1): 1–27.

Liebenberg, L. 1990. *The Art of Tracking and the Origin of Science*. Claremont, South Africa: David Philip.

Liebenberg, L. 2013. *The Origin of Science: On the Evolutionary Roots of Science and Its Implications for Self-Education and Citizen Science*. Cape Town: CyberTracker.

Love, J. R. B. 2009. *Kimberley People: Stone Age Bushmen of Today*. Darwin: Australian Aboriginal Culture Series.

Meggitt, M. 1965. *Desert People: A Study of the Walbiri Aborigines of Central Australia*. Sydney: Angus and Robinson.

Morrison, R. 1981. *A Field Guide to the Tracks and Traces of Australian Animals*. Adelaide: Rigby.

Naveh, D. 2016. "Social and Epistemological Dimensions of Learning among Nayaka Hunter-Gatherers." In *Social Learning and Innovation in Contemporary Hunter-Gatherers: Evolutionary and Ethnographic Perspectives*, edited by H. Terashima and B. Hewlett, 125–134. Dordrecht: Springer.

Shaw-Williams, K. 2014. "The Social Trackways Theory of the Evolution of Human Cognition." *Biological Theory* 9 (1): 16–26.

Shaw-Williams, K. 2017. "The Social Trackways Theory of the Evolution of Language." *Biological Theory* 12 (4): 195–210.

Sterelny, K. 2006. "The Evolution and Evolvability of Culture." *Mind & Language* 21 (2): 137–165.

Sterelny, K. 2012. *The Evolved Apprentice*. Cambridge, MA: MIT Press.

Sterelny, K., and P. Hiscock. (forthcoming). "Cumulative Culture, Archaeology, and the Zone of Latent Solutions." *Current Anthropology*.

Stout, D. 2002. "Skill and Cognition in Stone Tool Production: An Ethnographic Case Study from Irian Jaya." *Current Anthropology* 43 (5): 693–722.

Stout, D., and T. Chaminade. 2012. "Stone Tools, Language and the Brain in Human Evolution." *Philosophical Transactions of the Royal Society of London. Series B: Biological Sciences* 367:75–87.

Stout, D., E. Hecht, N. Khreisheh, B. Bradley, and T. Chaminade. 2015. "Cognitive Demands of Lower Paleolithic Toolmaking." *PLOS One* 10 (4): e0121804.

Tomasello, M. 1999. *The Cultural Origins of Human Cognition*. Cambridge, MA: Harvard University Press.

Tomasello, M. 2014. *A Natural History of Human Thinking*. Cambridge, MA: Harvard University Press.

Contributors

Rita Astuti, Department of Anthropology, London School of Economics, United Kingdom

Adam Howell Boyette, Department of Human Behavior, Ecology and Culture, Max Planck Institute for Evolutionary Anthropology, Germany

Blandine Bril, Écoles des Hautes Études en Sciences Sociales, France

Josep Call, School of Psychology and Neuroscience, University of St Andrews, United Kingdom

Mathieu Charbonneau, Africa Institute for Research in Economics and Social Sciences, Mohammed VI Polytechnic University, Morocco

Arianna Curioni, Vienna University of Technology (TU Wien), Austria

Nicola Cutting, School of Education, Language and Psychology, York St John University, United Kingdom

Bert De Munck, History Department, University of Antwerp, Belgium

György Gergely, Department of Cognitive Science, Central European University, Austria

Anne-Lise Goujon, CFEE (Centre Français des Etudes Ethiopiennes), Ethiopia

Ildikó Király, Department of Cognitive Psychology, Eötvös Loránd University, Hungary

Catherine Lara, IFEA (Institut Français des Etudes Andines), Peru

Sébastien Manem, UMR 8068 TEMPS, CNRS/Université Paris Nanterre/Université Paris 1 Panthéon-Sorbonne, France

Luke McEllin, Social Mind Center, Department of Cognitive Science, Central European University, Austria

Helena Miton, Santa Fe Institute, United States of America

Giulio Ongaro, Department of Anthropology, London School of Economics, United Kingdom

Sarah Pope-Caldwell, Department of Comparative Cultural Psychology, Max Planck Institute for Evolutionary Anthropology, Germany

Valentine Roux, UMR 7055 Préhistoire et Technologie, CNRS/Université Paris Nanterre, France

Manon Schweinfurth, School of Psychology and Neuroscience, University of St Andrews, United Kingdom

Dan Sperber, Department of Cognitive Science, Central European University, Austria

Kim Sterelny, College of Arts and Social Sciences, Australian National University, Australia

Dietrich Stout, Department of Anthropology, Emory University, United States of America

James W. A. Strachan, Department of Neurology, University Medical Center Hamburg-Eppendorf (UKE), Germany

Sadie Tenpas, School of Psychology and Neuroscience, University of St Andrews, United Kingdom

Index